T0297930

Spectral Theory of Large Dimensional Random Matrices and its Applications to Wireless Communications and Finance Statistics

Random Matrix Theory and its Applications

Spectral Theory of Large Dimensional Random Matrices and its Applications to Wireless Communications and Finance Statistics

Random Matrix Theory and its Applications

Zhidong Bai
Northeast Normal University, China &
National University of Singapore, Singapore

Zhaoben Fang
University of Science and Technology of China, China

Ying-Chang Liang
The Singapore Infocomm Research Institute, Singapore

Published by

World Scientific Publishing Co. Pte. Ltd.
5 Toh Tuck Link, Singapore 596224
USA office: 27 Warren Street, Suite 401-402, Hackensack, NJ 07601
UK office: 57 Shelton Street, Covent Garden, London WC2H 9HE

British Library Cataloguing-in-Publication Data
A catalogue record for this book is available from the British Library.

Spectral Theory of Large Dimensional Random Matrices and Its Applications to Wireless Communications and Finance Statistics

The work is originally published by University of Science and Technology of China Press in 2010.
This edition is published by World Scientific Company Pte Ltd by arrangement with University of Science and Technology of China Press, Anhui, China.

SPECTRAL THEORY OF LARGE DIMENSIONAL RANDOM MATRICES AND ITS APPLICATIONS TO WIRELESS COMMUNICATIONS AND FINANCE STATISTICS
Random Matrix Theory and Its Applications

ISBN 978-981-4579-05-6

Printed in Singapore

Dedicated to Fiftieth Anniversary of Alma Mater
University of Science and Technology of China

To my great teachers Professors Yongquan Yin
and Xiru Chen
my wife, Xicun Dan
and my sons, Li, and Steve Gang
and grandsons Yongji, Yonglin, and Yongbin

— Zhidong Bai

To my wife Guo Fujia
and my daughter Fang Min

— Zhaoben Fang

To
To my wife Shuo Gao
my son, Paul, and daughter, Wendy

— Liang Ying-Chang

Preface

Since the last three or four decades, computer science and computing facilities have been developed and adopted to almost every discipline. In consequence, data sets collected from experiments and/or natural phenomena have become large in both their sizes as well as dimensions. At the same time, the computing ability has become capable of dealing with extremely huge data sets.

For example, in the biological sciences, a DNA sequence can be as long as several billions. In finance research, the number of different stocks can be as large as tens of thousands. In wireless communications, the number of users can be several millions. In image processing, the pixels of a picture may be several thousands.

On the other hand, however, in statistics, classical limit theorems have been found to be seriously inadequate in aiding in the analysis of large dimensional data. All these are challenging classical statistics. Nowadays, an urgent need to statistics is to create new limiting theories that are applicable to large dimensional data analysis. Therefore, since last decade, the large dimensional data analysis has become a very hot topic in statistics and various disciplines where statistics is applicable.

Currently, spectral analysis of large dimensional random matrices (simply Random Matrix Theory (RMT)) is the only systematic theory that can be applied to some problems of large dimensional data analysis. The RMT dates back to the early development of Quantum Mechanics in the 1940's and 50's. In an attempt to explain the complex organizational structure of heavy nuclei, E. Wigner, Professor of Mathematical Physics at Princeton University, argued that one should not compute energy levels from Schrödinger's equation. Instead, one should imagine the complex nuclei system as a black box described by $n \times n$ Hamiltonian matrices with elements drawn from a probability distribution with only mild constraints dictated by symmetry considerations. Under these assumptions and a mild conditions imposed on the probability measure in the space of matrices, one finds the joint probability density of the n eigenvalues. Based on this consideration, Wigner established the well known semi-circular law. Since then, RMT has been developed into a big research area in mathematical physics and probability.

Due to needs of large dimensional data analysis, the number of researchers and publications on RMT has been rapidly increasing. As an evidence, it can be seen from the following statistics from Mathscinet database under keyword Random Matrix on 11 April 2008:

1955—1964	1965—1974	1975—1984	1985—1994	1995—2004	2005–04.2008
23	138	249	635	1205	493

Table 0.1. Publication numbers on RMT in 10 year periods since 1955

The purpose of this monograph is to introduce the basic concepts and results of RMT and some applications to Wireless Communications and Finance Statistics. The readers of this book would be graduate students and beginning researchers who are interested in RMT and/or its applications to their own research areas. As for the theorems in RMT, we only provide an outline of their proofs. The detailed proofs are referred to the book *Spectral Analysis of Large Dimensional Random Matrices* by Bai, Z. D. and Silverstein, J. W. (2006). A for the applications to Wireless communications and Finance Statistics, we are more emphasizing its formulation to illustrate how the RMT is applied to, rather than detailed mathematical proofs.

Special thanks go to Mr. Liuzhi Yin who contributed to the book by providing editing and extensive literature review.

Changchun, China
Hefei, China
Singapore

Bai Zhidong
Fang Zhaoben
Liang Ying-Chang
April 2008

Contents

1

Introduction

1.1 History of RMT and Current Development

The aim of this book is to introduce main results in the spectral theory of large dimensional random matrices (RM) and its rapidly spreading applications to many applied areas. As illustration, we briefly introduce some applications to the wireless communications and finance statistics.

In the past three or four decades, a significant and constant advancement in the world has been in the rapid development and wide application of computer science. Computing speed and storage capability have increased thousands folds. This has enabled one to collect, store and analyze data sets of huge size and very high dimension. These computational developments have had strong impact on every branch of science. For example, R. A. Fisher's resampling theory had been silent for more than three decades due to the lack of efficient random number generators, until Efron proposed his renowned bootstrap in the late 1970's; the minimum L_1 norm estimation had been ignored for centuries since it was proposed by Laplace, until Huber revived it and further extended it to robust estimation in the early 1970's. It is difficult to imagine that these advanced areas in statistics would have gotten such deep stages of development if there were no such assistance from the present day computer.

Although modern computer technology helps us in so many aspects, it also brings a new and urgent task to the statisticians. All classical limiting theorems employed in statistics are derived under the assumption that the dimension of data is fixed. However, it has been found that the large dimensionality would bring intolerable error when classical limiting theorems is employed to large dimensional statistical data analysis. Then, it is natural to ask whether there are any alternative theories can be applied to deal with large dimensional data. The theory of random matrix (RMT) has been found itself a powerful tool to deal some problems of large dimensional data.

1.1.1 A brief review of RMT

RMT traces back to the development of quantum mechanics (QM) in the 1940's and early 1950's. In QM, the energy levels of a system are described by eigenvalues of an Hermitian operator **A** on a Hilbert space, called the Hamiltonian. To avoid working with an infinite dimensional operator, it is common to approximate the system by discretization, amounting to a truncation, keeping only the part of the Hilbert space that is important to the problem under consideration. Hence, the limiting behavior of large dimensional random matrices attracts special interest among those working in QM and many laws were discovered during that time. For a more detailed review on applications of RMT in QM and other related areas, the reader is referred to the Books *Random Matrices* by Mehta (1991, 2004) and Bai and Silverstein (2006, 2009).

In the 1950's in an attempt to explain the complex organizational structure of heavy nuclei, E. P. Wigner, Jones Professor of Mathematical Physics at Princeton University, put forward a heuristic theory. Wigner argued that one should not try to solve the Schrödinger's equation which governs the n strongly interacting nucleons for two reasons firstly, it is computationally prohibitive - which perhaps remains true even today with the availability of modern high speed machines and, secondly the forces between the nucleons are not very well understood. Wigner's proposal is a pragmatic one: One should not compute from the Schrödinger's equation the energy levels, one should instead imagine the complex nuclei as a black box described by $n \times n$ Hamiltonian matrices with elements drawn from probability distribution with only mild constraint dictated by symmetry consideration.

Along with this idea, Wigner (1955, 1958) proved that the expected spectral distribution of a large dimensional Wigner matrix tends to the famous semicircular law. This work was generalized by Arnold (1967, 1971) and Grenander (1963) in various aspects. Bai and Yin (1988a) proved that the spectral distribution of a sample covariance matrix (suitably normalized) tends to the semicircular law when the dimension is relatively smaller than the sample size. Following the work of Marčenko and Pastur (1967) and Pastur (1972, 1973), the asymptotic theory of spectral analysis of large dimensional sample covariance matrices was developed by many researchers including Bai, Yin, and Krishnaiah (1986), Grenander and Silverstein (1977), Jonsson (1982), Wachter (1978), Yin (1986), and Yin and Krishnaiah (1983). Also, Bai, Yin, and Krishnaiah (1986, 1987), Silverstein (1985a), Wachter (1980), Yin (1986), and Yin and Krishnaiah (1983) investigated the limiting spectral distribution of the multivariate F-matrix, or more generally, of products of random matrices. In the early 1980's, major contributions on the existence of LSD and their explicit forms for certain classes of random matrices were made. In recent years, research on RMT is turning toward second order limiting theorems, such as the central limit theorem for linear spectral statistics, the limiting distributions of spectral spacings and extreme eigenvalues.

1.1.2 Spectral Analysis of Large Dimensional Random Matrices

Suppose $\mathbf{A}$ is an $m \times m$ matrix with eigenvalues $\lambda_j, j = 1, 2, \cdots, m$. If all these eigenvalues are real, e.g., if $\mathbf{A}$ is Hermitian, we can define a one-dimensional distribution function

$$F^{\mathbf{A}}(x) = \frac{1}{m}\#\{j \le m : \lambda_j \le x\}, \tag{1.1.1}$$

called the empirical spectral distribution (ESD) of the matrix $\mathbf{A}$. Here $\#E$ denotes the cardinality of the set E. If the eigenvalues λ_j's are not all real, we can define a two-dimensional empirical spectral distribution of the matrix $\mathbf{A}$:

$$F^{\mathbf{A}}(x, y) = \frac{1}{m}\#\{j \le m : \Re(\lambda_j) \le x, \Im(\lambda_j) \le y\}. \tag{1.1.2}$$

One of the main problems in RMT is to investigate the convergence of the sequence of empirical spectral distributions $\{F^{\mathbf{A}_n}\}$ for a given sequence of random matrices $\{\mathbf{A}_n\}$. The limit distribution F (possibly defective), which is usually nonrandom, is called the *Limiting Spectral Distribution* (LSD) of the sequence $\{\mathbf{A}_n\}$.

We are especially interested in sequences of random matrices with dimension (number of columns) tending to infinity, which refers to *the theory of large dimensional random matrices.*

The importance of ESD is due to the fact that many important statistics in multivariate analysis can be expressed as functionals of the ESD of some RM. We give now a few examples.

Example 1.1. *Let* $\mathbf{A}$ *be an* $n \times n$ *positive definite matrix. Then*

$$\det(\mathbf{A}) = \prod_{j=1}^{n} \lambda_j = \exp\left(n \int_0^{\infty} \log x F^{\mathbf{A}}(dx)\right).$$

Example 1.2. *Let the covariance matrix of a population have the form* $\boldsymbol{\Sigma} = \boldsymbol{\Sigma}_q + \sigma^2\mathbf{I}$, *where the dimension of* $\boldsymbol{\Sigma}$ *is* p *and the rank of* $\boldsymbol{\Sigma}_q$ *is* $q(< p)$. *Suppose* $\mathbf{S}$ *is the sample covariance matrix based on* n *iid. samples drawn from the population. Denote the eigenvalues of* $\mathbf{S}$ *by* $\sigma_1 \ge \sigma_2 \ge \cdots \ge \sigma_p$. *Then the test statistic for the hypothesis* H_0 : *rank*$(\boldsymbol{\Sigma}_q) = q$ *against* H_1 : *rank*$(\boldsymbol{\Sigma}_q) > q$ *is given by*

$$T = \frac{1}{p-q}\sum_{j=q+1}^{p} \sigma_j^2 - \left(\frac{1}{p-q}\sum_{j=q+1}^{p} \sigma_j\right)^2$$

$$= \frac{p}{p-q}\int_0^{\sigma_q} x^2 F^{\mathbf{S}}(dx) - \left(\frac{p}{p-q}\int_0^{\sigma_q} x F^{\mathbf{S}}(dx)\right)^2.$$

1.1.3 Limits of Extreme Eigenvalues

In applications of the asymptotic theorems of spectral analysis of large dimensional random matrices, two important problems arose after the LSD was found. The first is the bound on extreme eigenvalues; the second is the convergence rate of the ESD, with respect to sample size. For the first problem, the literature is extensive. The first success was due to Geman (1980), who proved that the largest eigenvalue of a sample covariance matrix converges almost surely to a limit under a growth condition on all the moments of the underlying distribution. Yin, Bai, and Krishnaiah (1988) proved the same result under the existence of the 4th moment, and Bai, Silverstein, and Yin (1988) proved that the existence of the 4th moment is also necessary for the existence of the limit. Bai and Yin (1988b) found the necessary and sufficient conditions for almost sure convergence of the largest eigenvalue of a Wigner matrix. By the symmetry between the largest and smallest eigenvalues of a Wigner matrix, the necessary and sufficient conditions for almost sure convergence of the smallest eigenvalue of a Wigner matrix was also found.

Comparing to almost sure convergence of the largest eigenvalue of a sample covariance matrix, a relatively harder problem is to find the limit of the smallest eigenvalue of a large dimensional sample covariance matrix. The first attempt was made in Yin, Bai, and Krishnaiah (1983), in which it was proved that the almost sure limit of the smallest eigenvalue of a Wishart matrix has a positive lower bound when the ratio of dimension to the degrees of freedom is less than 1/2. Silverstein (1984) modified the work to allowing the ratio less than 1. Silverstein (1985b) further proved that with probability one, the smallest eigenvalue of a Wishart matrix tends to the lower bound of the LSD when the ratio of dimension to the degrees of freedom is less than 1. However, Silverstein's approach strongly relies on the normality assumption on the underlying distribution and thus, it cannot be extended to the general case. The most current contribution was made in Bai and Yin (1993) in which it is proved that under the existence of the fourth moment of the underlying distribution, the smallest eigenvalue (when $p \le n$) or the $p-n+1$st smallest eigenvalue (when $p > n$) tends to $a(y) = \sigma^2(1-\sqrt{y})^2$, where $y = \lim(p/n) \in (0, \infty)$. Comparing to the case of the largest eigenvalues of a sample covariance matrix, the existence of the fourth moment seems to be necessary also for the problem of the smallest eigenvalue. However, this problem has not yet been solved.

1.1.4 Convergence Rate of ESD

The second problem, the convergence rate of the spectral distributions of large dimensional random matrices, is of practical interest, but has been open for decades. In finding the limits of both the LSD and the extreme eigenvalues of symmetric random matrices, a very useful and powerful method is the moment method which does not give any information about the rate of the convergence of the ESD to the LSD. The first success was made in Bai (1993a, b), in which a Berry-Esseen type inequality of the difference of two distributions was established in terms of their Stieltjes transforms. Applying this inequality, a convergence rate for the expected ESD of a large Wigner matrix was proved to be $O(n^{-1/4})$, that for the sample covariance matrix was shown to be $O(n^{-1/4})$ if the ratio of the dimension to the degrees of freedom is apart away from one, and to be $O(n^{-5/48})$, if the ratio is close to 1.

1.1.5 Circular Law

The most perplexing problem is the so-called circular law which conjectures that the spectral distribution of a non-symmetric random matrix, after suitable normalization, tends to the uniform distribution over the unit disc in the complex plane. The difficulty exists in that two most important tools used for symmetric matrices do not apply for non-symmetric matrices. Furthermore, certain truncation and centralization techniques cannot be used. The first known result was given in Mehta (1967) and in an unpublished paper of Silverstein (1984) which was reported in Hwang (1986). They considered the case where the entries of the matrix are iid standard complex normal. Their method uses the explicit expression of the joint density of the complex eigenvalues of the random matrix which was found by Ginibre (1965). The first attempt to prove this conjecture under some general conditions was made in Girko (1984a, b). However, his proofs have puzzled many who attempt to understand, without success, Girko's arguments. Recently, Edelman (1995) found the conditional joint distribution of complex eigenvalues of a random matrix whose entries are real normal $N(0,1)$ when the number of its real eigenvalues is given and proved that the expected spectral distribution of the real Gaussian matrix tends to the circular law. Under the existence of $4+\varepsilon$ moment and some smooth conditions, Bai (1997) proved the strong version of the circular law. The conjecture was finally proved by Tao and Vu (2010).

1.1.6 CLT of Linear Spectral Statistics

As mentioned above, functionals of the ESD of RM's are important in multivariate inference. Indeed, a parameter θ of the population can sometimes be expressed as

$$\theta = \int f(x)dF(x).$$

To make statistical inference on θ, one may use the integral

$$\hat{\theta} = \int f(x)dF_n(x),$$

which we call *linear spectral statistics* (LSS), as an estimator of θ, where $F_n(x)$ is the ESD of the RM computed from the data set. Further, one may want to know the limiting distribution of $\hat{\theta}$ through suitable normalization. In Bai and Silverstein (2004) the normalization has been found to be n, by showing the limiting distribution of the linear functional

$$X_n(f) = n \int f(t)d(F_n(t) - F(t))$$

to be Gaussian under certain assumptions.

The first work in this direction was done by D. Jonsson (1982) in which $f(t) = t^r$ and F_n is the ESD of normalized standard Wishart matrix. Further work was done by Johansson, K. (1998), Bai and Silverstein (2004), Bai and Yao (2004), Sinai and Soshnikov (1998), among others.

It would seem natural to pursue the properties of linear functionals by way of proving results on the process $G_n(t) = \alpha_n(F_n(t) - F(t))$ when viewed as a random element in $D[0, \infty)$, the metric space of functions with discontinuities of the first kind, along with the Skorokhod metric. Unfortunately, this is impossible. The work done in Bai and Silverstein (2004) shows that $G_n(t)$ cannot converge weakly to any non-trivial process for any choice of α_n. This fact appears to occur in other random matrix ensembles. When F_n is the empirical distribution of the angles of eigenvalues of an $n \times n$ Haar matrix, Diaconis and Evans (2001) proved that all finite dimensional distributions of $G_n(t)$ converge in distribution to independent Gaussian variables when $\alpha_n = n/\sqrt{\log n}$. This shows that with $\alpha_n = n/\sqrt{\log n}$, the process G_n cannot be tight in $D[0, \infty)$.

1.1.7 Limiting Distributions of Extreme Eigenvalues and Spacings

The first work on the limiting distributions was done by C. Tracy and H. Widom (1996) who found the expression for the largest eigenvalue of a Gaussian matrix when suitably normalized. Further, I. Johnstone (2001) found the limiting distribution of the largest eigenvalue of large Wishart matrix.

The work on spectrum spacing has a long history which dates back to M. L. Mehta (1960). Most of the work in these two directions assume the Gaussian (or generalized) distributions.

1.2 Applications to Wireless Communications

In recent years, there has been an increasing interest in applying RMT to wireless communications. One important scenario which can be well modeled using random matrix is direct-sequence code division multiple access (DS-CDMA) with random spreading codes. Let us consider a simplified input-output model for synchronous CDMA uplink:

$$\mathbf{y} = \sum_{i=1}^{K} s_i \mathbf{h}_i + \mathbf{w}, \tag{1.2.1}$$

where s_i and $\mathbf{h}_i \in \mathbb{C}^N$ are the transmitted symbol and spreading codes of user i, and $\mathbf{w} \sim \mathcal{CN}(0, \sigma^2\mathbf{I})$ is the additive complex Gaussian white noise. For brevity, we assume that the random variables $s_1, s_2, \cdots, s_K$ are independent and with $\mathsf{E}[s_i] = 0$ and $\mathsf{E}[|s_i|^2] = 1$.

Rewrite (1.2.1) as

$$\mathbf{y} = \mathbf{Hs} + \mathbf{w}, \tag{1.2.2}$$

where $\mathbf{H} = [\mathbf{h}_1, \cdots, \mathbf{h}_K]$ and $\mathbf{s} = [s_1, \cdots, s_K]^T$. Here $(\cdot)^T$ denotes transpose. The task of a receiver design is to estimate the transmitted data symbols, $s_1, \cdots, s_K$, from the received vector $\mathbf{y}$. Without loss of generality, we consider the problem of detecting the signal, s_1, from user 1.

One class of receivers is the linear receiver, which generates the signal estimate of user 1 as $\hat{s}_1 = \mathbf{c}_1^*\mathbf{y}$, where $\mathbf{c}_1 \in \mathbb{C}^N$ is the receiver weights, and $(\cdot)^*$ denotes transpose conjugate. The popular * denotes transpose conjugate. The popular linear receivers include matched filter, zero-forcing receiver and linear minimum mean-square-error (MMSE) receiver.

The performance of linear receivers depends on the signal-to-interference-plus-noise ratio (SINR) of the receiver output. Using MMSE receiver as an example, the SINR of the output is given by

$$SINR_1 = \mathbf{h}_1^*(\mathbf{H}_1\mathbf{H}_1^* + \sigma^2\mathbf{I})^{-1}\mathbf{h}_1,$$

where $\mathbf{H}_1$ is the matrix $\mathbf{H}$ with the first column being deleted.

Here, we are interested in a large scale system with both K (numbers of users) and N (spreading factor) going to infinity while their ratio being fixed ($\lim K/N = \alpha$). If the spreading codes can be modeled as i.i.d. random variables, the channel matrix $\mathbf{H}$ is then a large dimensional random matrix (LDRM). The RMT can then be used for analyzing the limiting SINR performance. In particular, if each of the elements in $\mathbf{H}$ is with mean zero and variance $1/N$, using the results from convergence of spectral distribution of large dimensional sample covariance matrix, we get

$$SINR_1 \to \beta^* = \int \frac{1}{x + \sigma^2} dF_\alpha(x)$$

almost surely, where $F_\alpha(x)$ is the Marćenko-Pastur law with index α. Interestingly, while the MMSE weights $\mathbf{c}_1$ depends on the random signature or channel responses, the SINR of the MMSE output tends to be a deterministic value for the limiting case.

1.3 Applications to Finance Statistics

Stephen A. Ross wrote the entry "Finance" for "The New Palgrave Dictionary of Money and Finance". He defined that "Finance is a subfield of economics distinguished by both its focus and its methodology. The primary focus of finance is the working of the capital assets. The methodology of finance is the use of close substitutes to price financial contracts and instruments." Ross described four important subjects in finance, including "efficient market", "return and risk", "option pricing theory", "corporate finance".

The four are just the main research areas of many Nobel Prize Laureates in Economics. In 1990 Nobel Prize was awarded to three scholars, who are Harry M. Markowitz for portfolio theory, William F. Sharp for Capital Asset Pricing Model (CAPM), and Metron M. Miller for MM proposition. And Black, Myron S. Scholes and Robert C. Merton won the Nobel Prize in Economics 1997 for their mathematical option pricing formula.

According to Markowitz's modern portfolio theory (1952), the portfolio with maximum expected return is not necessarily the one with minimum variance. For each investor there is a rate can gain expected return by taking on variance, or reduce variance by giving up expected return. His research analyzed mean-variances of portfolio to ensure the best portfolio selection for investors. But in the real world, the strategy cannot be applied easily because of so many securities.

In 1964, on the basis of Markowitz's work, William Sharpe brought up Capital Asset Pricing Model (CAPM). Mainly used to solve the relationship between market risk and expected return, CAPM is the backbone of modern price theory in financial market. CAPM simplified Markowitz's model, and its core concept is that the expected returns of portfolios and securities are related to risk premium (Sharpe, 1964). All securities are assumed to have the same market risk premiums, and expected returns are decided by risk factor beta.

CAPM is based on the following hypotheses: that all investors maximize their single period utilities; all investors are risk averse and rational; that information is free and immediately available; that investors may borrow at the same risk free rates; that returns are distributed normally; that there is no arbitrage opportunity; that there are perfectly efficient capital markets, and so on.

However, strict hypotheses place restrictions on its practical use. Therefore people tried to find another model as an alternative. In 1976 Ross published "The Arbitrage Theory of Capital Asset Pricing" in Journal of Economic Theory. He mentioned that market factor was not the only factor that effected returns of portfolio, and returns were jointly determined by several factors, such as property and human capital. This is the famous theory "Arbitrage Pricing Theory (APT)".

APT is based on some hypotheses including that (1) all investors are risk averse and rational; (2) there is perfect competition in the market; (3) market equilibrium must be consistent with no asymptotic arbitrage opportunity etc. APT is a multi factor model that assumes that assets returns are influenced by several independent systematic variables through a linear model. If returns of two stocks are affected by same factors, there is a correlation between their returns. According to APT, arbitrage opportunity will disappear very quickly, because investors will adjust their portfolios and take arbitrage opportunity until the price difference disappears.

Actually, APT is a generalized case of CAPM, while CAPM is the single factor form. APT provides an alternative method to comprehend the equilibrium between risk and return.

In financial derivatives markets, Black and Scholes put forward Black-Scholes (B-S) option pricing formula in 1973. From then on, derivatives had their own pricing model. B-S formula mainly studied the normal distributed option underlying price, while another important paper in 1976 written by Ross and John Cox discussed the option pricing model that introduced several jump and diffusion processes, and put forward risk-neutral concept. It was "The Valuation of Options for Alternative Stochastic Processes" in Journal of Financial Economics. Three years later Ross collaborated with Cox and Mark Rubinstein on "Option Pricing: A Simplified Approach", also known as Cox-Ross-Rubinstein Binomial options pricing model. The binomial option pricing model proceeds from the assumption that at each point in time specified the underlying asset price can only either increase or decrease. Using a binomial tree one can project all possible values of underlying at the option's expiration date, and from them, all possible final values for the option. The model is somewhat simple mathematically when compared to counterparts such as the Black-Scholes model, and is therefore widely used by all the Stock Exchanges.

Ross also brought his mathematical inspiration and financial intuition to describe interest rate term structure, which can be used to determine the time value of capital. In 1981 Ross, Cox and Ingersoll combined CAPM and stochastic process to study interest rate dynamics, and created Cox-Ingersoll-Ross (CIR) single factor model. Then it was generalized to two-factor model. That is to say, short-term and long-term random processes comprise the interest rate dynamics. CIR model had been validated for U.S coupon data.

Of course not all the persons greatly contributed to economics, like Ross, can be awarded Nobel Prize. They are definitely important in finance developmental history. For instance, Eugene Fama "Efficient Market Hypothesis (EMH)" is the hypotheses basis of those theories above. EMH indicates that, if the information won't be tortuous and can be fully reflected by securities price, the market can be considered efficient. The most important deduction of EMH is that any challenge to the market is unavailing, since all the probable information, public or private, is fully reflected by stocks price. So, there is no chance to get excess return. And for other examples, Michael Jensen's "agency costs theory", Clive Granger's study of causality and cointegration of economic factors, John Campbell's econometrics, and Darrel Duffie's dynamic asset pricing methods, are all influence the present financial markets and academia.

From the history of finance theories, we can see that the discussion about "value" exists all along, valuation of assets, financial derivatives, and also time. This discussion will continue and be further as long as the market is not disappearing.

When we consider applications of Large Dimensional Random matrix method to finance we have to discuss two main stream. One is the portfolio selection of Markowitz and related risk management issue relied on the risk measurement-covariance matrix. Another direction is the factor model rooted in CAPM and APT. We will discuss them for more details in Chapter 8 by examples in some important Financial Topic e.g. inflation forecasting and Financial crisis warning.

2

Limiting Spectral Distributions

2.1 Semicircular Law

A Wigner matrix is a symmetric (or Hermitian in the complex case) random matrix. Wigner matrices play an important role in nuclear physics. The reader is referred to Mehta (1991, 2004) for applications of Wigner matrices to that area. Here we mention that it has also a strong statistical meaning. Consider the limit of normalized Wishart matrix. Suppose that $\mathbf{x}_1, \cdots, \mathbf{x}_n$ are iid samples drawn from a p-dimensional multivariate normal population $N(\boldsymbol{\mu}, \mathbf{I}_p)$. Then, the sample covariance matrix is defined as

$$\mathbf{S}_n = \frac{1}{n-1}\sum_{i=1}^{n}(\mathbf{x}_i - \overline{\mathbf{x}})(\mathbf{x}_i - \overline{\mathbf{x}})',$$

where $\overline{\mathbf{x}} = \frac{1}{n}\sum_{i=1}^{n}\mathbf{x}_i$. When n tends to infinity, $\mathbf{S}_n \to \mathbf{I}_p$ and $\sqrt{n}\,(\mathbf{S}_n - \mathbf{I}_p) \to \sqrt{p}\mathbf{W}_p$. It can be seen that the entries above the main diagonal of $\sqrt{p}\mathbf{W}_p$ are iid $N(0,1)$ and the entries on the diagonal are iid $N(0,2)$. This matrix is called the (standard) Gaussian matrix or Wigner matrix.

A generalized definition of Wigner matrix only requires the matrix to be a Hermitian random matrix whose entries on or above the diagonal are independent. The study of spectral analysis of large dimensional Wigner matrix dates back to E. Wigner's (1958) famous **Semicircular Law**. He proved that the expected ESD of an $n \times n$ standard Gaussian matrix, normalized by $1/\sqrt{n}$, tends to the semicircular law F whose density is given by

$$F'(x) = \begin{cases} \frac{1}{2\pi}\sqrt{4-x^2} & \text{if } |x| \le 2, \\ 0 & \text{otherwise.} \end{cases} \tag{2.1.1}$$

This work was extended in various aspects. Grenander (1963) proved that $\|F^{\mathbf{W}_n} - F\| \to 0$ in probability. Further, this result was improved as in the sense of "almost sure" by Arnold (1967, 1971). Later on, this result was further generalized.

2.1.1 The iid Case

We first introduce the semicircle law for the iid case, that is, we shall prove the following theorem. For brevity of notation, we shall use $\mathbf{X}_n$ for an $n \times n$ Wigner matrix and save the notation $\mathbf{W}_n$ for the normalized Wigner matrix, *i.e.*, $\frac{1}{\sqrt{n}}\mathbf{X}_n$.

Theorem 2.1. *Suppose that $\mathbf{X}_n$ is an $n \times n$ Hermitian matrix whose diagonal entries are iid. real random variables and those above the diagonal are iid. complex random variables with variance $\sigma^2 = 1$. Then, with probability one, the ESD of $\mathbf{W}_n = \frac{1}{\sqrt{n}}\mathbf{X}_n$ tends to the semicircular law.*

The proof of this theorem may consist of the following steps:

Step 1. Removing the Diagonal Elements

Let $\widetilde{\mathbf{W}}_n$ be the matrix obtained from $\mathbf{W}_n$ by replacing the diagonal elements with zero. Let $N_n = \#\{|x_{ii}| \geq \sqrt[4]{n}\}$. Replace the diagonal elements of $\mathbf{W}_n$ by $\frac{1}{\sqrt{n}}x_{ii}I(|x_{ii}| < \sqrt[4]{n})$ and denote the resulting matrix by $\widehat{\mathbf{W}}_n$. Then, by the difference inequality in Bai (1999), we have

$$L^3(F^{\widehat{\mathbf{W}}_n}, F^{\widetilde{\mathbf{W}}_n}) \leq \frac{1}{n}\mathrm{tr}[(\widetilde{\mathbf{W}}_n - \widehat{\mathbf{W}}_n)^2] \leq \frac{1}{n^2}\sum_{i=1}^{n}|x_{ii}|^2 I(|x_{ii}| < \sqrt[4]{n}) \leq \frac{1}{\sqrt{n}},$$

where $L(F, G)$ denotes the Levy distance between the distribution functions F and G.

On the other hand, by the rank inequality in Bai (1999), we have

$$\left\|F^{\mathbf{W}_n} - F^{\widehat{\mathbf{W}}_n}\right\| \leq \frac{N_n}{n},$$

where $\|F\| = \sup_x |F(x)|$. Then, the proof of our assertion follows since $N_n/n \to 0$, almost surely, which can be easily proved by Bernstein inequality[1].

Step 2. Truncation

For any fixed positive constant C, truncate the variables at C and write $x_{ij(C)} = x_{ij}I(|x_{ij}| \leq C)$. Define a truncated Wigner matrix $\mathbf{W}_{n(C)}$ whose diagonal elements are zero and the off-diagonal elements are $\frac{1}{\sqrt{n}}x_{ij(C)}$. Then, by the truncation lemma and the law of large numbers, we have

$$\begin{aligned} L^3(F^{\mathbf{W}_n}, F^{\mathbf{W}_{n(C)}}) &\leq \frac{2}{n^2}\left(\sum_{1\leq i<j\leq n}|x_{ij}|^2 I(|x_{ij}| > C)\right) \\ &\to \mathrm{E}(|x_{12}|^2 I(|x_{12}| > C)). \end{aligned}$$

[1] Bernstein inequality states that if $X_1, \cdots, X_n$ are independent random variables with mean zero and uniformly bounded by b, then for any $\varepsilon > 0$, $P(|S_n| \geq \varepsilon) \leq 2\exp(-\varepsilon^2/[2(B_n^2 + b\varepsilon)])$, where $S_n = X_1 + \cdots + X_n$ and $B_n^2 = \mathrm{E}S_n^2$.

Note that the right-hand side of the above inequality can be made arbitrarily small by making C large. Therefore, in the proof of Theorem 2.1, we can assume that the entries of the matrix $\mathbf{X}_n$ are uniformly bounded.

Step 3. Centralization

Applying the rank inequality again, we have

$$\left\| F^{\mathbf{W}_{n(C)}} - F^{\mathbf{W}_{n(C)}-a\mathbf{1}\mathbf{1}'} \right\| \leq \frac{1}{n} \tag{2.1.2}$$

where, $a = \frac{1}{\sqrt{n}}\Re(\mathrm{E}(x_{12(C)}))$. We furthermore apply the difference lemma and obtain

$$L^3(F^{\mathbf{W}_{n(C)}-\Re(\mathrm{E}(\mathbf{W}_{n(C)}))}, F^{\mathbf{W}_{n(C)}-a\mathbf{1}\mathbf{1}'}) \leq \frac{|\Re(\mathrm{E}(x_{12(C)}))|^2}{n} \to 0. \tag{2.1.3}$$

This shows that we can assume that the real parts of the mean values of the off-diagonal elements are 0[2]. In the sequel, we proceed to remove the imaginary part of the mean values of the off-diagonal elements.

Before we treat the imaginary part, we introduce a lemma about eigenvalues of a skew-symmetric matrix.

Lemma 2.2. *Let $\mathbf{A}_n$ be an $n \times n$ skew-symmetric matrix whose elements above the diagonal are* 1 *and those below the diagonal are* -1. *Then, the eigenvalues of $\mathbf{A}_n$ are $\lambda_k = i\cot(\pi(2k-1)/2n)$, $k = 1, 2, \cdots, n$. The eigenvector associated with λ_k is $\mathbf{u}_k = \frac{1}{\sqrt{n}}(1, \rho_k, \cdots, \rho_k^{n-1})'$, where $\rho_k = (\lambda_k - 1)/(\lambda_k + 1) = \exp(-i\pi(2k-1)/n)$.*

The proof of this lemma can be found in Bai (1999).

Write $b = \mathrm{E}\Im(x_{12(C)})$. Then, $\mathrm{E}\Im(\mathbf{W}_{n(C)}) = ib\mathbf{A}_n$. By Lemma 2.2, the eigenvalues of the matrix $i\Im(\mathrm{E}(\mathbf{W}_{n(C)})) = ib\mathbf{A}_n$ are $ib\lambda_k = -n^{-1/2}b\cot(\pi(2k-1)/2n)$, $k = 1, \cdots, n$. If the spectral decomposition of $\mathbf{A}_n$ is $\mathbf{U}_n\mathbf{D}_n\mathbf{U}_n^*$, then we rewrite $i\Im(\mathrm{E}(\mathbf{W}_{n(C)})) = \mathbf{B}_1 + \mathbf{B}_2$ where $\mathbf{B}_j = -\frac{1}{\sqrt{n}}b\mathbf{U}_n\mathbf{D}_{nj}\mathbf{U}_n^*$, $j = 1, 2$, where $\mathbf{U}_n$ is a unitary matrix, $\mathbf{D}_n = \mathrm{diag}[\lambda_1, \cdots, \lambda_n]$ and

$$\mathbf{D}_{n1} = \mathbf{D}_n - \mathbf{D}_{n2} = \mathrm{diag}[0, \cdots, 0, \lambda_{[n^{3/4}]}, \lambda_{[n^{3/4}]+1}, \cdots, \lambda_{n-[n^{3/4}]}, 0, \cdots, 0].$$

For any $n \times n$ Hermitian matrix $\mathbf{C}$, by the difference inequality, we have

$$\begin{aligned} L^3(F^{\mathbf{C}}, F^{\mathbf{C}-\mathbf{B}_1}) &\leq \frac{1}{n^2} \sum_{n^{3/4} \leq k \leq n - n^{3/4}} \cot^2(\pi(2k-1)/2n) \\ &< \frac{2}{n\sin^2(n^{-1/4}\pi)} \to 0 \end{aligned} \tag{2.1.4}$$

and by the rank inequality

[2] Note that the diagonal elements of $\mathbf{W}_{nC} - a\mathbf{1}\mathbf{1}'$ may not be 0, but this doesn't matter. We can apply Step 1 again to remove it.

$$\|F^{\mathbf{C}} - F^{\mathbf{C}-\mathbf{B}_2}\| \leq \frac{2n^{3/4}}{n} \to 0. \tag{2.1.5}$$

Summing up estimations (2.1.2) — (2.1.5), we established the following centralization inequality

$$L(F^{\mathbf{W}_{n(C)}}, F^{\mathbf{W}_{n(C)}-\mathrm{E}(\mathbf{W}_{n(C)})}) = o(1). \tag{2.1.6}$$

Step 4. Rescaling

Write $\sigma^2(C) = \mathrm{Var}(x_{11(C)})$ and define $\widetilde{\mathbf{W}}_n = \sigma^{-1}(C)(\mathbf{W}_{n(C)} - \mathrm{E}(\mathbf{W}_{n(C)})$. Note that the off-diagonal entries of $\sqrt{n}\widetilde{\mathbf{W}}_n$ are $\widehat{x}_{kj} = (x_{kj(C)} - \mathrm{E}(x_{kj(C)}))/\sigma(C)$.

Applying the difference inequality, we obtain

$$\begin{aligned} L^3(F^{\widetilde{\mathbf{W}}_n}, F^{\mathbf{W}_{n(C)}-\mathrm{E}(\mathbf{W}_{n(C)})}) &\leq \frac{2(\sigma(C)-1)^2}{n^2\sigma^2(C)} \sum_{1\leq i<j\leq n} |x_{kj(C)} - \mathrm{E}(x_{kj(C)})|^2 \\ &\to (\sigma(C)-1)^2, \quad \text{a.s..} \end{aligned} \tag{2.1.7}$$

Note that $(\sigma(C)-1)^2$ can be made arbitrarily small if C is chosen large. Thus, we may assume that the variance of the off diagonal entries of $\mathbf{X}$ is 1.

Step 5. Proof of the Semicircle Law

Summing steps 1 – 4, we may assume the entries of $\mathbf{X}$ are uniformly bounded and mean 0. Then we can prove Theorem 2.1 by either the moment method or Stieltjes transform.

When applying the moment method, we first examine that the semicircle distribution satisfies Riesz condition or Carelman condition. Then show that the moments of the spectral distribution converge to the corresponding moments of the semicircular distribution almost surely. The k-th moment of the ESD of $\mathbf{W}_n$ is

$$\begin{aligned} \beta_k(\mathbf{W}_n) = \beta_k(F^{\mathbf{W}_n}) &= \int x^k \, dF^{\mathbf{W}_n}(x) \\ &= \frac{1}{n}\sum_{i=1}^{n} \lambda_i^k = \frac{1}{n}\mathrm{tr}(\mathbf{W}_n^k) = \frac{1}{n^{1+\frac{k}{2}}}\mathrm{tr}(\mathbf{X}_n^k) \\ &= \frac{1}{n^{1+\frac{k}{2}}} \sum_{\mathbf{i}} X(\mathbf{i}), \end{aligned} \tag{2.1.8}$$

where λ_i's are the eigenvalues of the matrix $\mathbf{W}_n$, $X(\mathbf{i}) = x_{i_1i_2}x_{i_2i_3}\cdots x_{i_ki_1}$, $\mathbf{i} = (i_1, \cdots, i_k)$ and the summation $\sum_{\mathbf{i}}$ runs over all possibilities that $\mathbf{i} \in \{1, \cdots, n\}^k$.

By applying the moment convergence theorem, the proof of the semicircular law for the iid case can be complete by showing

(1) $E[\beta_k(\mathbf{W}_n)]$ converges to the k-th moment β_k of the semicircle distribution, which are equal to $\beta_{2m-1} = 0$ and $\beta_{2m} = (2m)!/m!(m+1)!$.

(2) For each fixed k, $\sum_n \mathrm{Var}[\beta_k(\mathbf{W}_n)] < \infty$.

Step 5 can also be proved by Stieltjes transformation method. We will sketch it as follows:

Let $z = u + iv$ with $v > 0$ and $s(z)$ be the Stieltjes transform of the semicircular law. Then, we have

$$s(z) = \frac{1}{2\pi\sigma^2} \int_{-2\sigma}^{2\sigma} \frac{1}{x-z} \sqrt{4\sigma^2 - x^2} dx.$$

We shall frequently meet the square root of a complex number. For definiteness, throughout the book, the square-root of a complex number is specified as the one with positive imaginary part. By this convention, we have

$$\sqrt{z} = \text{sign}(\Im z) \frac{|z| + z}{\sqrt{2(|z| + \Re z)}} \tag{2.1.9}$$

or

$$\Re(\sqrt{z}) = \frac{1}{\sqrt{2}} \text{sign}(\Im z) \sqrt{|z| + \Re z} = \frac{\Im z}{\sqrt{2(|z| - \Re z)}}$$

and

$$\Im(\sqrt{z}) = \frac{1}{\sqrt{2}} \sqrt{|z| - \Re z} = \frac{|\Im z|}{\sqrt{2(|z| + \Re z)}}.$$

Under this convention, one can easily show that

$$s(z) = -\frac{1}{2\sigma^2}(z - \sqrt{z^2 - 4\sigma^2}).$$

By definition, the Stieltjes transform of $F^{\mathbf{W}_n}$ is given by

$$s_n(z) = \frac{1}{n} \text{tr}(\mathbf{W}_n - z\mathbf{I}_n)^{-1}. \tag{2.1.10}$$

We shall then proceed in our proof by taking the following 3 steps:

(i) For any fixed $z \in \mathbb{C}^+ = \{z, \Im(z) > 0\}$, $s_n(z) - \mathrm{E}s_n(z) \to 0$, a.s.
(ii) For any fixed $z \in \mathbb{C}^+$, $\mathrm{E}s_n(z) \to s(z)$, the Stieltjes transform of the semicircular law.
(iii) Outside a null set, $s_n(z) \to s(z)$, for every $z \in \mathbb{C}^+$.

Step 1. Almost sure convergence of the random part

For the first step, we show that for each fixed $z \in \mathbb{C}^+$,

$$s_n(z) - \mathrm{E}(s_n(z)) \to 0, \quad \text{a.s.} \tag{2.1.11}$$

We need the extended Burkholder inequalities.

Lemma 2.3. *Let $\{X_k\}$ be a complex martingale difference sequence with respect to the increasing σ-field $\{\mathcal{F}_k\}$. Then for $p > 1$,*

$$\mathrm{E}\left|\sum X_k\right|^p \le K_p \mathrm{E}\left(\sum |X_k|^2\right)^{p/2}.$$

Lemma 2.4. *Let $\{X_k\}$ be a complex martingale difference sequence with respect to the increasing σ-field $\mathcal{F}_k$, and let* E_k *denote conditional expectation w.r.t.* $\mathcal{F}_k$. *Then for $p \geq 2$,*

$$\mathrm{E}\left|\sum d_k\right|^p \leq K_p \left(\mathrm{E}\left(\sum \mathrm{E}_{k-1}|X_k|^2\right)^{p/2} + \mathrm{E}\sum |X_k|^p\right).$$

To proceed the proof of the almost sure convergence (2.1.11), we need a technique of martingale decomposition as follows: Denote by $\mathrm{E}_k(\cdot)$ conditional expectation with respect to the σ-field generated by the random variables $\{x_{ij}, i, j > k\}$, with the convention that $\mathrm{E}_n s_n(z) = \mathrm{E}s_n(z)$ and $\mathrm{E}_0 s_n(z) = s_n(z)$. Then, we have

$$s_n(z) - \mathrm{E}(s_n(z)) = \sum_{k=1}^{n}[\mathrm{E}_{k-1}(s_n(z)) - \mathrm{E}_k(s_n(z))] := \sum_{k=1}^{n} \gamma_k, \quad (2.1.12)$$

where, by the inverse formula of blocked matrix,

$$\begin{aligned}
\gamma_k &= \frac{1}{n}\left(\mathrm{E}_{k-1}\mathrm{tr}(\mathbf{W}_n - z\mathbf{I})^{-1} - \mathrm{E}_k\mathrm{tr}(\mathbf{W}_n - z\mathbf{I})^{-1}\right) \\
&= \frac{1}{n}\left(\mathrm{E}_{k-1}[\mathrm{tr}(\mathbf{W}_n - z\mathbf{I})^{-1} - \mathrm{tr}(\mathbf{W}_k - z\mathbf{I}_{n-1})^{-1}]\right. \\
&\quad \left. -\mathrm{E}_k[\mathrm{tr}(\mathbf{W}_n - z\mathbf{I})^{-1} - \mathrm{tr}(\mathbf{W}_k - z\mathbf{I}_{n-1})^{-1}]\right) \\
&= \frac{1}{n}\left(\mathrm{E}_{k-1}\frac{1 + \boldsymbol{\alpha}_k^*(\mathbf{W}_k - z\mathbf{I}_{n-1})^{-2}\boldsymbol{\alpha}_k}{-z - \boldsymbol{\alpha}_k^*(\mathbf{W}_k - z\mathbf{I}_{n-1})^{-1}\boldsymbol{\alpha}_k}\right. \\
&\quad \left. -\mathrm{E}_k\frac{1 + \boldsymbol{\alpha}_k^*(\mathbf{W}_k - z\mathbf{I}_{n-1})^{-2}\boldsymbol{\alpha}_k}{-z - \boldsymbol{\alpha}_k^*(\mathbf{W}_k - z\mathbf{I}_{n-1})^{-1}\boldsymbol{\alpha}_k}\right),
\end{aligned}$$

where $\mathbf{W}_k$ is the matrix obtained from $\mathbf{W}_n$ with the k-th row and column removed, $\boldsymbol{\alpha}_k$ is the k-th column of $\mathbf{W}_n$ with the k-th element removed.

Note that

$$\begin{aligned}
&|1 + \boldsymbol{\alpha}_k^*(\mathbf{W}_k - z\mathbf{I}_{n-1})^{-2}\boldsymbol{\alpha}_k| \\
&\leq 1 + \boldsymbol{\alpha}_k^*(\mathbf{W}_k - z\mathbf{I}_{n-1})^{-1}(\mathbf{W}_k - \bar{z}\mathbf{I}_{n-1})^{-1}\boldsymbol{\alpha}_k \\
&= v^{-1}\Im(z + \boldsymbol{\alpha}_k^*(\mathbf{W}_k - z\mathbf{I}_{n-1})^{-1}\boldsymbol{\alpha}_k)
\end{aligned}$$

which implies that

$$|\gamma_k| \leq 2/nv.$$

Noting that $\{\gamma_k\}$ forms a martingale difference sequence, applying Lemma 2.3 for $p = 4$, we have

$$\begin{aligned}
\mathrm{E}|s_n(z) - \mathrm{E}(s_n(z))|^4 &\leq K_4\mathrm{E}\left(\sum_{k=1}^{n}|\gamma_k|^2\right)^2 \\
&\leq K_4\left(\sum_{k=1}^{n}\frac{2}{n^2v^2}\right)^2 \leq \frac{4K_4}{n^2v^4}.
\end{aligned}$$

By Borel-Cantelli Lemma, we know that for each fixed $z \in \mathbb{C}^+$,

$$s_n(z) - \mathrm{E}(s_n(z)) \to 0, \text{ a.s.}$$

Step 2. Convergence of expected Stieltjes transform

Employ the formula inverse matrix, have

$$s_n(z) = \frac{1}{n}\mathrm{tr}(\mathbf{W}_n - z\mathbf{I}_n)^{-1}$$
$$= \frac{1}{n}\sum_{k=1}^{n} \frac{1}{-z - \boldsymbol{\alpha}_k^*(\mathbf{W}_k - z\mathbf{I}_{n-1})^{-1}\boldsymbol{\alpha}_k}. \tag{2.1.13}$$

Write $\varepsilon_k = \mathrm{E}s_n(z) - \boldsymbol{\alpha}_k^*(\mathbf{W}_k - z\mathbf{I}_{n-1})^{-1}\boldsymbol{\alpha}_k$. Then we have

$$\mathrm{E}s_n(z) = \frac{1}{n}\sum_{k=1}^{n} \frac{1}{-z - \mathrm{E}s_n(z) + \varepsilon_k}$$
$$= -\frac{1}{z + \mathrm{E}s_n(z)} + \frac{\delta_n}{z + \mathrm{E}s_n(z)}, \tag{2.1.14}$$

where

$$\delta_n = \frac{1}{n}\sum_{k=1}^{n} \mathrm{E}\left(\frac{\varepsilon_k}{-z - \mathrm{E}s_n(z) + \varepsilon_k}\right).$$

Solving equation (2.1.14), we obtain two solutions:

$$\frac{1}{2}(-z \pm \sqrt{z^2 - 4 + 4\delta_n}).$$

We show that only

$$\mathrm{E}s_n(z) = \frac{1}{2}(-z + \sqrt{z^2 - 4 + 4\delta_n}) \tag{2.1.15}$$

is the solution of $\mathrm{E}s_n(z)$ since the imaginary part of $\mathrm{E}s_n(z)$ is positive for all $z \in \mathbb{C}^+$.

From (2.1.15), the proof of this step follows by showing

$$\delta_n \to 0. \tag{2.1.16}$$

The details are omitted.

Step 3. Completion of the proof of Theorem 2.6

In this step, we need the Vitali's convergence theorem.

Lemma 2.5. *Let $f_1, f_2, \cdots$ be analytic in D, a connected open set of $\mathbb{C}$, satisfying $|f_n(z)| \le M$ for every n and z in D, and $f_n(z)$ converges, as $n \to \infty$ for each z in a subset of D having a limit point in D. Then there exists a function f, analytic in D for which $f_n(z) \to f(z)$ and $f_n'(z) \to f'(z)$ for all $z \in D$. Moreover, on any set bounded by a contour interior to D the convergence is uniform and $\{f_n'(z)\}$ is uniformly bounded.*

The conclusions on $\{f_n\}$ are from Vitali's convergence theorem (see Titchmarsh (1939), p. 168). Those on $\{f'_n\}$ can be found in Bai and Silverstein (2004).

By Steps 1 and 2, for any fixed $z \in \mathbb{C}^+$, we have

$$s_n(z) \to s(z), \quad \text{a.s.}$$

where $s(z)$ is the Stieltjes transform of the standard semicircular law. That is, for each $z \in \mathbb{C}^+$, there exists a null set N_z (i.e., $P(N_z) = 0$) such that

$$s_n(z,\omega) \to s(z), \text{ for all } \omega \in N_z^c.$$

Now, let $\mathbb{C}_0^+ = \{z_m\}$ be a countable dense subset of $\mathbb{C}^+$ (e.g., all z of rational real and imaginary parts) and let $N = \cup N_{z_m}$. Then

$$s_n(z,\omega) \to s(z), \text{ for all } \omega \in N^c \text{ and } z \in \mathbb{C}_0^+.$$

Let $\mathbb{C}_m^+ = \{z \in \mathbb{C}^+, \Im z > 1/m, \ |z| \le m\}$. When $z \in \mathbb{C}_m^+$, we have $|s_n(z)| \le m$. Applying Lemma 2.5, we have

$$s_n(z,\omega) \to s(z), \text{ for all } \omega \in N^c \text{ and } z \in \mathbb{C}_m^+.$$

Since the above convergence holds for every m, we conclude that

$$s_n(z,\omega) \to s(z), \text{ for all } \omega \in N^c \text{ and } z \in \mathbb{C}^+.$$

Applying the property of Stieltjes transform, we conclude that

$$F^{\mathbf{W}_n} \xrightarrow{w} F, \text{ a.s.}$$

2.1.2 Independent but not Identically Distributed

Sometimes, it is of practical interest to consider the case where for each n, the entries above or on the diagonal of $\mathbf{W}_n$ are independent complex random variables with mean zero and variance σ^2 (for simplicity we assume $\sigma = 1$ in the sequel), but may depend on n. For this case, we present the following theorem.

Theorem 2.6. *Suppose that $\mathbf{W}_n = \frac{1}{\sqrt{n}}\mathbf{X}_n$ is a Wigner matrix, the entries above or on the diagonal of $\mathbf{X}_n$ are independent but may be dependent on n and may not be necessarily identically distributed. Assume that all the entries of $\mathbf{X}_n$ are of mean zero and variance 1 and satisfy the following condition. For any constant $\eta > 0$,*

$$\lim_{n\to\infty} \frac{1}{n^2} \sum_{jk} \mathrm{E}|x_{jk}^{(n)}|^2 I(|x_{jk}^{(n)}| \ge \eta\sqrt{n}) = 0, \tag{2.1.17}$$

Then, the ESD of $\mathbf{W}_n$ converges to the semicircular law almost surely.

The proof consists of similar steps as that of Theorem 2.1.

Step 1. Truncation

Note that condition (2.1.17) is equivalent to: for any $\eta > 0$,

$$\lim_{n\to\infty} \frac{1}{\eta^2 n^2} \sum_{jk} \mathrm{E}|x_{jk}^{(n)}|^2 I(|x_{jk}^{(n)}| \geq \eta\sqrt{n}) = 0. \tag{2.1.18}$$

Thus, one can select a sequence $\eta_n \downarrow 0$ such that (2.1.18) remains true when η is replaced by η_n. Define $\widetilde{\mathbf{W}}_n = \frac{1}{\sqrt{n}} n(x_{ij}^{(n)} I(|x_{ij}^{(n)}| \leq \eta_n\sqrt{n})$. By using the rank inequality, one has

$$\begin{aligned} \|F^{\mathbf{W}_n} - F^{\widetilde{\mathbf{W}}_n}\| &\leq \frac{1}{n}\mathrm{rank}(\mathbf{W}_n - \mathbf{W}_{n(\eta_n\sqrt{n})}) \\ &\leq \frac{2}{n} \sum_{1\leq i\leq j\leq n} I(|x_{ij}^{(n)}| \geq \eta_n\sqrt{n}). \end{aligned} \tag{2.1.19}$$

By condition (2.1.18), we have

$$\begin{aligned} &\mathrm{E}\left(\frac{1}{n} \sum_{1\leq i\leq j\leq n} I(|x_{ij}^{(n)}| \geq \eta_n\sqrt{n})\right) \\ &\leq \frac{2}{\eta_n^2 n^2} \sum_{jk} \mathrm{E}|x_{ij}^{(n)}|^2 I(|x_{ij}^{(n)}| \geq \eta_n\sqrt{n}) = o(1), \end{aligned}$$

and

$$\begin{aligned} &\mathrm{Var}\left(\frac{1}{n} \sum_{1\leq i\leq j\leq n} I(|x_{ij}^{(n)}| \geq \eta_n\sqrt{n})\right) \\ &\leq \frac{4}{\eta_n^2 n^3} \sum_{jk} \mathrm{E}|x_{ij}^{(n)}|^2 I(|x_{ij}^{(n)}| \geq \eta_n\sqrt{n}) = o(1/n). \end{aligned}$$

Then, applying Bernstein inequality, for all small $\varepsilon > 0$ and large n, we have

$$\mathrm{P}\left(\frac{1}{n} \sum_{1\leq i\leq j\leq n} I(|x_{ij}^{(n)}| \geq \eta_n\sqrt{n}) \geq \varepsilon\right) \leq 2e^{-\varepsilon n}, \tag{2.1.20}$$

which is summable. Thus, by (2.1.19) and (2.1.20), to prove that with probability one, $F^{\mathbf{W}_n}$ converges to the semicircular law, it suffices to show that with probability one, $F^{\widetilde{\mathbf{W}}_n}$ converges to the semicircular law.

Step 2. Removing diagonal elements

Let $\widehat{\mathbf{W}}_n$ be the matrix $\widetilde{\mathbf{W}}_n$ with diagonal elements replaced by 0. Then, by difference inequality, we have

$$L^3\left(F^{\widetilde{\mathbf{W}}_n}, F^{\widehat{\mathbf{W}}_n}\right) \le \frac{1}{n^2}\sum_{k=1}^{n}|x_{kk}^{(n)}|^2 I(|x_{kk}^{(n)}| \le \eta_n\sqrt{n}) \le \eta_n^2 \to 0.$$

Step 3. Centralization

By difference inequality, it follows that

$$\begin{aligned}
&L^3\left(F^{\widehat{\mathbf{W}}_n}, F^{\widehat{\mathbf{W}}_n - \mathrm{E}\widehat{\mathbf{W}}_n}\right) \\
&\le \frac{1}{n^2}\sum_{i\neq j}|\mathrm{E}(x_{ij}^{(n)} I(|x_{ij}^{(n)}| \le \eta_n\sqrt{n}))|^2 \\
&\le \frac{1}{n^3\eta_n^2}\sum_{ij}\mathrm{E}|x_{jk}^{(n)}|^2 I(|x_{jk}^{(n)}| \ge \eta_n\sqrt{n}) \to 0.
\end{aligned} \tag{2.1.21}$$

Step 4. Rescaling

Write $\widetilde{\mathbf{W}}_n = \frac{1}{\sqrt{n}}\widetilde{\mathbf{X}}_n = \frac{1}{\sqrt{n}}(\tilde{x}_{ij})$ where, for $i > j$,

$$\tilde{x}_{ij} = \begin{cases} \frac{x_{ij}^{(n)} I(|x_{ij}^{(n)}| \le \eta_n\sqrt{n}) - \mathrm{E}(x_{ij}^{(n)} I(|x_{ij}^{(n)}| \le \eta_n\sqrt{n}))}{\sigma_{ij}} & \text{if } \sigma_{ij} > \frac{1}{2}, \\ y_{ij} & \text{otherwise} \end{cases}$$

where $\sigma_{ij}^2 = \mathrm{E}|x_{ij}^{(n)} I(|x_{ij}^{(n)}| \le \eta_n\sqrt{n}) - \mathrm{E}(x_{ij}^{(n)} I(|x_{ij}^{(n)}| \le \eta_n\sqrt{n}))|^2$ and y_{ij}'s are i.i.d. random variables taking values ± 1 with probability $\frac{1}{2}$ and independent of all $x_{ij}^{(n)}$'s.

By the difference inequality, it follows that

$$L^3\left(F^{\widetilde{\mathbf{W}}_n}, F^{\widehat{\mathbf{W}}_n - \mathrm{E}\widehat{\mathbf{W}}_n}\right) \le R_1 + R_2 \tag{2.1.22}$$

where

$$R_1 \le \frac{4}{n^2}\sum_{\substack{i\neq j \\ \sigma_{ij}^2 > \frac{1}{2}}}(1-\delta_{ij})^2|x_{ij}^{(n)} I(|x_{ij}^{(n)}| \le \eta_n\sqrt{n}) - \mathrm{E}(x_{ij}^{(n)} I(|x_{ij}^{(n)}| \le \eta_n\sqrt{n}))|^2$$

and

$$R_2 \le \frac{1}{n^2}\sum_{\substack{i\neq j, \\ \sigma_{ij}^2 < \frac{1}{2}}}|z_{ij}|^2.$$

Note that

$$\begin{aligned}
\mathrm{E}R_1 &\le \mathrm{E}\Big(\frac{4}{n^2}\sum_{i\neq j}(1-\delta_{ij})^2|x_{ij}^{(n)} I(|x_{ij}^{(n)}| \le \eta_n\sqrt{n}) - \mathrm{E}(x_{ij}^{(n)} I(|x_{ij}^{(n)}| \le \eta_n\sqrt{n}))|^2\Big) \\
&\le \frac{4}{n^2\eta_n^2}\sum_{ij}(1-\sigma_{ij})^2 \le \frac{4}{n^2\eta_n^2}\sum_{ij}(1-\sigma_{ij}^2) \\
&\le \frac{4}{n^2\eta_n^2}\sum_{ij}[\mathrm{E}|x_{jk}^{(n)}|^2 I(|x_{jk}^{(n)}| \ge \eta_n\sqrt{n}) + \mathrm{E}^2|x_{jk}^{(n)}|I(|x_{jk}^{(n)}| \ge \eta_n\sqrt{n})] \to 0.
\end{aligned}$$

Also, we have[3]

$$
\begin{aligned}
&\mathrm{E}|R_1 - \mathrm{E}R_1|^4 \\
&\leq \frac{C}{n^8}\left[\sum_{i\neq j}\mathrm{E}|x_{ij}^{(n)}|^8 I(|x_{ij}^{(n)}| \leq \eta_n\sqrt{n}) + \left(\sum_{i\neq j}\mathrm{E}|x_{ij}^{(n)}|^4 I(|x_{ij}^{(n)}| \leq \eta_n\sqrt{n})\right)^2\right] \\
&\leq Cn^{-2}[n^{-1}\eta_n^6 + \eta_n^4],
\end{aligned}
$$

which is summable. Similarly, noting that

$$\#\{(i,j); \sigma_{ij}^2 < \frac{1}{2}\} = o(n^2),$$

one can similarly show that

$$\mathrm{E}R_2 \to 0$$

and

$$\mathrm{E}|R_2 - \mathrm{E}R_2|^2 \leq Cn^{-2}[n^{-1}\eta_n^6 + \eta_n^4].$$

From the estimates above, we conclude that

$$L\left(F^{\widetilde{\mathbf{W}}_n}, F^{\widehat{\mathbf{W}}_n - \mathrm{E}\widehat{\mathbf{W}}_n}\right) \to 0, \text{ a.s.}$$

Step 5. Proof of Theorem 2.6 by MCT

Up to here, we have proved that we may truncate, centralize and rescale the entries of the Wigner matrix at $\eta_n\sqrt{n}$ and remove the diagonal elements without changing the LSD. These four steps are almost the same as those we did for the iid case.

Now, we assume that the variables are truncated at $\eta_n\sqrt{n}$ and then centralized and rescaled.

Again for simplicity, the truncated and centralized variables are still denoted by x_{ij}, We assume:
(i) The variables $\{x_{ij}, 1 \leq i < j \leq n\}$ are independent and $x_{ii} = 0$;
(ii) $\mathrm{E}(x_{ij}) = 0$ and $\mathrm{Var}(x_{ij}) = 1$;
and
(iii) $|x_{ij}| \leq \eta_n\sqrt{n}$.

Similar to what we did in the last section, in order to prove Theorem 2.6, we need to show that
(1) $E[\beta_k(\mathbf{W}_n)]$ converges to the k-th moment β_k of the semicircle distribution;
(2) For each fixed k, $\sum_n \mathrm{E}|\beta_k(\mathbf{W}_n) - \mathrm{E}(\beta_k(\mathbf{W}_n))|^4 < \infty$.

The details of the proof of Theorem 2.6 are omitted since they are similar to those of Theorem 2.1.

Remark 2.7. *In Girko's book (1990), it is stated that condition (2.1.18) is necessary and sufficient for the conclusion of Theorem 2.6.*

---

[3] Here we use the elementary inequality $\mathrm{E}|\sum X_i|^{2k} \leq C_k(\sum \mathrm{E}|X_i|^{2k} + (\sum \mathrm{E}|X_i|^2)^k)$ for some constant C_k if the X_i's are independent with zero means.

2.2 Marčenko-Pastur Law

The sample covariance matrix is one of the most important random matrices in multivariate statistical inference since many test statistics are defined by its eigenvalues or functionals. Its definition is as follows. Suppose that $\{x_{jk}, j, k = 1, 2, \cdots\}$ is a double array of iid complex random variables with mean zero and variance σ^2. Write $\mathbf{x}_j = (x_{1j}, \cdots, x_{pj})'$ and $\mathbf{X} = (\mathbf{x}_1, \cdots, \mathbf{x}_n)$. The sample covariance matrix is defined by

$$\mathbf{S} = \frac{1}{n-1}\sum_{k=1}^{n}(\mathbf{x}_k - \bar{\mathbf{x}})(\mathbf{x}_k - \bar{\mathbf{x}})^*,$$

where $\bar{\mathbf{x}} = \frac{1}{n}\sum \mathbf{x}_j$.

However, in most cases of spectral analysis of large dimensional random matrices, the sample covariance matrix is simply defined as

$$\mathbf{S} = \frac{1}{n}\sum_{k=1}^{n}\mathbf{x}_k\mathbf{x}_k^* = \frac{1}{n}\mathbf{X}\mathbf{X}^*, \tag{2.2.1}$$

because the $\bar{\mathbf{x}}\bar{\mathbf{x}}^*$ is a rank 1 matrix and hence the removal of $\bar{\mathbf{x}}$ does not affect the LSD due to the rank inequality.

In spectral analysis of large dimensional sample covariance matrices, it is usual to assume that the dimension p tends to infinity proportionally to the degrees of freedom n, namely, $p/n \to y \in (0, \infty)$.

The first success in finding the limiting spectral distribution of the large sample covariance matrix $\mathbf{S}_n$ (named as the Marčenko-Pastur (MP) law by some authors) is due to Marčenko and Pastur (1967). Succeeding work was done in Bai and Yin (1988a), Grenander and Silverstein (1977), Jonsson (1982), Silverstein (1995), Wachter (1978), and Yin (1986). When the entries of $\mathbf{X}$ are not independent, Yin and Krishnaiah (1985) investigated the limiting spectral distribution of $\mathbf{S}$ when the underlying distribution is isotropic. The theorem in the next section is a consequence of a result in Yin (1986) where the real case is considered.

2.2.1 MP Law for iid Case

The MP law $F_y(x)$ has a density function

$$p_y(x) = \begin{cases} \frac{1}{2\pi x y \sigma^2}\sqrt{(b-x)(x-a)} & \text{if } a \le x \le b, \\ 0 & \text{otherwise} \end{cases} \tag{2.2.1}$$

and has a point mass $1 - 1/y$ at the origin if $y > 1$, where $a = \sigma^2(1-\sqrt{y})^2$ and $b = \sigma^2(1+\sqrt{y})^2$. Here, the constant y is the dimension to sample size ratio index and σ^2 is the scale parameter. If $\sigma^2 = 1$, the MP law is said to be the standard MP law.

The moments $\beta_k = \beta_k(y, \sigma^2) = \int_a^b x^k p_y(x)dx$ of MP law can be written in a more explicit form as:

$$\beta_k = \sigma^{2k} \sum_{r=0}^{k-1} \frac{1}{r+1} \binom{k}{r} \binom{k-1}{r} y^r.$$

First, we introduce

Theorem 2.8. *Suppose that* $\{x_{ij}\}$ *are iid. real random variables with mean 0 and variance* σ^2*. Also assume that* $p/n \to y \in (0, \infty)$*. Then, with probability one,* $F^{\mathbf{S}}$ *tends to the MP law which is defined in* (2.2.1).

Yin (1986) considered existence of the LSD of the sequence of random matrices $\mathbf{S}_n \mathbf{T}_n$ where $\mathbf{T}_n$ is a positive definite random matrix and is independent of $\mathbf{S}_n$. When $\mathbf{T}_n = \mathbf{I}_p$, Yin's result reduces to Theorem 2.8.

The result of Theorem 2.8 can be easily extended to the complex random sample covariance matrix.

Theorem 2.9. *Suppose that* $\{x_{ij}\}$ *are iid. complex random variables with variance* σ^2*. Also assume that* $p/n \to y \in (0, \infty)$*. Then, with probability one,* $F^{\mathbf{S}}$ *tends to a limiting distribution same as described in Theorem* 2.8.

The proofs will be separated into several steps. Note that the MP law varies with the scale parameter σ^2. Therefore, in the proof we shall assume that $\sigma^2 = 1$, without loss of generality.

In most work in multivariate statistics, it is assumed that the means of the entries of $\mathbf{X}_n$ are zero. The centralization technique can be employed to remove the common mean of the entries of $\mathbf{X}_n$ without altering the LSD of sample covariance matrices.

Step 1. Truncation, Centralization and Rescaling

Let C be a positive number and define

$$\begin{aligned}
\hat{x}_{ij} &= x_{ij} I(|x_{ij}| \le C), \\
\tilde{x}_{ij} &= \hat{x}_{ij} - \mathrm{E}(\hat{x}_{11}), \\
\widehat{\mathbf{x}}_i &= (\hat{x}_{i1}, \cdots, \hat{x}_{ip})', \\
\widetilde{\mathbf{x}}_i &= (\tilde{x}_{i1}, \cdots, \tilde{x}_{ip})', \\
\widehat{\mathbf{S}}_n &= \frac{1}{n} \sum_{i=1}^n \widehat{\mathbf{x}}_i \widehat{\mathbf{x}}_i^* = \frac{1}{n} \widehat{\mathbf{X}} \widehat{\mathbf{X}}^*, \\
\widetilde{\mathbf{S}}_n &= \frac{1}{n} \sum_{i=1}^n \widetilde{\mathbf{x}}_i \widetilde{\mathbf{x}}_i^* = \frac{1}{n} \widetilde{\mathbf{X}} \widetilde{\mathbf{X}}^*.
\end{aligned}$$

Write the ESD's of $\widehat{\mathbf{S}}_n$ and $\widetilde{\mathbf{S}}_n$ as $F^{\widehat{\mathbf{S}}_n}$ and $F^{\widetilde{\mathbf{S}}_n}$, respectively. By difference inequality for sample covariance matrix (see Bai (1999) or Bai and Silverstein (2006, 2009)) and the strong law of large numbers, we have

$$\begin{aligned}
L^4(F^{\mathbf{S}}, F^{\widehat{\mathbf{S}}_n}) &\le \left(\frac{2}{np}\sum_{i,j}(|x_{ij}^2| + |\hat{x}_{ij}^2|)\right)\left(\frac{1}{np}\sum_{i,j}(|x_{ij} - \hat{x}_{ij}|^2)\right) \\
&\le \left(\frac{4}{np}\sum_{i,j}|x_{ij}^2|\right)\left(\frac{1}{np}\sum_{i,j}(|x_{ij}^2|I(|x_{ij}| > C))\right) \\
&\to 4\mathrm{E}(|x_{ij}^2|I(|x_{ij}| > C)), \text{ a.s..}
\end{aligned} \tag{2.2.2}$$

Note that the right hand side of (2.2.2) can be made arbitrarily small by choosing C large enough.

Also, by rank inequality for sample covariance (see Bai (1999) or Bai and Silverstein (2006)), we obtain

$$||F^{\widehat{\mathbf{S}}_n} - F^{\widetilde{\mathbf{S}}_n}|| \le \frac{1}{p}\mathrm{rank}(\mathrm{E}\widehat{\mathbf{X}}) = \frac{1}{p}. \tag{2.2.3}$$

Write $\tilde{\sigma}^2 = \mathrm{E}(|\tilde{x}_{jk}|^2) \to 1$, as $C \to \infty$. Applying difference inequality for sample covariance matrix, we obtain

$$\begin{aligned}
L^4(F^{\widetilde{\mathbf{S}}_n}, F^{\tilde{\sigma}^{-2}\widetilde{\mathbf{S}}_n}) &\le 2\left(\frac{1+\tilde{\sigma}^2}{np\tilde{\sigma}^2}\sum_{i,j}|\tilde{x}_{ij}|^2\right)\left(\frac{1-\tilde{\sigma}^2}{np\tilde{\sigma}^2}\sum_{i,j}|\tilde{x}_{ij(c)}|^2\right) \\
&\to 2(1-\tilde{\sigma}^4), \text{ a.s..}
\end{aligned} \tag{2.2.4}$$

Note that the right-hand side of the above inequality can be made arbitrarily small by choosing C large. Combining (2.2.2), (2.2.3) and (2.2.4), in the proof of Theorem 2.9, we may assume that the variables x_{jk} are uniformly bounded, mean zero and variance 1. For abbreviation, in proofs given in the next step, we still use $\mathbf{S}_n$, $\mathbf{X}_n$ for the matrices associated with the truncated variables.

Step 2. Completion of the proof for Theorem 2.9

If one wants to employ the moment approach, he needs to note

$$\begin{aligned}
\beta_k(\mathbf{S}_n) &= \int x^k F^{\mathbf{S}_n}(dx) \\
&= p^{-1}n^{-k}\sum_{\{i_1,\cdots,i_k\}}\sum_{\{j_1,\cdots,j_k\}} x_{i_1j_1}\bar{x}_{i_2j_1}x_{i_2j_2}\cdots x_{i_kj_k}\bar{x}_{i_1j_k},
\end{aligned}$$

where the summation runs over for the indices in $\mathbf{i} = (i_1, \cdots, i_k)$ run over $1, 2, \cdots, p$ and the indices in $\mathbf{j} = (j_1, \cdots, j_k)$ run over $1, 2, \cdots, n$.

To complete the proof of the almost sure convergence of the ESD of $\mathbf{S}_n$, we need only show the following two assertions:

$$\mathrm{E}(\beta_k(\mathbf{S}_n)) = \sum_{r=0}^{k-1}\frac{y_n^r}{r+1}\binom{k}{r}\binom{k-1}{r} + O(n^{-1}) \tag{2.2.5}$$

and

$$\mathrm{Var}(\beta_k(\mathbf{S}_n)) = O(n^{-2}), \tag{2.2.6}$$

where $y_n = p/n$.

Details of the proofs are omitted and referred to Bai and Silverstein (2006, 2009).

Remark 2.10. *The existence of the second moment of the entries is obviously necessary and sufficient for the Marčenko-Pastur law since the limiting distribution involves the parameter σ^2.*

2.2.2 Generalization to the Non-iid Case

Sometimes it is of practical interest to consider the case where the entries of $\mathbf{X}_n$ depend on n and for each n they are independent but not necessarily identically distributed. As in Section 2.1, we shall briefly present a proof of the following theorem.

Theorem 2.11. *Suppose that for each n, the entries of $\mathbf{X}$ are independent complex variables, with a common mean μ and variance σ^2. Assume that $p/n \to y \in (0, \infty)$ and that for any $\eta > 0$,*

$$\frac{1}{\eta^2 np} \sum_{jk} \mathrm{E}(|x_{jk}^{(n)}|^2 I(|x_{jk}^{(n)}| \geq \eta\sqrt{n})) \to 0. \tag{2.2.7}$$

Then, with probability one, $F^{\mathbf{S}}$ tends to the Marčenko-Pastur law with ratio index y and scale index σ^2.

Sketch of the proof. We shall only give an outline of the proof of this theorem. The details are left to the reader. Without loss of generality, we assume that $\mu = 0$ and $\sigma^2 = 1$. Similar to what we did in the proof of Theorem 2.6, we may select a sequence $\eta_n \downarrow 0$ such that condition (2.2.7) holds true when η is replaced by some $\eta_n \downarrow 0$. In the sequel, once condition (2.2.7) is used, we always mean this condition with η replaced by η_n.

Applying the rank inequality and the Bernstein inequality, by condition (2.2.7), we may truncate the variables $x_{ij}^{(n)}$ at $\eta_n\sqrt{n}$. Then, applying difference inequality, by condition (2.2.7), we may renormalize the truncated variables to have mean 0 and variances 1. Thus, in the rest of the proof, we shall drop the superscript (n) from the variables for brevity. We further assume that

$$\begin{aligned} &1)\ |x_{ij}| < \eta_n\sqrt{n}, \\ &2)\ \mathrm{E}(x_{ij}) = 0 \ \text{ and } \ \mathrm{Var}(x_{ij}) = 1. \end{aligned} \tag{2.2.8}$$

By similar arguments as given in the proof of Theorem 2.6, one can show the following two assertions:

$$\mathrm{E}(\beta_k(\mathbf{S}_n)) = \sum_{r=0}^{k-1} \frac{y_n^r}{r+1}\binom{k}{r}\binom{k-1}{r} + o(1); \tag{2.2.9}$$

and

$$\mathrm{E}\left|\beta_k(\mathbf{S}_n) - \mathrm{E}\left(\beta_k(\mathbf{S}_n)\right)\right|^4 = O(n^{-2}). \tag{2.2.10}$$

The proof of Theorem 2.11 is then complete.

2.2.3 Proof of Theorem 2.11 by Stieltjes Transform

As an illustration of applying Stieltjes transforms to sample covariance matrices, we give a proof of Theorem 2.11 in this section. Using the same approach of truncating, centralizing and rescaling as we did in the last section, we may assume the additional conditions given in (2.2.8).

Let $z = u + iv$ with $v > 0$ and $s(z)$ be the Stieltjes transform of the MP law. When $y < 1$, we have

$$s(z) = \int_a^b \frac{1}{x-z}\frac{1}{2\pi xy\sigma^2}\sqrt{(b-x)(x-a)}dx,$$

where $a = \sigma^2(1-\sqrt{y})^2$ and $b = \sigma^2(1+\sqrt{y})^2$.

By elementary calculation, one has

$$s(z) = \frac{\sigma^2(1-y) - z + \sqrt{(\sigma^2 - z - y\sigma^2)^2 - 4yz}}{2yz\sigma^2}. \tag{2.2.11}$$

When $y > 1$, since the MP law has also a point mass $1 - 1/y$ at zero, $s(z)$ equals the above integral plus $-(y-1)/yz$. Finally, one finds that expression (2.2.11) still holds. When $y = 1$, the above expression is still true by continuity in y.

Let the Stieltjes transform of the ESD of $\mathbf{S}_n$ be denoted by $s_n(z)$. Define

$$s_n(z) = \frac{1}{p}\mathrm{tr}(\mathbf{S}_n - z\mathbf{I}_p)^{-1}.$$

As we did in Section 2.1, we shall complete the proof by the following 3 steps:

(i) For any fixed $z \in \mathbb{C}^+$, $s_n(z) - \mathrm{E}s_n(z) \to 0$, a.s.
(ii) For any fixed $z \in \mathbb{C}^+$, $\mathrm{E}s_n(z) \to s(z)$, the Stieltjes transform of the MP law.
(iii) Except a null set, $s_n(z) \to s(z)$ for every $z \in \mathbb{C}^+$.

The proof of Steps (i) and (iii) are similar to what we did in Section 2.1 and thus is omitted. We now sketch Step (ii).

Step (ii) Convergence of $\mathrm{E}s_n(z)$

We will show that

$$\mathrm{E}s_n(z) \to s(z), \tag{2.2.12}$$

where $s(z)$ is defined in (2.2.11) with $\sigma^2 = 1$.

By the inverse matrix formula or Theorem 11.4 of Bai and Silverstein (2006), we have

$$s_n(z) = \frac{1}{p}\sum_{k=1}^{p} \frac{1}{\frac{1}{n}\boldsymbol{\alpha}_k'\overline{\boldsymbol{\alpha}}_k - z - \frac{1}{n^2}\boldsymbol{\alpha}_k'\mathbf{X}_k^*(\frac{1}{n}\mathbf{X}_k\mathbf{X}_k^* - z\mathbf{I}_{p-1})^{-1}\mathbf{X}_k\overline{\boldsymbol{\alpha}}_k}, \quad (2.2.13)$$

where $\mathbf{X}_k$ is the matrix obtained from $\mathbf{X}$ with the k-th row removed, $\boldsymbol{\alpha}_k'$ $(n \times 1)$ is the k-th row of $\mathbf{X}$.

Set

$$\varepsilon_k = \frac{1}{n}\boldsymbol{\alpha}_k'\overline{\boldsymbol{\alpha}}_k - 1 - \frac{1}{n^2}\boldsymbol{\alpha}_k'\mathbf{X}_k^*(\frac{1}{n}\mathbf{X}_k\mathbf{X}_k^* - z\mathbf{I}_{p-1})^{-1}\mathbf{X}_k\overline{\boldsymbol{\alpha}}_k + y_n + y_n z \mathrm{E}s_n(z), \quad (2.2.14)$$

where $y_n = p/n$. Then, by (2.2.13), we have

$$\mathrm{E}s_n(z) = \frac{1+\delta_n}{1 - z - y_n - y_n z\mathrm{E}s_n(z)}, \quad (2.2.15)$$

where

$$\delta_n = -\frac{1}{p}\sum_{k=1}^{p} \mathrm{E}\Big(\frac{\varepsilon_k}{1 - z - y_n - y_n z\mathrm{E}s_n(z) + \varepsilon_k}\Big). \quad (2.2.16)$$

Solving $\mathrm{E}s_n(z)$ from equation (2.2.15), we get the solution

$$\mathrm{E}s_n(z) = \frac{1}{2y_n z}\Big(1 - z - y_n + \sqrt{(1 - z - y_n)^2 - 4y_n z(1+\delta_n)}\Big).$$

Comparing with (2.2.11), it suffices to show that

$$\delta_n \to 0. \quad (2.2.17)$$

The details is omitted and refers to Bai and Silverstein (2006).

2.3 LSD of Products

In this section, we shall show some results on the existence of the LSD of a product of two random matrices, one of them is a sample covariance matrix and the other is an arbitrary Hermitian matrix. This topic is related to two areas: The first is the study of LSD of a multivariate F-matrix which is a product of a sample covariance matrix and the inverse of another sample covariance matrix, independent of each other. The second is the investigation of the LSD of a sample covariance matrix when the population covariance matrix is not a multiple of the identity.

Pioneering work was done by Wachter (1980), who considered the limiting distribution of the solutions to the following equation

$$\det(\mathbf{X}_{1,n_1}\mathbf{X}'_{1,n_1} - \lambda\mathbf{X}_{2,n_2}\mathbf{X}'_{2,n_2}) = 0 \tag{2.3.1}$$

where $\mathbf{X}_{j,n_j}$ is a $p \times n_j$ matrix whose entries are iid $N(0,1)$ and $\mathbf{X}_{1,n_1}$ is independent of $\mathbf{X}_{2,n_2}$. When $\mathbf{X}_{2,n_2}\mathbf{X}'_{2,n_2}$ is of full rank, the solutions to (2.3.1) are n_2/n_1 times of the eigenvalues of the multivariate F matrix $(\frac{1}{n_1}\mathbf{X}_{1,n_1}\mathbf{X}'_{1,n_1})(\frac{1}{n_2}\mathbf{X}_{2,n_2}\mathbf{X}'_{2,n_2})^{-1}$.

Yin and Krishnaiah (1983) established the existence of the LSD of the matrix sequence $\{\mathbf{S}_n\mathbf{T}_n\}$, where $\mathbf{S}_n$ is a standard Wishart matrix of dimension p and degrees of freedom n with $p/n \to y \in (0,\infty)$, $\mathbf{T}_n$ is a positive definite matrix satisfying $\beta_k(\mathbf{T}_n) \to H_k$ and the sequence H_k satisfy the Carleman condition. In Yin (1986), this result was generalized to the case where the sample covariance matrix is formed based on iid real random variables of mean zero and variance one. Using the result of Yin and Krishnaiah (1983), Yin, Bai and Krishnaiah (1983) showed the existence of the LSD of the multivariate F-matrix. The explicit form of the LSD of multivariate F-matrices was derived in Bai, Yin, and Krishnaiah (1987) and Silverstein (1985). Under the same structure, Bai, Yin, and Krishnaiah (1986) established the existence of the LSD when the underlying distribution of $\mathbf{S}_n$ is isotropic.

Some further extensions were done in Silverstein (1995) and Silverstein and Bai (1995). In this section, we shall introduce some recent developments in this direction.

2.3.1 Existence of the ESD of $\mathbf{S}_n\mathbf{T}_n$

Here we present the following results.

Theorem 2.12. *Suppose that the entries of $\mathbf{X}_n$ $(p \times n)$ are independent complex random variables satisfying (2.2.7) and that $\mathbf{T}_n$ is a sequence of Hermitian matrices independent of $\mathbf{X}_n$ and that the ESD of $\mathbf{T}_n$ tends to a nonrandom limit F^T in some sense (in pr. or a.s.). If $\lim(p/n) \to y \in (0,\infty)$, then the ESD of the product $\mathbf{S}_n\mathbf{T}_n$ tends to a non-random limit in probability or almost surely (accordingly), where $\mathbf{S}_n = \frac{1}{n}\mathbf{X}_n\mathbf{X}_n^*$.*

Remark 2.13. *Note that the eigenvalues of the product matrix $\mathbf{S}_n\mathbf{T}_n$ are all real although it is not symmetric, because the whole set of eigenvalues is the same as that of the symmetric matrix $\mathbf{S}_n^{1/2}\mathbf{T}_n\mathbf{S}_n^{1/2}$.*

This theorem contains Yin's result as a special case. In Yin (1986), the entries of $\mathbf{X}$ are assumed to be real and iid with mean zero and variance one and the matrix $\mathbf{T}_n$ being real and positive definite and satisfying, for each fixed k,

$$\frac{1}{p}\mathrm{tr}(\mathbf{T}_n^k) \to H_k \quad \text{(in pr. or a.s.)} \tag{2.3.2}$$

while the constant sequence $\{H_k\}$ satisfies the Carleman condition.

In Silverstein and Bai (1995), the following theorem is proved.

Theorem 2.14. *Suppose that the entries of* $\mathbf{X}_n$ $(n \times p)$ *are complex random variables which are independent for each* n *and identically distributed for all* n *and satisfy* $\mathrm{E}(|x_{1\,1} - \mathrm{E}(x_{1\,1})|^2) = 1$. *Also, assume that* $\mathbf{T}_n = \mathrm{diag}(\tau_1, \ldots, \tau_p)$, τ_i *real, and the empirical distribution function of* $\{\tau_1, \ldots, \tau_p\}$ *converges almost surely to a probability distribution function* H *as* $n \to \infty$. *The entries of both* $\mathbf{X}_n$ *and* $\mathbf{T}_n$ *may depend on* n, *which is suppressed for brevity. Set* $\mathbf{B}_n = \mathbf{A}_n + \frac{1}{n}\mathbf{X}_n\mathbf{T}_n\mathbf{X}_n^*$, *where* $\mathbf{A}_n$ *is Hermitian* $n \times n$ *satisfying* $F^{\mathbf{A}_n} \to F^A$ *almost surely, where* F^A *is a distribution function (possibly defective) on the real line. Assume also that* $\mathbf{X}_n$, $\mathbf{T}_n$, *and* $\mathbf{A}_n$ *are independent. When* $p = p(n)$ *with* $p/n \to y > 0$ *as* $n \to \infty$, *then, almost surely,* $F^{\mathbf{B}_n}$, *the ESD of the eigenvalues of* $\mathbf{B}_n$, *converges vaguely, as* $n \to \infty$, *to a (non-random) d.f.* F, *whose Stieltjes transform* $s(z)$ *is given by*

$$s(z) = s_A\left(z - y\int \frac{\tau dH(\tau)}{1 + \tau s(z)}\right), \tag{2.3.3}$$

where z *is a complex number with a positive imaginary part and* s_A *is the Stieltjes transform of* F_A.

Remark 2.15. *Note that Theorem 2.14 is more general than Yin's result in the sense that there is no requirement on the moment convergence of the ESD of* $\mathbf{T}_n$ *as well as there is no requirement on the positive definiteness of the matrix* $\mathbf{T}_n$. *Also, it allows a perturbation matrix* $\mathbf{A}_n$ *involved in* $\frac{1}{n}\mathbf{X}_n^*\mathbf{T}_n\mathbf{X}_n$. *However, it is more restrictive than Yin's result in the sense that it requires the matrix* $\mathbf{T}_n$ *to be diagonal. An extension of Yin's work on another direction is made in Silverstein* (1995) *which will be introduced later. Weak convergence of* (2.3.3) *was established in Marčenko-Pastur (1967) under higher moment conditions than assumed in Theorem* 2.12, *but with mild dependence between the entries of* $\mathbf{X}_n$.

The proof of Theorem 2.14 uses the Stieltjes transform which will be given in Section 2.3.6.

2.3.2 Truncation of the ESD of $\mathbf{T}_n$

For brevity, we shall suppress the superscript n from the x-variables.

Step 1. Reducing to the case where $\mathbf{T}_n$'s are non-random

If the ESD of $\mathbf{T}_n$ converges to a limit F^T almost surely, we may consider LSD of $\mathbf{S}_n\mathbf{T}_n$ conditioning on given all $\mathbf{T}_n$ and hence we may assume that $\mathbf{T}_n$ is non-random. Then the final result follows by the Fubini Theorem. If the convergence is in probability, then we may use the subsequence method or use the strong representation theorem (see Skorokhod (1956) or Dudley (1985))[4]. The strong representation theorem says that there is a probability space on which we can define a sequence of random matrices $(\widetilde{\mathbf{X}}_n, \widetilde{\mathbf{T}}_n)$ such that for each n, the joint distribution of $(\widetilde{\mathbf{X}}_n, \widetilde{\mathbf{T}}_n)$ is identical to that of $(\mathbf{X}_n, \mathbf{T}_n)$ and the ESD of $\widetilde{\mathbf{T}}_n$ converges to F^T almost surely. Therefore, to prove theorem 2.12, it suffices to show it for the case of a.s. convergence.

Now, suppose that $\mathbf{T}_n$ are nonrandom and that the ESD of $\mathbf{T}_n$ convergence to F^T.

Step 2. Truncation of the ESD of $\mathbf{T}_n$

Suppose that the spectral decomposition of $\mathbf{T}_n$ is $\sum_{i=1}^p \lambda_{in}\mathbf{u}_i\mathbf{u}_i^*$. Define a matrix $\widetilde{\mathbf{T}}_n = \sum_{i=1}^p \tilde{\lambda}_{in}\mathbf{u}_i\mathbf{u}_i^*$, where $\tilde{\lambda}_{in} = \lambda_{in}$ or zero in accordance with $|\lambda_{in}| \le \tau_0$ or not, where τ_0 is pre-chosen constant such that both $\pm\tau_0$ are continuity points of F^T. Then, the ESD of $\widetilde{\mathbf{T}}_n$ converges to the limit

$$F_{T,\tau_0}(x) = \int_{-\infty}^{x} I_{[-\tau_0,\tau_0]}(u)F^T(du) + (F^T(-\tau_0) + 1 - F^T(\tau_0))I_{[0,\infty)}(x),$$

and (2.3.2) is true for $\widetilde{\mathbf{T}}_n$ with $\tilde{H}_k = \int_{|x|\le\tau_0} x^k dF^T(x)$.

Applying the rank inequality, we obtain

$$\left\| F^{\mathbf{S}_n\mathbf{T}_n} - F^{\mathbf{S}_n\widetilde{\mathbf{T}}_n} \right\| \le \frac{1}{p}\text{rank}(\mathbf{T}_n - \widetilde{\mathbf{T}}_n) \to F^T(-\tau_0) + 1 - F^T(\tau_0) \quad (2.3.4)$$

as $n \to \infty$. Note that the right-hand side of the above inequality can be made arbitrarily small if τ_0 is large enough.

Therefore, we may assume that the eigenvalues of $\mathbf{T}_n$ are bounded by a constant, say τ_0.

2.3.3 Truncation, Centralization and Rescaling of the X-variables

Following the truncation technique used in Section 2.2.2, let $\widetilde{\mathbf{X}}_n$ and $\widetilde{\mathbf{S}}_n$ denote the sample matrix and the sample covariance matrix defined by the truncated variables at the truncation location $\eta_n\sqrt{n}$. Note that $\mathbf{S}_n\mathbf{T}_n$ and $\frac{1}{n}\mathbf{X}_n^*\mathbf{T}_n\mathbf{X}_n$ have the same set of nonzero eigenvalues, so do the matrices $\widetilde{\mathbf{S}}_n\mathbf{T}_n$ and $\frac{1}{n}\widetilde{\mathbf{X}}_n^*\mathbf{T}_n\widetilde{\mathbf{X}}_n$. Thus,

$$\begin{aligned} &\|F^{\mathbf{S}_n\mathbf{T}_n} - F^{\widetilde{\mathbf{S}}_n\mathbf{T}_n}\| \\ &= \frac{n}{p}\|F^{\frac{1}{n}\mathbf{X}_n^*\mathbf{T}_n\mathbf{X}_n} - F^{\frac{1}{n}\widetilde{\mathbf{X}}_n^*\mathbf{T}_n\widetilde{\mathbf{X}}_n}\| = \frac{n}{p}\|F^{\mathbf{X}_n^*\mathbf{T}_n\mathbf{X}_n} - F^{\widetilde{\mathbf{X}}_n^*\mathbf{T}_n\widetilde{\mathbf{X}}_n}\|. \end{aligned}$$

[4] in an unpublished work by Bai and Liang (1985), Skorokhod's results was generalized to *Suppose that μ_n is a probability measure defined on a Polish space (i.e., a complete and separable metric space) S_n and φ_n is a measurable mapping from S_n to another Polish space S_0. If $\mu_n\varphi_n^{-1}$ tends to μ_0 weakly, where μ_0 is a probability measure defined on the space S_0, then there exists a probability space $(\Omega, \mathcal{F}, P)$ on which we have random mappings $X_n : \Omega \mapsto S_n$, such that μ_n is the distribution of X_n and $\varphi_n(X_n) \to X_0$ almost surely.* Skorokhod's result is the special case where all S_n is identical to S_0 and all $\varphi_n(x) = x$.

Then, by the rank inequality, for any $\varepsilon > 0$, we have

$$\begin{aligned}
&\mathrm{P}\left(\left\|F^{\mathbf{S}_n\mathbf{T}_n} - F^{\widetilde{\mathbf{S}}_n\mathbf{T}_n}\right\| \geq \varepsilon\right) \\
&= \mathrm{P}\left(\left\|F^{\mathbf{X}_n^*\mathbf{T}_n\mathbf{X}_n} - F^{\widetilde{\mathbf{X}}_n^*\mathbf{T}_n\widetilde{\mathbf{X}}_n}\right\| \geq \varepsilon p/n\right) \\
&\leq \mathrm{P}(\mathrm{rank}(\mathbf{X}_n^*\mathbf{T}_n\mathbf{X}_n - \widetilde{\mathbf{X}}_n^*\mathbf{T}_n\widetilde{\mathbf{X}}_n) \geq \varepsilon p) \\
&\leq \mathrm{P}(2\mathrm{rank}(\mathbf{X}_n - \widetilde{\mathbf{X}}_n) \geq \varepsilon p) \\
&\leq \mathrm{P}\bigg(\sum_{ij} I_{\{|x_{ij}|\geq \eta_n\sqrt{n}\}} \geq \varepsilon p/2\bigg).
\end{aligned}$$

From the condition (2.2.7), one can easily see that

$$\mathrm{E}\bigg(\sum_{ij} I_{\{|x_{ij}|\geq\eta_n\sqrt{n}\}}\bigg) \leq \frac{1}{\eta_n^2 n}\sum_{ij}\mathrm{E}|x_{ij}|^2 I_{\{|x_{ij}|\geq\eta_n\sqrt{n}\}} = o(p)$$

and

$$\mathrm{Var}\bigg(\sum_{ij} I_{\{|x_{ij}|\geq\eta_n\sqrt{n}\}}\bigg) \leq \frac{1}{\eta_n^2 n}\sum_{ij}\mathrm{E}|x_{ij}|^2 I_{\{|x_{ij}|\geq\eta_n\sqrt{n}\}} = o(p).$$

Then, applying Bernstein's inequality one obtains

$$\mathrm{P}\bigg(\left\|F^{\mathbf{S}_n\mathbf{T}_n} - F^{\widetilde{\mathbf{S}}_n\mathbf{T}_n}\right\| \geq \varepsilon\bigg) \leq 2\exp\left(-\frac{1}{8}\varepsilon^2 p\right) \tag{2.3.5}$$

which is summable. By the Borel-Cantelli lemma we conclude that with probability 1,

$$\left\|F^{\mathbf{S}_n\mathbf{T}_n} - F^{\widetilde{\mathbf{S}}_n\mathbf{T}_n}\right\| \to 0. \tag{2.3.6}$$

We may do the centralization and rescaling of the X-variables by the same way given in Section 2.2.2. We leave the details to the reader.

2.3.4 Sketch of the Proof of Theorem 2.12

Therefore, the proof of Theorem 2.12 can be done under the following additional conditions:

$$\begin{aligned}
&\|\mathbf{T}_n\| \leq \tau_0; \\
&|x_{jk}| \leq \eta_n\sqrt{n}; \\
&\mathrm{E}(x_{jk}) = 0, \\
&\mathrm{E}|x_{jk}|^2 = 1.
\end{aligned} \tag{2.3.7}$$

Now, we will proceed in the proof of Theorem 2.12 by applying the MCT under the above additional conditions. We need to show the convergence of the spectral moments of $\mathbf{S}_n\mathbf{T}_n$. We have

$$\beta_k(\mathbf{S}_n\mathbf{T}_n) = \frac{1}{p}\mathrm{E}[(\mathbf{S}_n\mathbf{T}_n)^k]$$
$$= p^{-1}n^{-k}\sum x_{i_1j_1}\overline{x}_{i_2j_1}t_{i_2i_3}x_{i_3j_2}\cdots x_{i_{2k-1}j_k}\overline{x}_{i_{2k}j_k}t_{i_{2k}i_1}, \qquad (2.3.8)$$

where $Q(\mathbf{i},\mathbf{j})$ is the Q-graph defined by $i_1,\cdots,i_{2k}$ take values in $\{1,\cdots,p\}$ and $j_1,\cdots,j_k$ run over $1,\cdots,n$.

By tedious but elementary combinatorial argument one can prove the theorem by the following steps:

$$\mathrm{E}\beta_k(\mathbf{S}_n\mathbf{T}_n) \to \beta_k^{st} = \sum_{s=1}^{k} y^{k-s} \sum_{\substack{i_1+\cdots+i_s=k-s+1\\ i_1+\cdots+si_s=k}} \frac{k!}{s!}\prod_{m=1}^{s}\frac{H_m^{i_m}}{i_m!}; \qquad (2.3.9)$$

$$\mathrm{E}|\beta_k(\mathbf{S}_n\mathbf{T}_n) - \mathrm{E}\beta_k(\mathbf{S}_n\mathbf{T}_n)|^4 = O(n^{-2}), \qquad (2.3.10)$$

β_k^{st} satisfy the Carleman condition.

The details of the proof are omitted.

2.3.5 LSD of F Matrix

We have seen in the last section that the moments of the LSD F^{st} of $\mathbf{S}_n\mathbf{T}_n$ is given by

$$\beta_k^{st} = \sum_{s=1}^{k} y^{k-s} \sum_{\substack{i_1+\cdots+i_s=k-s+1\\ i_1+\cdots+si_s=k}} \frac{k!}{s!}\prod_{m=1}^{s}\frac{H_m^{i_m}}{i_m!}$$

For $k \ge 1$, β_k^{st} is the coefficient of z^k in the Taylor expansion of

$$\sum_{s=0}^{k} y^{k-s}\frac{k!}{s!(k-s+1)!}\left(\sum_{\ell=1}^{\infty} z^\ell H_\ell\right)^{k-s+1} + \frac{1}{y(k+1)}$$
$$= \frac{1}{y(k+1)}\left[1 + y\sum_{\ell=1}^{\infty} z^\ell H_\ell\right]^{k+1} \qquad (2.3.1)$$

where H_k is the moments of the LSD H of $\mathbf{T}_n$. Therefore, β_k^{st} can be written as

$$\beta_k^{st} = \frac{1}{2\pi i y(k+1)}\oint_{|\zeta|=\rho} \zeta^{-k-1}\left[1 + y\sum_{\ell=1}^{\infty}\zeta^\ell H_\ell\right]^{k+1} d\zeta,$$

for any $\rho \in (0, 1/\tau_0)$, which guarantees the convergence of the series $\sum \zeta^\ell H_\ell$.

Using the above expression, we can construct a generating function of β_k^{st} as follows: for all small z with $|z| < 1/\tau_0 b$, where $b = (1+\sqrt{y})^2$,

$$g(z) - 1 = \sum_{k=1}^{\infty} z^k\beta_k^{st} = \frac{1}{2\pi i y}\oint_{|\zeta|=\rho}\sum_{k=1}^{\infty}\frac{1}{k+1}z^k\zeta^{-1-k}\left(1 + y\sum_{\ell=1}^{\infty}\zeta^\ell H_\ell\right)^{k+1} d\zeta$$

$$= \frac{1}{2\pi i y} \oint_{|\zeta|=\rho} \left[-\zeta^{-1} - y \sum_{\ell=1}^{\infty} \zeta^{\ell-1} H_\ell - \frac{1}{z} \log(1 - z\zeta^{-1} - zy \sum_{\ell=1}^{\infty} \zeta^{\ell-1} H_\ell) \right] d\zeta$$

$$= -\frac{1}{y} - \frac{1}{2\pi i y z} \oint_{|\zeta|=\rho} \log \left(1 - z\zeta^{-1} - zy \sum_{\ell=1}^{\infty} \zeta^{\ell-1} H_\ell \right) d\zeta.$$

The exchange of summation and integral is justified provided that $|z| < \rho/(1+y\sum \rho^\ell |H_\ell|)$. Therefore, we have

$$g(z) = 1 - \frac{1}{y} - \frac{1}{2\pi i y z} \oint_{|\zeta|=\rho} \log \left(1 - z\zeta^{-1} - zy \sum_{\ell=1}^{\infty} \zeta^{\ell-1} H_\ell \right) d\zeta. \tag{2.3.2}$$

Let $s_F(z)$ and $s_H(z)$ denote the Stieltjes transforms of F^{st} and H respectively. It is easy to verify that

$$-\frac{1}{z} s_F(\frac{1}{z}) = 1 + \sum_{k=1}^{\infty} z^k \beta_k^{st},$$

$$-\frac{1}{z} s_H(\frac{1}{z}) = 1 + \sum_{k=1}^{\infty} z^k H_k.$$

Then, from (2.3.2) it follows that

$$\frac{1}{z} s_F(\frac{1}{z}) = \frac{1}{y} - 1 + \frac{1}{2\pi i y z} \oint_{|\zeta|=\rho} \log(1 - z\zeta^{-1} + \zeta^{-1} zy + \zeta^{-2} zy s_H(\frac{1}{\zeta})) d\zeta. \tag{2.3.3}$$

Now, let us use this formula to derive the LSD of general multivariate F matrices. A multivariate F matrix is defined as a product of $\mathbf{S}_n$ with the inverse of another covariance matrix, i.e. $\mathbf{T}_n$ is the inverse of another covariance matrix with dimension p and degrees of freedom n_2. To guarantee the existence of the inverse matrix, we assume that $p/n_2 \to y' \in (0, 1)$. In this case, it is easy to verify that H will have a density function

$$H'(x) = \begin{cases} \frac{1}{2\pi y' x^2} \sqrt{(xb' - 1)(1 - a'x)} & \text{if } \frac{1}{b'} < x < \frac{1}{a'}, \\ 0 & \text{otherwise,} \end{cases}$$

where $a' = (1 - \sqrt{y'})^2$ and $b' = (1 + \sqrt{y'})^2$. Noting that the k-th moment of H is the $-k$-th moment of the Marčenko-Pastur law with index y', one can verify that

$$\zeta^{-1} s_H(\frac{1}{\zeta}) = -\zeta s_{y'}(\zeta) - 1,$$

where $s_{y'}$ is the Stieltjes transform of the MP law with index y'. Thus,

$$s_F(z) = \frac{1}{yz} - \frac{1}{z} + \frac{1}{2\pi i y} \oint_{|\zeta|=\rho} \log(z - \zeta^{-1} - y s_{y'}(\zeta)) d\zeta. \tag{2.3.4}$$

By (2.2.11), we have

$$s_{y'}(\zeta) = \frac{1 - y' - \zeta + \sqrt{(1 + y' - \zeta)^2 - 4y'}}{2y'\zeta}. \tag{2.3.5}$$

By integration by parts, we have

$$\begin{aligned} &\frac{1}{2\pi i y} \oint_{|\zeta|=\rho} \log(z - \zeta^{-1} - y s_{y'}(\zeta)) d\zeta \\ &= -\frac{1}{2\pi i y} \oint_{|\zeta|=\rho} \zeta \frac{\zeta^{-2} - y s'_{y'}(\zeta)}{z - \zeta^{-1} - y s_{y'}(\zeta)} d\zeta \\ &= -\frac{1}{2\pi i y} \oint_{|\zeta|=\rho} \frac{1 - y\zeta^2 s'_{y'}(\zeta)}{z\zeta - 1 - y\zeta s_{y'}(\zeta)} d\zeta. \end{aligned} \tag{2.3.6}$$

For easy evaluation of the integral, we make a variable change from ζ to s. Note that $s_{y'}$ is a solution of the equation (see (2.2.15) with $\delta = 0$),

$$s = \frac{1}{1 - \zeta - y' - \zeta y' s}. \tag{2.3.7}$$

From this, we have

$$\begin{aligned} \zeta &= \frac{s - sy' - 1}{s + s^2 y'}, \\ \frac{ds}{d\zeta} &= \frac{s + s^2 y'}{1 - y' - \zeta - 2sy'\zeta} = \frac{s^2(1 + sy')^2}{1 + 2sy' - s^2 y'(1 - y')}. \end{aligned}$$

Note that when ζ runs along $\zeta = \rho$ anticlockwise, s will also run along a contour $\mathcal{C}$ anticlockwise. Therefore,

$$\begin{aligned} &-\frac{1}{2\pi i y} \oint_{|\zeta|=\rho} \frac{1 - y\zeta^2 (ds_{y'}(\zeta)/d\zeta)}{z\zeta - 1 - y\zeta s_{y'}(\zeta)} d\zeta \\ &= -\frac{1}{2\pi i y} \oint_{\mathcal{C}} \frac{1 + 2sy' - s^2 y'(1 - y') - y(s - sy' - 1)^2}{s(1 + sy')[z(s - sy' - 1) - s(1 + sy') - ys(s - sy' - 1)]} ds \\ &= -\frac{1}{2\pi i y} \oint_{\mathcal{C}} \frac{(y' + y - yy')(1 - y')s^2 - 2s(y' + y - yy') - 1 + y}{(s + s^2 y')[(y' + y - yy')s^2 + s((1 - y) - z(1 - y')) + z]} ds. \end{aligned}$$

The integrand has 4 poles at $s = 0, -1/y'$ and

$$\begin{aligned} s_{1,2} &= \frac{-(1 - y) + z(1 - y') \pm \sqrt{((1 - y) + z(1 - y'))^2 - 4z}}{2(y + y' - yy')} \\ &= \frac{2z}{-(1 - y) + z(1 - y') \mp \sqrt{((1 - y) + z(1 - y'))^2 - 4z}}. \end{aligned}$$

We need to decide which pole is located inside the contour $\mathcal{C}$. From (2.3.7), it is easy to see that when ρ is small, for all $|\zeta| \le \rho$, $s_{y'}(\zeta)$ is close to $\frac{1}{1-y}$, that is, the contour $\mathcal{C}$ and its inner region are around $\frac{1}{1-y}$. Hence, 0 and $-1/y'$ are not inside the contour $\mathcal{C}$.

Let $z = u + iv$ with large u and $v > 0$. Then we have

$$\Im(((1-y)+z(1-y'))^2 - 4z) = 2v[(1-y)(u(1-y')+(1-y)) - 2] > 0.$$

By the convention for square root of complex numbers, both real and imaginary parts of $\sqrt{((1-y)+z(1-y'))^2-4z}$ are positive. Therefore, $|s_1| > |s_2|$ and s_1 may take very large values. Also, s_2 will stay around $1/(1-y')$. We conclude that only s_2 is the pole inside the contour $\mathcal{C}$ for all z with large real part and positive imaginary part.

Now, let us compute the residue at s_2. By using $s_1s_2 = z/(y+y'-yy')$, the residue is given by

$$\begin{aligned} R &= \frac{(y'+y-yy')(1-y')s_2^2 - 2s_2(y'+y-yy') - 1 + y}{(s_2+s_2^2y')(y'+y-yy')(s_2-s_1)} \\ &= \frac{(1-y')zs_2s_1^{-1} - 2zs_1^{-1} - 1 + y}{(zs_1^{-1} + zs_2s_1^{-1}y')(s_2-s_1)} \\ &= \frac{z(1-y')s_2 - 2z - (1-y)s_1}{z(1+s_2y')(s_2-s_1)} \\ &= \frac{[(1-y+z-zy') - \sqrt{((1-y)+z(1-y'))^2-4z}](y+y'-yy')}{z(2y+y'-yy'+zy'(1-y') - y'\sqrt{((1-y)+z(1-y'))^2-4z})}. \end{aligned}$$

Multiplying both the numerator and denominator by $2y+y'-yy'+zy'(1-y')+y'\sqrt{((1-y)+z(1-y'))^2-4z}$, after simplification, we obtain

$$R = \frac{y(1-y+z-zy')+2y'z) - y\sqrt{((1-y)+z(1-y'))^2-4z}}{2z(yz+y')}.$$

So, for all large $z \in \mathbb{C}^+$,

$$\begin{aligned} s_F(z) &= \frac{1}{zy} - \frac{1}{z} - \frac{y(z(1-y')+1-y)+2zy' - y\sqrt{((1-y)+z(1-y'))^2-4z}}{2zy(y+zy')} \\ &= \frac{1-y-z(1+y')+\sqrt{((1-y)+z(1-y'))^2-4z}}{2z(y+zy')}. \end{aligned}$$

Since $s_F(z)$ is analytic on $\mathbb{C}^+$, the above identity is true for all $z \in \mathbb{C}^+$. Now, letting $z \downarrow x + i0$, $\pi^{-1}\Im s_F(z)$ tends to the density function of the LSD of multivariate F matrices. Thus, the density function of the LSD of multivariate F matrices is

$$\begin{cases} \frac{\sqrt{4x-((1-y)+x(1-y'))^2}}{2\pi x(y+y'x)} & \text{when } 4x-((1-y)+x(1-y'))^2 > 0, \\ 0 & \text{otherwise.} \end{cases}$$

Or equivalently

$$\begin{cases} \frac{(1-y')\sqrt{(b-x)(x-a)}}{2\pi x(y+y'x)} & \text{when } a < x < b, \\ 0 & \text{otherwise,} \end{cases}$$

where $a, b = \left(\frac{1\mp\sqrt{y+y'-yy'}}{1-y'}\right)^2$. Now we determine the possible atom at 0. When $z = u + iv \to 0$ with $v > 0$, we have

$$\Im((1-y+z(1-y'))^2 - 4z) = 2v[(1-y+u(1-y'))(1-y') - 2] < 0$$

Hence $\Re(\sqrt{(1-y+z(1-y'))^2 - 4z}) < 0$. Thus $\sqrt{(1-y+z(1-y'))^2 - 4z} \to -|1-y|$. Consequently,

$$\begin{aligned} F(\{0\}) &= -\lim_{z\to 0} z s_F(z) = 1 - \frac{1}{y} + \frac{1-y+|1-y|}{2y} \\ &= \begin{cases} 1 - \frac{1}{y} & \text{if } y > 1 \\ 0 & \text{otherwise.} \end{cases} \end{aligned}$$

This conclusion coincides with the intuitive observation that the matrix $\mathbf{S}_n\mathbf{T}_n$ has $p - n$ 0 eigenvalues.

2.3.6 Sketch of the Proof of Theorem 2.14

In this section, we shall sketch the proof of Theorem 2.14 by using Stieltjes transforms. Instead, we shall prove it under a weaker condition that the entries of $\mathbf{X}_n$ satisfy (2.2.7).

We omit the truncation and centralization steps and only briefing the proof by Stieltjes Transform.

Let

$$s_n(z) = \frac{1}{n}\text{tr}(\mathbf{B}_n - z\mathbf{I})^{-1}.$$

We first make an additional assumption that F^A, the LSD of the sequence $\mathbf{A}_n$, is not a zero measure. For otherwise, $s_A(z) \equiv 0$ for any $z \in \mathbb{C}^+$. In this case, the proof of (2.3.3) reduces to showing that $s_n(z) \to 0$. In this case, except $o(n)$ eigenvalues of $\mathbf{A}_n$, all other eigenvalues of $\mathbf{A}_n$ will tend to $\pm$ infinity. By the relations between eigenvalues of two Hermitian matrices and their sum, we have

$$s_{i+j+1}(\mathbf{A}_n) \le s_{i+1}(\mathbf{B}_n) + s_{j+1}(-\frac{1}{n}\mathbf{X}_n\mathbf{T}_n\mathbf{X}_n^*). \tag{2.3.8}$$

Note that

$$s_{j+1}(-\frac{1}{n}\mathbf{X}_n\mathbf{T}_n\mathbf{X}_n^*) \le \tau_0\lambda_{j+1}(\frac{1}{n}\mathbf{X}_n\mathbf{X}_n^*).$$

Let j_0 be an integer such that $\lambda_{j_0+1}(\frac{1}{n}\mathbf{X}_n\mathbf{X}_n^*) < b + 1 \le \lambda_{j_0}(\frac{1}{n}\mathbf{X}_n\mathbf{X}_n^*)$. Since the ESD of $\frac{1}{n}\mathbf{X}_n\mathbf{X}_n^*$ converges a.s. to the Marčenko-Pastur law with spectrum bounded in $[0, b]$, it follows that $j_0 = o(n)$. For any $M > 0$, define ν_0 be such that $s_{n-\nu_0+2}(\mathbf{A}_n) < M \le s_{n-\nu_0+1}(\mathbf{A}_n)$. By the assumption that F^A is a zero measure, we conclude that $\nu_0 = o(n)$. Define $i_0 = n - \nu_0 - j_0$. By (2.3.8), we have

$$s_{i_0+1}(\mathbf{B}_n) \geq s_{n-\nu_0+1} - s_{j_0+1}(-\frac{1}{n}\mathbf{X}_n\mathbf{T}_n\mathbf{X}_n^*) \geq M - b - 1.$$

This shows that except $o(n)$ eigenvalues of $\mathbf{B}_n$, the absolute values of all its other eigenvalues will be larger than $M - b - 1$. By the arbitrariness of M, we conclude that except for $o(n)$ eigenvalues of $\mathbf{B}_n$, all its other eigenvalues tend to infinity. This shows that $F^{\mathbf{B}_n}$ tends to a zero measure or $s_n(z) \to 0$, a.s., so that (2.3.3) holds trivially.

Now, we assume that $F^A \neq 0$. We claim that for any subsequence n', with probability 1, $F^{\mathbf{B}_{n'}} \not\to 0$. For otherwise, using the same arguments given above, one may show that $F^{\mathbf{A}_{n'}} \to 0 = F^A$, a contradiction. This shows that with probability 1, there is a constant m such that

$$\inf_n F^{\mathbf{B}_n}([-m, m]) > 0,$$

which simply implies that

$$\delta = \inf_n |\mathrm{E}s_n(z)| \geq \inf_n \mathrm{E}\Im(s_n(z)) \geq \mathrm{E}\inf_n \int \frac{vdF^{\mathbf{B}_n}(x)}{(x-u)^2+v^2} > 0. \tag{2.3.9}$$

Now, we shall complete the proof of Theorem 2.14 by showing

(a) $s_n(z) - \mathrm{E}s_n(z) \to 0$, a.s. (2.3.10)

(b) $\mathrm{E}s_n(z) \to s(z)$, (2.3.11)

which satisfies (2.3.3).

(c) The equation (2.3.3) has a unique solution in $\mathbb{C}^+$. (2.3.12)

Step 1. Proof of (2.3.10)

Let $\mathbf{x}_k$ denote the k-th column of $\mathbf{X}_n$ and set

$$\mathbf{q}_k = \frac{1}{\sqrt{n}}\mathbf{x}_k,$$
$$\mathbf{B}_{k,n} = \mathbf{B}_n - \tau_k\mathbf{q}_k\mathbf{q}_k^*.$$

Write E_k to denote the conditional expectation given $\mathbf{x}_{k+1}, \cdots, \mathbf{x}_p$. With this notation, we have $s_n(z) = \mathrm{E}_0(s_n(z))$ and $\mathrm{E}(s_n(z)) = \mathrm{E}_p(s_n(z))$. Therefore, we have

$$\begin{aligned}
s_n(z) - \mathrm{E}(s_n(z)) &= \sum_{k=1}^{p}[\mathrm{E}_{k-1}(s_n(z)) - \mathrm{E}_k(s_n(z))] \\
&= \frac{1}{n}\sum_{k=1}^{p}[\mathrm{E}_{k-1} - \mathrm{E}_k](\mathrm{tr}(\mathbf{B}_n - z\mathbf{I})^{-1}) - \mathrm{tr}(\mathbf{B}_{k,n} - z\mathbf{I})^{-1}) \\
&= \frac{1}{n}\sum_{k=1}^{p}[\mathrm{E}_{k-1} - \mathrm{E}_k]\gamma_k,
\end{aligned}$$

where

$$\gamma_k = \frac{\tau_k \mathbf{q}_k^*(\mathbf{B}_{k,n} - z\mathbf{I})^{-2}\mathbf{q}_k}{1 + \tau_k \mathbf{q}_k^*(\mathbf{B}_{k,n} - z\mathbf{I})^{-1}\mathbf{q}_k}.$$

By elementary calculation, we have

$$|\gamma_k| \le \frac{|\tau_k \mathbf{q}_k^*(\mathbf{B}_{k,n} - z\mathbf{I})^{-2}\mathbf{q}_k|}{|\Im(1 + \tau_k \mathbf{q}_k^*(\mathbf{B}_{k,n} - z\mathbf{I})^{-1}\mathbf{q}_k)|} \le v^{-1}. \tag{2.3.13}$$

Note that $\{[\mathrm{E}_{k-1} - \mathrm{E}_k]\gamma_k\}$ forms a bounded martingale difference sequence. By applying Burkholder's inequality (see Lemma 2.3), one can easily show that for any $\ell > 1$,

$$\begin{aligned} \mathrm{E}|s_n(z) - \mathrm{E}s_n(z)|^\ell &\le K_p n^{-\ell} \mathrm{E}\left(\sum_{k=1}^n |(\mathrm{E}_{k-1} - \mathrm{E}_k)\gamma_k|^2\right)^{\ell/2} \\ &\le K_\ell (2/v)^\ell n^{-\ell/2}. \end{aligned} \tag{2.3.14}$$

From this with $p > 2$, it follows easily that

$$\frac{1}{n}\sum_{k=1}^p [\mathrm{E}_{k-1} - \mathrm{E}_k]\gamma_k \to 0, \text{ a.s.}$$

Then, what to be shown follows.

Step 2. Proof of (2.3.11)

Write

$$x = x_n = \frac{1}{n}\sum_{k=1}^p \frac{\tau_k}{1 + \tau_k \mathrm{E}s_n(z)}.$$

It is easy to verify that $\Im x \le 0$. Write

$$\mathbf{B}_n - z\mathbf{I} = \mathbf{A}_n - (z - x)\mathbf{I} + \sum_{k=1}^p \tau_k \mathbf{q}_k \mathbf{q}_k^* - x\mathbf{I}.$$

Then, we have

$$\begin{aligned} &(\mathbf{A}_n - (z - x)\mathbf{I})^{-1} - (\mathbf{B}_n - z\mathbf{I})^{-1} \\ &= (\mathbf{A}_n - (z - x)\mathbf{I})^{-1}\left(\sum_{k=1}^p \tau_k \mathbf{q}_k \mathbf{q}_k^* - x\mathbf{I}\right)(\mathbf{B}_n - z\mathbf{I})^{-1}. \end{aligned}$$

From this and the definition of Stieltjes transform of ESD of random matrices, using the formula $\mathbf{q}_k^*(\mathbf{B}_n - z\mathbf{I})^{-1} = \frac{\mathbf{q}_k^*(\mathbf{B}_{k,n} - z\mathbf{I})^{-1}}{1 + \tau_k \mathbf{q}_k^*(\mathbf{B}_{k,n} - z\mathbf{I})^{-1}\mathbf{q}_k}$, we have

$$s_{A_n}(z - x) - s_n(z) = \frac{1}{n}\mathrm{tr}(\mathbf{A}_n - (z - x)\mathbf{I})^{-1}\left(\sum_{k=1}^p \tau_k \mathbf{q}_k \mathbf{q}_k^* - x\mathbf{I}\right)(\mathbf{B}_n - z\mathbf{I})^{-1}$$

$$= \frac{1}{n}\mathrm{tr}(\mathbf{A}_n - (z-x)\mathbf{I})^{-1}\sum_{k=1}^{p}\tau_k\mathbf{q}_k\mathbf{q}_k^*(\mathbf{B}_n - z\mathbf{I})^{-1}$$
$$-\frac{x}{n}\mathrm{tr}(\mathbf{A}_n - (z-x)\mathbf{I})^{-1}(\mathbf{B}_n - z\mathbf{I})^{-1}$$
$$= \frac{1}{n}\sum_{k=1}^{p}\frac{\tau_k d_k}{1+\tau_k \mathrm{E}s_n(z)},$$

where

$$d_k = \frac{1+\tau_k \mathrm{E}s_n(z)}{1+\tau_k\mathbf{q}_k^*(\mathbf{B}_{k,n}-z\mathbf{I})^{-1}\mathbf{q}_k}\mathbf{q}_k^*(\mathbf{B}_{k,n}-z\mathbf{I})^{-1}(\mathbf{A}_n-(z-x)\mathbf{I})^{-1}\mathbf{q}_k$$
$$-\frac{1}{n}\mathrm{tr}(\mathbf{B}-z\mathbf{I})^{-1}(\mathbf{A}_n-(z-x)\mathbf{I})^{-1}.$$

Write $d_k = d_{k1} + d_{k2} + d_{k3}$, where

$$d_{k1} = \frac{1}{n}\mathrm{tr}(\mathbf{B}_{k,n}-z\mathbf{I})^{-1}(\mathbf{A}_n-(z-x)\mathbf{I})^{-1} - \frac{1}{n}\mathrm{tr}(\mathbf{B}_n-z\mathbf{I})^{-1}(\mathbf{A}_n-(z-x)\mathbf{I})^{-1},$$
$$d_{k2} = \mathbf{q}_k^*(\mathbf{B}_{k,n}-z\mathbf{I})^{-1}(\mathbf{A}_n-(z-x)\mathbf{I})^{-1}\mathbf{q}_k - \frac{1}{n}\mathrm{tr}(\mathbf{B}_{k,n}-z\mathbf{I})^{-1}(\mathbf{A}_n-(z-x)\mathbf{I})^{-1},$$
$$d_{k3} = \frac{\tau_k(\mathrm{E}s_n(z)-\mathbf{q}_k^*(\mathbf{B}_{k,n}-z\mathbf{I})^{-1}\mathbf{q}_k)(\mathbf{q}_k^*(\mathbf{B}_{k,n}-z\mathbf{I})^{-1}(\mathbf{A}_n-(z-x)\mathbf{I})^{-1}\mathbf{q}_k)}{1+\tau_k\mathbf{q}_k^*(\mathbf{B}_{k,n}-z\mathbf{I})^{-1}\mathbf{q}_k}.$$

Noting that $\|(\mathbf{A}_n - (z-x)\mathbf{I})^{-1}\| \le v^{-1}$, we have

$$|d_{k1}| = \frac{1}{n}\left|-\frac{\tau_k\mathbf{q}_k^*(\mathbf{B}_{k,n}-z\mathbf{I})^{-1}(\mathbf{A}_n-(z-x)\mathbf{I})^{-1}(\mathbf{B}_{k,n}-z\mathbf{I})^{-1}\mathbf{q}_k}{1+\tau_k\mathbf{q}_k^*(\mathbf{B}_{k,n}-z\mathbf{I})^{-1}\mathbf{q}_k}\right|$$
$$\le n^{-1}v^{-1}\frac{|\tau_k|\mathbf{q}_k^*(\mathbf{B}_{k,n}-z\mathbf{I})^{-1}(\mathbf{B}_{k,n}-\bar{z}\mathbf{I})^{-1}\mathbf{q}_k}{|\Im(1+\tau_k\mathbf{q}_k^*(\mathbf{B}_{k,n}-z\mathbf{I})^{-1}\mathbf{q}_k)|}$$
$$\le \frac{1}{nv^2}.$$

Therefore, by (2.3.9), we obtain

$$\frac{1}{n}\sum_{k=1}^{p}\frac{|\tau_k d_{k1}|}{|1+\tau_k\mathrm{E}s_n(z)|} \le \frac{1}{nv^2\delta} \to 0.$$

Obviously, $\mathrm{E}d_{k2} = 0$.

To estimate $\mathrm{E}d_{k3}$, we first show that

$$\left|\frac{\tau_k(\mathbf{q}_k^*(\mathbf{B}_{k,n}-z\mathbf{I})^{-1}(\mathbf{A}_n-(z-x)\mathbf{I})^{-1}\mathbf{q}_k)}{1+\tau_k\mathbf{q}_k^*(\mathbf{B}_{k,n}-z\mathbf{I})^{-1}\mathbf{q}_k}\right| \le 2\tau_0 v^{-2}\|\mathbf{q}_k\|^2. \tag{2.3.15}$$

One can consider $\mathbf{q}_k^*(\mathbf{B}_{k,n} - z\mathbf{I})^{-1}\mathbf{q}_k/\|\mathbf{q}_k\|^2$ as the Stieltjes transform of a distribution. Thus, by Cauchy-Schwarz inequality[5], we have

[5] Suppose $s(z) = \int \frac{dF(x)}{x-z}$ with $z = u + iv$, then

$$|\Re(\mathbf{q}_k^*(\mathbf{B}_{k,n}-z\mathbf{I})^{-1}\mathbf{q}_k)| \le v^{-1/2}\|\mathbf{q}_k\|\sqrt{\Im(\mathbf{q}_k^*(\mathbf{B}_{k,n}-z\mathbf{I})^{-1}\mathbf{q}_k)}.$$

Thus, if

$$\tau_0 v^{-1/2}\|\mathbf{q}_k\|\sqrt{\Im(\mathbf{q}_k^*(\mathbf{B}_{k,n}-z\mathbf{I})^{-1}\mathbf{q}_k)} \le 1/2,$$

then

$$|1+\tau_k(\mathbf{q}_k^*(\mathbf{B}_{k,n}-z\mathbf{I})^{-1}\mathbf{q}_k)| \ge 1-\tau_0|\Re(\mathbf{q}_k^*(\mathbf{B}_{k,n}-z\mathbf{I})^{-1}\mathbf{q}_k)| \ge 1/2.$$

Hence,

$$\left|\frac{\tau_k(\mathbf{q}_k^*(\mathbf{B}_{k,n}-z\mathbf{I})^{-1}(\mathbf{A}_n-(z-x)\mathbf{I})^{-1}\mathbf{q}_k)}{1+\tau_k\mathbf{q}_k^*(\mathbf{B}_{k,n}-z\mathbf{I})^{-1}\mathbf{q}_k}\right| \le 2\tau_0 v^{-2}\|\mathbf{q}_k\|^2.$$

Otherwise, we have

$$\begin{aligned}
&\left|\frac{\tau_k(\mathbf{q}_k^*(\mathbf{B}_{k,n}-z\mathbf{I})^{-1}(\mathbf{A}_n-(z-x)\mathbf{I})^{-1}\mathbf{q}_k)}{1+\tau_k\mathbf{q}_k^*(\mathbf{B}_{k,n}-z\mathbf{I})^{-1}\mathbf{q}_k}\right| \\
&\le \frac{|\tau_k|\|(\mathbf{B}_{k,n}-z\mathbf{I})^{-1}\mathbf{q}_k\|\|(\mathbf{A}_n-(z-x)\mathbf{I})^{-1}\mathbf{q}_k)\|}{|\Im(1+\tau_k\mathbf{q}_k^*(\mathbf{B}_{k,n}-z\mathbf{I})^{-1}\mathbf{q}_k)|} \\
&= \frac{\|(\mathbf{A}_n-(z-x)\mathbf{I})^{-1}\mathbf{q}_k\|}{\sqrt{v\Im(\mathbf{q}_k^*(\mathbf{B}_{k,n}-z\mathbf{I})^{-1}\mathbf{q}_k)}} \\
&\le 2\tau_0 v^{-2}\|\mathbf{q}_k\|^2.
\end{aligned}$$

Therefore, for some constant C,

$$|\mathrm{E}d_{k3}|^2 \le C\mathrm{E}|\mathrm{E}s_n(z)-\mathbf{q}_k^*(\mathbf{B}_{k,n}-z\mathbf{I})^{-1}\mathbf{q}_k|^2\mathrm{E}\|\mathbf{q}_k\|^4. \tag{2.3.16}$$

At first, we have

$$\begin{aligned}
\mathrm{E}\|\mathbf{q}_k\|^4 &= \frac{1}{n^2}\mathrm{E}\left(\sum_{i=1}^n |x_{ik}|^2\right)^2 \\
&= \frac{1}{n^2}\left[\sum_{i=1}^n \mathrm{E}|x_{ik}|^4 + \sum_{i\ne j}\mathrm{E}|x_{ik}|^2\mathrm{E}|x_{jk}|^2\right] \\
&\le \frac{1}{n^2}[n^2\eta_n^2+n(n-1)] \le 1+\eta_n^2.
\end{aligned}$$

To complete the proof of the convergence of $\mathrm{E}s_n(z)$, we need to show that

---

$$\begin{aligned}
\Re(s(z)) &= \int \frac{(x-u)dF(x)}{(x-u)^2+v^2} \le \left(\int \frac{(x-u)^2dF(x)}{(x-u)^2+v^2}\int\frac{dF(x)}{(x-u)^2+v^2}\right)^{1/2} \\
&\le v^{-1/2}\sqrt{\Im(s(z))}.
\end{aligned}$$

$$\frac{1}{n}\sum_{k=1}^{p}(\mathrm{E}|\mathrm{E}s_n(z)-\mathbf{q}_k^*(\mathbf{B}_{k,n}-z\mathbf{I})^{-1}\mathbf{q}_k|^2)^{1/2}\to 0. \tag{2.3.17}$$

Write $(\mathbf{B}_{k,n}-z\mathbf{I})^{-1}=(b_{ij})$. Then, we have

$$\begin{aligned}
&\mathrm{E}\left|\mathbf{q}_k^*(\mathbf{B}_{k,n}-z\mathbf{I})^{-1}\mathbf{q}_k-\frac{1}{n}\sum_{i=1}^{n}\sigma_{ik}^2 b_{ii}\right|^2\\
&\le\frac{1}{n^2}\left[\sum_{i=1}^{n}\mathrm{E}|x_{ik}^2-\sigma_{ik}^2|^2+2\sum_{i\neq j}\mathrm{E}|x_{ik}^2|\mathrm{E}|x_{jk}^2||b_{ij}|^2\right]\\
&\le v^{-2}\eta_n^2+\frac{2}{n^2}\mathrm{tr}((\mathbf{B}_{k,n}-u\mathbf{I})^2+v^2\mathbf{I})^{-1}\\
&\le v^{-2}[\eta_n^2+n^{-1}]\to 0.
\end{aligned}$$

By noticing $1-\sigma_{ik}^2\ge 0$,

$$\left|\frac{1}{n}\sum_{i=1}^{n}(\sigma_{ik}^2-1)b_{ii}\right|\le\frac{1}{nv}\sum_{i=1}^{n}(1-\sigma_{ik}^2),$$

$$\left|\frac{1}{n}\mathrm{tr}(\mathbf{B}_{k,n}-z\mathbf{I})^{-1}-s_n(z)\right|\le 1/nv.$$

By Step 1, we have

$$\mathrm{E}|s_n(z)-\mathrm{E}(s_n(z))|^2\le\frac{1}{nv^2}.$$

Then, (2.3.17) follows from the above estimates.

Up to the present, we have proved that for any $z\in\mathbb{C}^+$,

$$s_{A_n}(z-x)-\mathrm{E}s_n(z)\to 0.$$

For any subsequence n' such that $\mathrm{E}s_n(z)$ tends to a limit, say s, by assumption of the theorem, we have

$$x=x_{n'}=\frac{1}{n}\sum_{k=1}^{p}\frac{\tau_k}{1+\tau_k\mathrm{E}s_{n'}(z)}\to y\int\frac{\tau dH(\tau)}{1+\tau s}.$$

Therefore, s will satisfy (2.3.3). We have thus proved (2.3.11) if equation (2.3.3) has a unique solution $s\in\mathbb{C}^+$, which is done in the next step.

Step 3. Uniqueness of solution of (2.3.3)

If F^A is a zero measure, the unique solution is obviously $s(z)=0$. Now, suppose that $F^A\neq 0$ and we have two solutions s_1, $s_2\in\mathbb{C}^+$ of equation (2.3.3) for a common $z\in\mathbb{C}^+$, that is,

$$s_j=\int\frac{dF^A(\lambda)}{\lambda-z+y\int\frac{\tau dH(\tau)}{1+\tau s_j}}, \tag{2.3.18}$$

from which we obtain

$$s_1 - s_2$$
$$= y \int \frac{(s_1 - s_2)\lambda^2 dH(\tau)}{(1+\tau s_1)(1+\tau s_2)} \int \frac{dF^A(\lambda)}{(\lambda - z + y\int \frac{\tau dH(\tau)}{1+\tau s_1})(\lambda - z + y\int \frac{\tau dH(\tau)}{1+\tau s_2})}.$$

If $s_1 \neq s_2$, then

$$\int \frac{y\int \frac{\lambda^2 dH(\tau)}{(1+\tau s_1)(1+\tau s_2)} dF^A(\lambda)}{(\lambda - z + y\int \frac{\tau dH(\tau)}{1+\tau s_1})(\lambda - z + y\int \frac{\tau dH(\tau)}{1+\tau s_2})} = 1.$$

By Cauchy-Schwarz Inequality, we have

$$1 \leq \left(\int \frac{y\int \frac{\lambda^2 dH(\tau)}{|1+\tau s_1|^2} dF^A(\lambda)}{|\lambda - z + y\int \frac{\tau dH(\tau)}{1+\tau s_1}|^2} \int \frac{y\int \frac{\lambda^2 dH(\tau)}{|1+\tau s_2|^2} dF^A(\lambda)}{|\lambda - z + y\int \frac{\tau dH(\tau)}{1+\tau s_2}|^2} \right)^{1/2}.$$

From (2.3.18), we have

$$\Im s_j = \int \frac{v + y\Im s_j \int \frac{\tau^2 dH(\tau)}{|1+\tau s_j|^2} dF^A(\lambda)}{|\lambda - z + y\int \frac{\tau dH(\tau)}{1+\tau s_j}|^2} > \int \frac{y\Im s_j \int \frac{\tau^2 dH(\tau)}{|1+\tau s_j|^2} dF^A(\lambda)}{|\lambda - z + y\int \frac{\tau dH(\tau)}{1+\tau s_j}|^2}$$

which implies that for both $j = 1$ and 2,

$$1 > \int \frac{y\int \frac{\tau^2 dH(\tau)}{|1+\tau s_j|^2} dF^A(\lambda)}{|\lambda - z + y\int \frac{\tau dH(\tau)}{1+\tau s_j}|^2}.$$

The inequality is strict even if F^A is a zero measure, which leads to a contradiction. The contradiction proves that $s_1 = s_2$ and hence equation (2.3.3) has at most one solution. The existence of solution to (2.3.3) has been seen in Step 2. The proof of this theorem is then complete.

2.3.7 When T is a Wigner Matrix

Bai, Jin and Miao (2007) derived the explicit form of LSD of $\mathbf{S}_n\mathbf{W}$ when $\mathbf{S}_n$ is independent of $\mathbf{W}$, where $\mathbf{W}$ is a Wigner matrix.

A generalized definition of Wigner matrix only requires the matrix to be a Hermitian random matrix whose entries on or above the diagonal are independent.

Suppose that $\mathbf{W}_n = n^{-1/2}\mathbf{Y}_n$ is a Wigner matrix, the entries above or on the diagonal of $\mathbf{Y}_n$ are independent but may be dependent of n and may not be identically distributed. Assume that all the entries of $\mathbf{Y}_n$ are of mean zero and variance 1 and satisfy the following condition. For any constant $\eta > 0$

$$\lim_{n\to\infty}\frac{1}{n^2\eta^2}\sum_{jk}\mathrm{E}|y_{jk}^{(n)}|^2 I(|y_{jk}^{(n)}|\geq\eta\sqrt{n})=0. \tag{2.3.19}$$

Suppose that the entries of $\mathbf{X}_n$ $(p\times n)$ are independent complex random variables satisfying (2.2.7), and $\mathbf{S}_n=\frac{1}{n}\mathbf{X}_n\mathbf{X}_n^*$.

Theorem 2.16. *Under above conditions, we have*

$$\beta_k^{sw}=\begin{cases}\sum_{j=1}^{k/2}\binom{k}{j-1}\binom{k}{k/2-j}2y^{j-1}/k & \text{if } k \text{ is even,}\\ 0 & \text{if } k \text{ is odd,}\end{cases} \tag{2.3.20}$$

where β_k^{sw} *is the moments of the LSD of* $\mathbf{S}_n\mathbf{W}_n$.

Furthermore, the density function of the LSD of $\mathbf{S}_n\mathbf{W}_n$ *is*

$$\begin{cases}\frac{1}{2\pi}\sqrt{\frac{R}{x^2}}\sqrt{2+R+\frac{2}{y}-\frac{2}{y}\sqrt{\frac{x^2}{R}}} & \text{if } |x|<a,\\ \\ 0 & \text{otherwise,}\end{cases} \tag{2.3.21}$$

where

$$a=\sqrt{\frac{2(1+14y+y^2)^{3/2}+72y(1+y)-2(1+y)^3}{27y}},$$

and

$$R=\frac{-2(1+y)}{3y}+\frac{2\sqrt{1+14y+y^2}}{3y}\cos\frac{\varphi}{3},$$

where $\cos\varphi=\frac{-72y(1+y)+2(1+y)^3+27yx^2}{2^{1/3}\sqrt{1+14y+y^2}}$, $\varphi\in[0,\pi]$.

And the point mass at 0 is:

$$F(\{0\})=\begin{cases}1-\frac{1}{y} & \text{if } y>1,\\ 0 & \text{otherwise.}\end{cases}$$

That is, the matrix S_nW_n *has* $p-n$ *0 eigenvalues.*

2.4 Hadamard Product

In nuclear physics, since the particles move in a very high velocity in a small range, many exciting states in very short time periods cannot be observed. Generally, if a real physical system is not of full connectivity, the random matrix describing the interactions between the particles in the system will have a large proportion of zero elements. In this case, a sparse random matrix provides a more natural and relevant description of the system. Indeed, in neural network theory, the number of neurons in one person's brain is probably of several orders of magnitude larger than that of the dendrites connected with one individual neuron (see Grenander and Silverstein (1977)). Sparse random matrices are adopted in modeling these partially connected systems in neural network theory. A sparse or dilute matrix is a random matrix in which some entries will be replaced by 0 if not observed. Sometimes a large portion of the entries of the interesting random matrix can be 0's. Due to its special application background, the sparse matrix has received special attention in quantum mechanics, atomic physics, neural networks, and many other areas, such as linear algebra, neural networks, algorithms and computing, finance modeling, electrical engineering, bio-interactions and theoretical physics.

A sparse matrix can be expressed by a Hadamard product. If $\mathbf{B}_m = [b_{ij}]$ and $\mathbf{D}_m = [d_{ij}]$ are two $m \times m$ matrices, then the Hadamard product $\mathbf{A}_p = (A_{ij})$ with $A_{ij} = b_{ij}d_{ij}$ is denoted by

$$\mathbf{A}_p = \mathbf{B}_m \circ \mathbf{D}_m.$$

The matrix $\mathbf{A}_p$ is sparse if the elements d_{ij} of $\mathbf{D}_m$ take values 0 and 1 with $\sum_{i=1}^m P(d_{ij} = 1) = p = o(m)$. The index p usually stands for the level of sparseness; *i.e.*, after performing the Hadamard product, the resulting matrix will have p nonzero elements per row on the average.

It is commonly assumed that the matrix $\mathbf{D}_m$ is symmetric and its entries $\{d_{ij} : i \le j\}$ are independent Bernoulli trials with $P(d_{ij} = 1) = p_{ij}$ and independent of the entries of the matrix $\mathbf{B}_m$. In Kohrunzhy and Rodgers (1997, 1998), it is assumed that

$$p_{ij} = \frac{\alpha}{m^\beta},$$

with $0 \le \beta \le 1$, $0 < \frac{\alpha}{m^\beta} < 1$, and the entries of $\mathbf{B}_m$ are centralized elements of a sample covariance matrix. More precisely, suppose $\mathbf{X}_{m,n} = [x_{ij} : i = 1, 2, \ldots, m, j = 1, 2, \ldots, n]$ is an $m \times n$ matrix with independent entries of mean 0 and variance σ^2. Let the sample covariance matrix of $\mathbf{X}_{m,n}$ be defined as $\mathbf{S}_n = \frac{1}{n}\mathbf{X}_{m,n}\mathbf{X}_{m,n}^*$. Then, define $\mathbf{B}_m = \sqrt{n/p}(\mathbf{S}_n - \sigma^2\mathbf{I}_m)$ and $\mathbf{A}_p = \mathbf{B}_m \circ \mathbf{D}_m$.

Bai and Zhang (2006) considered a kind of Hadamard product of a normalized sample covariance matrix with a sparsing matrix (whose entries are not necessarily Bernoulli trials) and show the weak and strong convergence to the semicircle law of the ESD of this kind of Hadamard product. To this end, they made the following assumptions. In what follows, the entries of $\mathbf{D}_m$ and $\mathbf{X}_{m,n}$ are allowed to depend on n. For brevity, the dependence on n is suppressed.

Assumptions on $\mathbf{D}_m$.

(D1) $\mathbf{D}_m = [d_{ij}]$ is $m \times m$ Hermitian with $\{d_{ij} : i \le j\}$ independent complex random variables.

(D2) $\max_j \left|\sum_{i=1}^m p_{ij} - p\right| = o(p)$, where $p_{ij} = \mathrm{E}|d_{ij}^2|$.

(D3.1) For some $\delta \in [0, 1/2]$, there exists a constant $C_1 > 0$ such that $\max_j \sum_i \mathrm{E}|d_{ij}| \leq C_1 m^\delta p^{1-\delta}$.

(D3.2) For each $k > 2$ there is a constant C_k such that

$$\mathrm{E}|d_{ij}|^k \leq C_k p_{ij}. \tag{2.4.22}$$

We remark here that condition (D3.2) implies that p_{ij} are uniformly bounded. In fact, $p_{ij} = \mathrm{E}|d_{ij}|^2 \leq (\mathrm{E}|d_{ij}|^4)^{1/2} \leq C_4$. Combining this fact with condition (D2), we indeed have $p \leq Km$, for some constant $K > 0$. In view of this relation between p and m, we notice that if condition (D3.1) holds for some $\delta_0 \in [0, 1/2]$, then it must hold for every $\delta \geq \delta_0$, $\delta \in [0, 1/2]$. Therefore, we clarify here and in what follows that when we say condition (D3.1) holds for some $\delta^* \in [0, 1/2]$ we are referring to δ^* as the smallest value in $[0, 1/2]$ for which condition (D3.1) holds. In this sense, for any $0 \leq \delta_1 \leq \delta_2 \leq 1/2$, we say we have a stronger sparseness in the case when condition (D3.1) holds for δ_1 than in the case when condition (D3.1) holds for δ_2. Also note that in the case of the weakest sparseness, i.e., $\delta = 1/2$, condition (D3.1) is a direct consequence of Hölder's inequality and condition (D2); that is, no additional assumption is imposed on the first moments of the sparsing factors in this case.

Assumptions on $\mathbf{X}_{m,n}$.

(X1) $\mathbf{X}_{m,n} = [x_{ij}]$ is $m \times n$ consisting of independent random variables with $\mathrm{E}x_{ij} = 0$, $\mathrm{E}|x_{ij}|^2 = \sigma^2$.

(X2.1) For any $\eta > 0$,

$$\frac{1}{mn} \sum_{ij} \mathrm{E}|x_{ij}^2| I[|x_{ij}| > \eta \sqrt[4]{np}] \to 0.$$

(X2.2) For any $\eta > 0$,

$$\sum_{u=1}^{\infty} \frac{1}{mn} \sum_{ij} \mathrm{E}|x_{ij}^2| I[|x_{ij}| > \eta \sqrt[4]{np}] < \infty,$$

where u may take $[p]$, m, or n.

(X3) For any $\eta > 0$,

$$\frac{1}{m} \sum_{i=1}^{m} P\left(\left| \sum_{k=1}^{n} (|x_{ik}|^2 - \sigma^2) d_{ii} \right| > \eta \sqrt{np} \right) \to 0. \tag{2.4.23}$$

We shall present the following theorem.

Theorem 2.17. (i) *Conditions* (2.4.22) *and* (2.4.23) *hold.*

(ii) *The entries of* $\mathbf{D}_m$ *are independent of those of* $X_{m,n}$.

(iii) $p/n \to 0$ *and* $p \to \infty$.

(iv) *Condition* (D3.1) *holds for* $\delta = 1/2$ *and* $m/n \to 0$, *or condition* (D3.1) *holds for some* $\delta \in (0, 1/2)$ *and* $m \leq Kn$ *for some constant* K, *or condition* (D3.1) *holds for* $\delta = 0$ *and there is no restriction between* m *and* n.

Then, the ESD $F^{\mathbf{A}_p}$ converges weakly to the semicircle law with scale parameter σ^2 as $[p] \to \infty$, where $\mathbf{A}_p = \frac{1}{\sqrt{np}}(\mathbf{X}_{m,n}\mathbf{X}_{m,n}^* - \sigma^2 n\mathbf{I}_m) \circ \mathbf{D}_m$. The convergence is in the sense of in probability (i.p.) if condition (X2.1) is assumed, and the convergence is in the sense of a.s. for $[p] \to \infty$ or $m \to \infty$ if condition (X2.2) is assumed for $u = [p]$ or $u = m$, respectively.

Remark 2.18. Theorem 2.17 covers all well-known results on sparse matrices, since the Bernoulli trials satisfy conditions (D3.1) (with $\delta = 0$) and (D3.2) obviously. The new contribution of Theorem 2.17 is to allow the sparsing factors d_{ij} to be very non-homogenous. Consider the following example. Let $\mathbf{D}_m = [d_{ij}]$ be symmetric. Let $m = kL$ with L fixed, and let for all $(\ell - 1)k < i, j \le \ell k$ with $\ell \le L$,

$$P(d_{ij} = 1) = p/k = 1 - P(d_{ij} = 0),$$

and for all other indices i, j, $d_{ij} \equiv 0$. Then conditions (D1), (D2), (D3.1), and (D3.2) are true whenever $p \le k$.

Remark 2.19. In the case of the strongest sparseness, condition (D3.1) seems not to allow the d_{ij}'s to take large values. In fact, it is not the case. For example, consider that

$$d_{ij} = c_n^{-1}|z_{ij}|I(|z_{ij}| > c_n),$$

where z_{ij} are i.i.d. $N(0,1)$ subject to the condition $d_{ij} = d_{ji}$ and c_n is a positive constant uniquely solving the equation $\mathrm{E}z_{ij}^2 I(|z_{ij}| > c_n) = c_n^2 p/m$. Then obviously d_{ij} can take very large values, and $\mathbf{D}_m$ is symmetric with

$$\sum_{i=1}^m p_{ij} = c_n^{-2}\sum_{i=1}^m \mathrm{E}z_{ij}^2 I(|z_{ij}| > c_n) = p,$$

i.e., conditions (D1) and (D2) are satisfied.

Now we show that condition (D3.1) holds for $\delta = 0$ if $p/m \to 0$. In fact, we can see that if $p/m \to 0$, then $c_n \to \infty$ and consequently

$$Ez_{ij}^2 I(|z_{ij}| > c_n) \simeq 2c_n\varphi(c_n),$$

which implies that

$$\frac{p}{m} \simeq \frac{2}{c_n}\varphi(c_n),$$

where the notation "$\simeq$" is used to represent the relation that the two quantities on its two sides have a ratio tending to 1 as $n \to \infty$, while $\varphi(\cdot)$ is the density function of standard normal variables.

Therefore, we get

$$\sum_{i=1}^m \mathrm{E}|d_{ij}| = mc_n^{-1}\mathrm{E}|z_{ij}|I(|z_{ij}| > c_n) \simeq 2mc_n^{-1}\phi(c_n) \simeq p,$$

and

$$\begin{aligned}\mathrm{E}|d_{ij}|^k &= c_n^{-k}\mathrm{E}|z_{ij}|^k I(|z_{ij}| > c_n)\\ &\simeq 2c_n^{-k}c_n^{k-1}\phi(c_n) = 2c_n^{-1}\phi(c_n)\\ &\simeq p/m = \mathrm{E}|d_{ij}|^2,\end{aligned}$$

which implies that condition (D3.1) holds for $\delta = 0$, and that condition (D3.2) holds.

Remark 2.20. However, if condition (D3.1) is assumed for $\delta = 0$, then sometimes it may happen that the d_{ij}'s are not allowed to take small values with large probabilities. For example,

$$d_{ij} = \sqrt{p/m} \text{ with probability } 1.$$

Then obviously (D3.1) holds for and only for $\delta = 1/2$, i.e., the weakest sparseness. For this case, condition (D3.1) holds automatically when condition (D3.2) is assumed. Condition (D3.2), assuming that higher moments of the d_{ij}'s are not larger than a multiple of their second moments, is not seriously restrictive because the d_{ij}'s are usually small random variables.

Nonetheless, if condition (D3.1) is assumed for $\delta = 1/2$, the condition does allow the first moments of d_{ij}'s to be much larger than their second moments, e.g., $d_{ij} = c\sqrt{p/m}|z_{ij}|$, where z_{ij} are i.i.d. random variables subject to the restriction $d_{ij} = d_{ji}$ and c makes $\mathrm{E}d_{ij}^2 = p/m$. As a price, however, we need to require m to have a smaller order than that of n, that is, $m/n \to 0$. The requirement is necessary in some sense. In Section 2.4.2, we will present a counterexample showing that the semicircle law fails to hold if this requirement is not satisfied.

Remark 2.21. Note that p may not be an integer, and it may increase very slowly as n increases. Thus, the limit for $p \to \infty$ may not be true for almost sure convergence. So, we consider the limit when the integer part of p tends to infinity. However, if we consider the convergence in probability, Theorem 2.17 is true for $p \to \infty$.

Remark 2.22. Conditions (D2) and (D3.2) imply that $p \leq Km$; that is, the order of p cannot be larger than that of m. In the theorem, it is assumed that $p/n \to 0$; that is, p also has a lower order than n. This is essential. However, the relation between m and n can be arbitrary if condition (D3.1) holds for $\delta = 0$. It is important to remind the readers that the statement "the relation between m and n can be arbitrary" only says that there are examples with $m/n \to \infty$ as well as examples with $m/n \to 0$, for which the results of Theorem 2.17 are applicable equally well once condition (D3.1) is satisfied with $\delta = 0$. For example, if the d_{ij}'s (subject to the condition $d_{ij} = d_{ji}$) are the Bernoulli trials defined by $P(d_{ij} = 1) = p/m = 1 - P(d_{ij} = 0)$ for any i, j, then $E|d_{ij}| = E|d_{ij}|^2 = E|d_{ij}|^k$ for any $k > 2$. This implies conditions (D1),

(D2), (D3.1) (with $\delta = 0$), and (D3.2) always hold. But no matter $m/n \to 0$, $m/n \le K < \infty$, or $m/n \to \infty$, Theorem 2.17 always holds, provided that $p/n \to 0$.

Remark 2.23. From the proof of the theorem, one can see that the almost sure convergence is true for $m \to \infty$ in all places except the part of the truncation on the entries of $\mathbf{X}_{m,n}$ which was guaranteed by condition (X2.2). Thus, if condition (X2.2) holds for $u = m$, then the almost sure convergence is true in the sense of $m \to \infty$. Sometimes, it may be of interest to consider the almost sure convergence in the sense of $n \to \infty$. Examining the proof given in the next sections, one can find that to guarantee the almost sure convergence for $n \to \infty$, the removal of the diagonal elements of the matrix requires $m/\log n \to \infty$; the truncation on the entries of $\mathbf{X}_{m,n}$ requires condition (X2.2) to be true for $u = n$. As for Proposition 2.28, one may modify the conclusion of (II) on page 51 as

$$E|M_k - EM_k|^{2\mu} = O(m^{-\mu})$$

for any fixed integer μ, where M_k is defined on Page 50. Thus, if $m \ge n^{\varepsilon}$ for some positive constant ε, then the almost sure convergence of the ESD of the matrix after the truncation and centralization is true for $n \to \infty$. We therefore see that the conclusion of Theorem 2.17 can be strengthened to the almost sure convergence as $n \to \infty$ under the additional assumptions that, for some small positive constant ε, $m \ge n^{\varepsilon}$ and condition (X2.2) holds for $u = n$.

Remark 2.24. In Theorem 2.17, if $p = m$ and $d_{ij} \equiv 1$ for all i and j and the entries of $X_{m,n}$ are i.i.d., then the model considered in Theorem 2.17 reduces to that of Bai and Yin (1988), where it is assumed that the fourth moment of x_{ij} is finite. It can be easily verified that the conditions of Theorem 2.17 are satisfied under Bai and Yin's assumption. Thus, Theorem 2.17 contains Bai and Yin's result as a special case.

2.4.1 Truncation and Centralization

In this section, we simply outline the truncation and centralization techniques to A_p.

Step 1. Removal of the diagonal elements of $\mathbf{A}_p$ For any $\varepsilon > 0$, denote by $\widehat{\mathbf{A}}_p$ the matrix obtained from A_p by replacing its diagonal elements whose absolute values are greater than ε by 0 and denote by $\widetilde{\mathbf{A}}_p$ the matrix obtained from $\mathbf{A}_p$ by replacing all its diagonal elements by 0.

Proposition 2.25. *Under the assumptions of Theorem* 2.17,

$$\|F^{\widehat{A}_p} - F^{A_p}\| \to 0, \ a.s.$$

and

$$L^3(F^{\widehat{A}_p}, F^{\widetilde{A}_p}) \le \varepsilon^2.$$

Proof. The second conclusion of the proposition is a trivial consequence of the difference inequality. As for the first conclusion, by the rank inequality,

$$\|F^{\widehat{A}_p} - F^{A_p}\| \leq \frac{1}{m}\sum_{i=1}^{m} I\left[\left|\frac{1}{\sqrt{np}}\sum_{k=1}^{n}(|x_{ik}|^2 - \sigma^2)d_{ii}\right| > \varepsilon\right].$$

By condition (X3) in (2.4.23), we have

$$\sum_{i=1}^{m} P\left(\left|\frac{1}{\sqrt{np}}\sum_{k=1}^{n}(|x_{ik}|^2 - \sigma^2)d_{ii}\right| > \varepsilon\right) = o(m).$$

It then turns out that the first conclusion is a consequence of the Bernstein's inequality and the Borel-Cantelli Lemma.

Combining the two conclusions in Proposition 2.25, we have shown that

$$L(F^{A_p}, F^{\widetilde{A}_p}) \to 0, \text{ a.s.}$$

Hence, in what follows, we can assume that the diagonal elements are 0; i.e., assume $d_{ii} = 0$ for all $i = 1, \ldots, m$.

Step 2. Truncation and centralization of the entries of $\mathbf{X}_{m,n}$

Note that condition (X2.1) in (2.4.23) guarantees the existence of $\eta_n \downarrow 0$ such that

$$\frac{1}{mn\eta_n^2}\sum_{ij} \mathrm{E}|x_{ij}|^2 I(|x_{ij}| > \eta_n \sqrt[4]{np}) \to 0.$$

Similarly, if condition (X2.2) holds, there exists $\eta_n \downarrow 0$ such that

$$\sum_{u=1}^{\infty} \frac{1}{mn\eta_n^2}\sum_{ij} \mathrm{E}|x_{ij}|^2 I[|x_{ij}| > \eta_n \sqrt[4]{np}] < \infty,$$

for u takes $[p]$, m, or n. In the subsequent truncation procedure, we shall not distinguish under whichever condition the sequence $\{\eta_n\}$ is defined. The reader should remember that whatever condition is used, the $\{\eta_n\}$ is defined by that condition.

Define $\tilde{x}_{ij} = x_{ij}I[|x_{ij}| \leq \eta_n \sqrt[4]{np}] - \mathrm{E}x_{ij}I[|x_{ij}| \leq \eta_n \sqrt[4]{np}]$ and $\hat{x}_{ij} = x_{ij} - \tilde{x}_{ij}$. Also, define $\widetilde{\mathbf{B}}_m$ with $\widetilde{\mathbf{B}}_{ij} = \frac{1}{\sqrt{np}}\sum_{k=1}^{n} \tilde{x}_{ik}\bar{\tilde{x}}_{jk}$ $(i \neq j)$, and denote its Hadamard product with $\mathbf{D}_m$ by $\widetilde{\mathbf{A}}_p$. Then we have the following proposition.

Proposition 2.26. *Under condition* (X2.1) *in* (2.4.23) *and the other assumptions of Theorem* 2.17,

$$L(F^{\widetilde{A}_p},\ F^{A_p}) \to 0, \ i.p.$$

If condition (X2.1) *is strengthened to* (X2.2), *then*

$$L(F^{\widetilde{A}_p},\ F^{A_p}) \to 0, \ a.s. \text{ as } \ u \to \infty,$$

where $u = [p]$, m, or n in accordance with the choice of u in condition (X2.2).

The proof of a routine application of difference inequality and simple argument of limiting theorems and hence omitted.

From the above two propositions, we are allowed to make the following additional assumptions:

$$\begin{aligned} &\text{(i)} \quad d_{ii} = 0. \\ &\text{(ii)} \quad \mathrm{E}x_{ij} = 0, \quad |x_{ij}| \le \eta_n \sqrt[4]{np}. \end{aligned} \tag{2.4.24}$$

Note that, we shall no longer have $\mathrm{E}|x_{ij}|^2 = \sigma^2$ after the truncation and centralization on the $X_{m,n}$ entries. Write $\mathrm{E}|x_{ij}|^2 = \sigma_{ij}^2$. We need the following proposition.

Proposition 2.27. *Under the assumptions of Theorem* 2.17,
(a) $\max_j |\sum_i \mathrm{E}|d_{ij}|^2 - p| = o(p)$,
(b) *for any* i, j, $\sigma_{ij}^2 \le \sigma^2$, *and* $\frac{1}{mn}\sum_{ij}\sigma_{ij}^2 \to \sigma^2$.

The proof of this proposition is trivial and thus omitted.

2.4.2 Outlines of Proof of the theorem

In the last section, we have shown that to prove Theorem 2.17, it suffices to do it under the additional conditions (i), (ii) in (2.4.24) and (a), (b) in Proposition 2.27. In the present section, we shall prove the following proposition.

Proposition 2.28. *Suppose that the assumptions of Theorem* 2.17 *and the additional conditions* (i), (ii) *in* (2.4.24) *and* (a), (b) *in Proposition* 2.27 *hold. Then with probability one as* $m \to \infty$, *the empirical spectral distribution* $F^{A_p}(x)$ *of* $\mathbf{A}_p$ *converges weakly to the semicircle law with scale parameter* σ^2.

Proof. Recall the definition of the empirical k-th moment, we have

$$M_k = \frac{1}{mn^{k/2}p^{k/2}} \sum_{(\mathbf{i},\mathbf{j})\in\mathcal{I}} d_{(\mathbf{i},\mathbf{j})} X_{(\mathbf{i},\mathbf{j})},$$

where

$$d_{(\mathbf{i},\mathbf{j})} = d_{i_1 i_2} \cdots d_{i_k i_1},$$

$$X_{(\mathbf{i},\mathbf{j})} = x_{i_1 j_1}\overline{x}_{i_2 j_1} x_{i_2 j_2}\overline{x}_{i_3 j_2} \cdots \overline{x}_{i_k j_{k-1}} x_{i_k j_k}\overline{x}_{i_1 j_k}.$$

The moment convergence theorem requires to show that $M_k \to m_k$ a.s. and that the sequence $\{m_k\}$ satisfies the Carleman condition $\sum_{k=1}^{\infty} m_{2k}^{-1/2k} = \infty$, where

$$m_k = \begin{cases} \frac{\sigma^{4s}(2s)!}{s!(s+1)!} & \text{if } k = 2s, \\ 0 & \text{if } k = 2s+1. \end{cases}$$

By noting that $m_{2k} \leq \sigma^{4k} 2^{2k}$, it is easy to see that the Carleman condition holds. Thus, to complete the proof of the proposition, by using the Borel–Cantelli lemma, we only need to prove

$$(\mathrm{I}) : \mathrm{E}(M_k) = m_k + o(1),$$

and

$$(\mathrm{II}) : \mathrm{E}|M_k - \mathrm{E}M_k|^4 = O\left(\frac{1}{m^2}\right).$$

The assertions (I) and (II) can be proved under conditions of Proposition 2.28. The details are referred to Bai and Zhang (2006).

Before concluding this section, we present two examples. The first example is to show, when condition (D3.1) is assumed for $\delta = 1/2$, to ensure the convergence of the semicircle law of F^{A_p}, it is necessary to require $m/n \to 0$.

Example 2.29. Let $\mathbf{D}_m = [d_{ij}]$ consist of $d_{ii} = 0$ and $d_{ij} = \sqrt{p/m}$ for $i \neq j$. Let $\mathbf{X}_{m,n} = [x_{ij}]$ consist of i.i.d. standard normal random variables. Now assume $m/n \to c > 0$ and $p/n \to 0$. Then conditions (D2), (D3.1), and (D3.2) hold. Specifically, $1/2$ is the smallest parameter in $[0, 1/2]$ such that condition (D3.1) is satisfied by $\mathbf{A}_p$.

Consider the kth moment of F^{A_p}. By evaluating the leading term in $\mathrm{E}M_k$, we have

$$\mathrm{E}M_k \to m_k = \sum_{s=1}^{[\frac{k}{2}]} c^{k/2-s} \mu_s,$$

where μ_s is a number depending only on s and satisfying

$$\mu_s \leq \frac{1}{k} \binom{k}{s} \binom{k}{s-1}.$$

This shows that the Carleman condition is satisfied.

Also, one can easily show that

$$\mathrm{E}(M_k - \mathrm{E}M_k)^{2\mu} = O(m^{-2\mu}).$$

Therefore, with probability one, F^{A_p} converges to a nonrandom limiting distribution, say F. It is easy to verify that when $k = 3$,

$$m_3 = \sqrt{c}.$$

Since the third moment of F is not 0, F is not the semicircle law. That is, we have shown with probability one that F^{A_p} converges weakly to a limiting spectral distribution which is however not the semicircle law.

The next example is to show for the case when condition (D3.1) is assumed for $\delta \in (0, 1/2)$, that the condition m/n is bounded is also necessary for the convergence to the semicircle law.

Example 2.30. Let $\mathbf{D}_m = [d_{ij}]$ be defined as in Example 2.29. We assume the same conditions $m/n \to c > 0$ and $p/n \to 0$. Now we define $\tilde{\mathbf{D}}_h = \mathbf{D}_m \otimes \mathbf{I}_h$ and $\tilde{\mathbf{B}}_h = \frac{1}{\sqrt{np}}(\mathbf{X}_{mh,n}\mathbf{X}^*_{mh,n} - \sigma^2 n\mathbf{I}_{mh})$, where "$\otimes$" denotes the Kronecker product of matrices, $h = [m^\eta]$ with $\eta > 0$, and $\mathbf{X}_{mh,n}$ is $mh \times n$ consisting of i.i.d. standard normal random variables.

Let $\tilde{A}_p = \tilde{\mathbf{B}}_h \circ \tilde{\mathbf{D}}_h$. Then $\tilde{\mathbf{A}}_p = \text{diag}[\mathbf{A}_{1,m}, \ldots, \mathbf{A}_{h,m}]$, where

$$\mathbf{A}_{i,m} = \mathbf{B}_{ii} \circ \mathbf{D}_m, \ i = 1, \ldots, h,$$

and $\mathbf{B}_{ii}$ is the ith $m \times m$ major submatrix of $\tilde{\mathbf{B}}_h$.

Note that $\mathbf{A}_{1,m}, \ldots, \mathbf{A}_{h,m}$ are independent with the same distribution as $\mathbf{A}_p$ defined in Example 2.29. Denote by $\tilde{M}_k$, $M_{i,k}$ and M_k, the kth moment of $\tilde{\mathbf{A}}_p$, $\mathbf{A}_{i,m}$ and $\mathbf{A}_p$, respectively. Then it follows that $\mathrm{E}M_{i,k} = \mathrm{E}M_k$ and $\mathrm{E}(M_{i,k} - \mathrm{E}M_{i,k})^{2\mu} = \mathrm{E}(M_k - \mathrm{E}M_k)^{2\mu}$. Since $F^{\tilde{A}_p} = \frac{1}{h}\sum_{i=1}^h F^{A_{i,m}}$ so that $\tilde{M}_k = \frac{1}{h}\sum_{i=1}^h M_{i,k}$, we get $\mathrm{E}\tilde{M}_k = \mathrm{E}M_k$ and $\mathrm{E}(\tilde{M}_k - \mathrm{E}\tilde{M}_k)^{2\mu} \le \mathrm{E}(M_k - \mathrm{E}M_k)^{2\mu}$. By the results we proved in Example 2.29, it follows with probability one that $F^{\tilde{A}_p}$ converges weakly but the limiting spectral distribution is not the semicircle law.

Let us now check the validity of the assumptions of Theorem 2.17 for $\tilde{\mathbf{A}}_p$. Conditions (D1), (D2), and (D3.2) hold for $\tilde{\mathbf{A}}_p$ automatically by definition. We now show that for any $\delta \in (0, 1/2)$ by choosing $\eta > 0$ such that $2\delta(1+\eta) = 1$, condition (D3.1) is satisfied by $\tilde{\mathbf{A}}_p$ for the given δ. To see this, note that the dimension of $\tilde{\mathbf{A}}_p$ is mh, and so we have

$$\sum_i \mathrm{E}d_{ij} \le \sqrt{mp} \le \frac{\sqrt{m}}{(mh)^\delta}(mh)^\delta p^{1-\delta} \le C_1(mh)^\delta p^{1-\delta}.$$

By requiring $p = O(\log m)$, we can further see for any $\delta_0 < \delta$ that

$$\left(\sum_i \mathrm{E}d_{ij}\right) / \left((mh)^{\delta_0} p^{1-\delta_0}\right) \ge \frac{1}{2} m^{\frac{1}{2}(1-\delta_0/\delta)} p^{\delta_0 - \frac{1}{2}} \to \infty,$$

which confirms that δ is the smallest parameter in $(0, 1/2)$ such that condition (D3.1) is satisfied by $\tilde{\mathbf{A}}_p$. Noticing that $mh/n \to \infty$, we see that $\tilde{\mathbf{A}}_p$ satisfies all assumptions of Theorem 2.17 except only the condition that in case of $\delta \in (0, 1/2)$ the ratio between the vector dimension and the sample size should be bounded. We achieved our target.

2.5 Circular Law

This is a famous conjecture which has been open for more than half a century. Its final solution is given by Tao and Vu (2010). The conjecture is stated as follows. Suppose that $\mathbf{X}_n$ is an $n \times n$ matrix with entries x_{kj} where $\{x_{kj},\ k, j =$

$1, 2, \cdots\}$ forms an infinite double array of iid complex random variables of mean zero and variance one. Using the complex eigenvalues $\lambda_1, \lambda_2, \cdots, \lambda_n$ of $\frac{1}{\sqrt{n}}\mathbf{X}_n$, we can construct a two-dimensional empirical distribution by

$$\mu_n(x, y) = \frac{1}{n}\#\{i \le n : \Re(\lambda_k) \le x,\ \Im(\lambda_k) \le y\},$$

which is called the empirical spectral distribution of the matrix $\frac{1}{\sqrt{n}}\mathbf{X}_n$.

Since early 1950's, it has been conjectured that the distribution $\mu_n(x, y)$ converges to the so-called circular law, *i.e.*, the uniform distribution over the unit disk in the complex plane. The first answer was given by Mehta (1967) for the case where x_{ij} are iid complex normal variables. He used the joint density function of the eigenvalues of the matrix $\frac{1}{\sqrt{n}}\mathbf{X}_n$ which was derived by Genibre (1965). The joint density is

$$p(\lambda_1, \cdots, \lambda_n) = c_n \prod_{i<j} |\lambda_i - \lambda_j|^2 \prod_{i=1}^{n} e^{-n|\lambda_i|^2},$$

where λ_i, $i \le n$, are the complex eigenvalues of the matrix $\frac{1}{\sqrt{n}}\mathbf{X}_n$ and c_n is a normalizing constant.

Partial answers under more general assumptions are made in Girko (1984) and Bai (1997). The problem under the only condition of finite second moment is still open. For details of the history of this problem, the reader is referred to Bai (1997). Recently, the final answer is given by Tao and Vu (2010).

2.5.1 Failure of Techniques Dealing with Hermitian Matrices

In this section, we show that the methodologies to deal with Hermitian matrices do not apply to non-Hermitian matrices.

1. Failure of truncation method

It has been seen in previous chapters that a small change to all entries or large change to a small number of entries of Hermitian matrices will cause a small change in their empirical spectral distributions, and thus the truncation technique has played an important role in the spectral theory of large dimensional hermitian matrices. However, it is not the case for non-Hermitian matrices. See the following example:

Example 2.31. *Consider the following two $n \times n$ matrices*

$$\mathbf{A} = \begin{pmatrix} 0 & 1 & 0 & 0 & \cdots & 0 \\ 0 & 0 & 1 & 0 & \cdots & 0 \\ 0 & 0 & 0 & 1 & \cdots & 0 \\ \vdots & \vdots & \vdots & \vdots & & \vdots \\ 0 & 0 & 0 & 0 & \cdots & 1 \\ 0 & 0 & 0 & 0 & \cdots & 0 \end{pmatrix} \quad \textit{and} \quad \mathbf{B} = \begin{pmatrix} 0 & 1 & 0 & 0 & \cdots & 0 \\ 0 & 0 & 1 & 0 & \cdots & 0 \\ 0 & 0 & 0 & 1 & \cdots & 0 \\ \vdots & \vdots & \vdots & \vdots & & \vdots \\ 0 & 0 & 0 & 0 & \cdots & 1 \\ \frac{1}{n^3} & 0 & 0 & 0 & \cdots & 0 \end{pmatrix}$$

It is easy to see that all the n eigenvalues of $\mathbf{A}$ are 0 while those of $\mathbf{B}$ are

$$\lambda_k = n^{-3/n} e^{2k\pi/n}, \quad k = 0, 1, \cdots, n-1.$$

When n is large, $|\lambda_k| = n^{-3/n} \sim 1$. This example shows that the empirical spectral distribution of $\mathbf{A}$ and $\mathbf{B}$ are very different although they only have one different entry which is as small as n^{-3}. Therefore, the truncation technique does not apply.

2. Failure of moment method

Although the moment method has successively been used as a powerful tool in establishing the spectral theory of Hermitian (symmetric) large matrices, it fails to apply to non-symmetric matrices. The reason can be seen from the fact that for any $k \geq 1$,

$$\mathrm{E}Z^k = 0, \tag{2.5.1}$$

if Z denotes a complex random variable uniformly distributed over any disk with center 0. That is, even we had proved

$$\frac{1}{n}\mathrm{tr}(\frac{1}{\sqrt{n}}\mathbf{X}_n)^k \to 0, \quad \text{a.s.}, \tag{2.5.2}$$

we could not claim that the spectral distribution of $\frac{1}{\sqrt{n}}\mathbf{X}_n$ tends to the circular law because (2.5.1) does not uniquely determine the distribution of Z.

3. Difficulty of the method of Stieltjes transform

When $|z| > 1$, by Theorem 3.12, we may write

$$s_n(z) =: \frac{1}{n}\mathrm{tr}\left(\frac{1}{\sqrt{n}}\mathbf{X}_n - z\right)^{-1} = -\frac{1}{z}\left(1 + \sum_{k=1}^{\infty}\frac{1}{z^k}\left(\frac{1}{n}\mathrm{tr}\left(\frac{1}{\sqrt{n}}\mathbf{X}_n^k\right)\right)\right). \tag{2.5.3}$$

By (2.5.2), we should have

$$s_n(z) \to -\frac{1}{z}, \quad \text{a.s.} \tag{2.5.4}$$

The limit is the same as the Stieltjes transform of any uniform distribution over a disk with center 0 and radius $\rho \leq 1$. Although the Stieltjes transform of $\frac{1}{\sqrt{n}}\mathbf{X}_n$ can uniquely determine all eigenvalues of $\frac{1}{\sqrt{n}}\mathbf{X}_n$, even only by values at z with $|z| > 1$, limit (2.5.4) cannot determine convergence to circular law.

Therefore, to establish the circular law by Stieltjes transforms, one has to consider the convergence of $s_n(z)$ for z with $|z| \leq 1$. Unlike the case for Hermitian matrices, the Stieltjes transform $s_n(z)$ is not bounded for $|z| \leq 1$, even impossible to any bound depending on n. This leads to serious difficulties in the mathematical analysis on $s_n(z)$.

2.5.2 Revisit of Stieltjes Transformation

Despite the difficulty shown in last subsection, one usable piece of information is that the function $s_n(z)$ uniquely determines all eigenvalues of $\frac{1}{\sqrt{n}}\mathbf{X}_n$. We make some modification on it so that the resulting version is easier to deal with.

Denote the eigenvalues of $\frac{1}{\sqrt{n}}\mathbf{X}_n$ by

$$\lambda_k = \lambda_{kr} + i\lambda_{ki}, \ k = 1, 2, \cdots, n$$

and $z = s + it$. Then,

$$s_n(z) = \frac{1}{n}\sum_{k=1}^{n} \frac{1}{\lambda_k - z}. \tag{2.5.5}$$

Because $s_n(z)$ is analytic at all z except the n eigenvalues, the real (or imaginary) part can also determine the eigenvalues of $\frac{1}{\sqrt{n}}\mathbf{X}_n$. Write $s_n(z) = s_{nr}(z) + is_{ni}(z)$. Then we have

$$\begin{aligned} s_{nr}(z) &= \frac{1}{n}\sum_{k=1}^{n} \frac{\lambda_{kr} - s}{|\lambda_k - z|^2} \\ &= -\frac{1}{2n}\sum_{k=1}^{n} \frac{\partial}{\partial s}\log(|\lambda_k - z|^2) \\ &= -\frac{1}{2}\frac{\partial}{\partial s}\int_0^{\infty} \log x \nu_n(dx, z), \end{aligned} \tag{2.5.6}$$

where $\nu_n(\cdot, z)$ is the ESD of the Hermitian matrix $\mathbf{H}_n = (\frac{1}{\sqrt{n}}\mathbf{X}_n - z\mathbf{I})(\frac{1}{\sqrt{n}}\mathbf{X}_n - z\mathbf{I})^*$.

Now, let us consider the Fourier transform of the function $s_{nr}(z)$. We have

$$\begin{aligned} &-2\iint s_{nr}(z)e^{i(us+vt)}dtds \\ &= \iint e^{i(us+vt)}\frac{\partial}{\partial s}\int_0^{\infty} \log x \nu_n(dx, z)dtds \\ &= \frac{2}{n}\sum_{k=1}^{n}\iint \frac{s - \lambda_{kr}}{(\lambda_{kr} - s)^2 + (\lambda_{ki} - t)^2}e^{i(us+vt)}dtds \\ &= \frac{2}{n}\sum_{k=1}^{n}\iint \frac{s}{s^2 + t^2}e^{i(us+vt)+i(u\lambda_{kr}+v\lambda_{ki})}dtds. \end{aligned} \tag{2.5.7}$$

We note here that in (2.5.7) and throughout the following, when integration with respect to s and t is performed on unbounded domains, it is iterated, first with respect to t, then with respect to s. Fubini's theorem cannot be applied, since $s/(s^2 + t^2)$ is not integrable on $\mathbb{R}^2$ (although it is integrable on bounded subsets of the plane).

Recalling the characteristic function of the Cauchy distribution, we have

$$\frac{1}{\pi}\int \frac{|s|}{s^2+t^2}e^{itv}dt = e^{-|sv|}.$$

Therefore,

$$\begin{aligned}
&\int\Big[\int \frac{s}{s^2+t^2}e^{i(us+vt)}dt\Big]ds\\
&= \pi\int \operatorname{sgn}(s)e^{ius-|sv|}ds\\
&= 2i\pi\int_0^\infty \sin(su)e^{-|v|s}ds\\
&= \frac{2i\pi u}{u^2+v^2}.
\end{aligned}$$

Substituting the above into (2.5.7), we have

$$\begin{aligned}
&\iint e^{i(us+vt)}\frac{\partial}{\partial s}\int_0^\infty \log x\nu_n(dx,z)dtds\\
&= \frac{4i\pi u}{u^2+v^2}\iint \frac{1}{n}\sum_{k=1}^n e^{i(u\lambda_{kr}+t\lambda_{ki})}.
\end{aligned}$$

Therefore, we have established the following lemma.

Lemma 2.32. *For any $uv \neq 0$, we have*

$$\begin{aligned}
c_n(u,v) &=: \iint e^{iux+ivy}\mu_n(dx,dy)\\
&= \frac{u^2+v^2}{4iu\pi}\iint \frac{\partial}{\partial s}\Big[\int_0^\infty \ln x\nu_n(dx,z)\Big]e^{ius+ivt}dtds, \qquad (2.5.8)
\end{aligned}$$

where $z = s+it$, $i=\sqrt{-1}$ and

$$\mu_n(x,y) = \frac{1}{n}\#\{k\le n : \lambda_{kr}\le x, \lambda_{ki}\le y\},$$

the empirical spectral distribution of $\frac{1}{\sqrt{n}}\mathbf{X}_n$.

If we assume the 4-th moment of the underlying distribution is finite, then by Theorem 3.12, with probability 1, the family of distributions $\mu_n(x,y)$ is tight. In fact, one can prove that the family of distributions $\mu_n(x,y)$ is also tight under only the finiteness of second moment. Therefore, to prove the circular law, applying Lemma 2.32, one needs only show that the right-hand side of (2.5.8) converges to its counterpart generated by the circular law.

Note that the function $\ln x$ is not bounded at both infinity and zero. Therefore, the convergence of the right-hand side of (2.5.8) cannot simply reduce to the convergence of ν_n. In view of Theorem 3.5, the upper limit of the inner integral does not pose a serious problem, since the support of ν_n is bounded from the right by $(2+\varepsilon+|z|)^2$ under the assumption of finite 4-th moment. The most difficult part is in dealing with the lower limit of the integral.

2.5.3 A Partial Answer to the Circular Law

We shall prove the following theorem by using Lemma 2.32.

Theorem 2.33. *Suppose that the underlying distribution of elements of $\mathbf{X}_n$ have finite $(2+\eta)$-th moment and that the joint distribution of the real and imaginary part of the entries has a bounded density. Then, with probability one, the ESD $\mu_n(x,y)$ of $\frac{1}{\sqrt{n}}\mathbf{X}_n$ tends to the uniform distribution over the unit disc in the complex plane.*

The proof of the theorem needs to go through the next steps:

1. Reduction of the range of integration. We need to reduce the range of integration to a finite rectangle, so that the dominated convergence theorem is applicable. As will be seen, the proof of the circular law reduces to showing that for every large $A > 0$ and small $\varepsilon > 0$,

$$\begin{aligned} &\iint_T \left[\frac{\partial}{\partial s}\int_0^\infty \ln x\nu_n(dx,z)\right] e^{ius+ivt}dsdt \\ &\to \iint_T \left[\frac{\partial}{\partial s}\int_0^\infty \ln x\nu(dx,z)\right] e^{ius+ivt}dsdt, \end{aligned} \tag{2.5.9}$$

where $T = \{(s,t);\ |s| \le A, |t| \le A^3, |\sqrt{s^2+t^2}-1| \ge \varepsilon\}$ and $\nu(x,z)$ is the LSD of the sequence of matrices $\mathbf{H}_n = (\frac{1}{\sqrt{n}}\mathbf{X}_n - z\mathbf{I})(\frac{1}{\sqrt{n}}\mathbf{X}_n - z\mathbf{I})^*$ which determines the circular law. The rest of the proof will be divided into the following steps.

2. Identifying the limiting spectrum $\nu(\cdot,z)$ of $\nu_n(\cdot,z)$ and showing that it determines the circular law.

3. Establishing a convergence rate of $\nu_n(x,z)$ to $\nu(x,z)$ uniformly in every bounded region of z.

Then, we will be able to apply the convergence rate to establish (2.5.9). As argued earlier, it is sufficient to show the following.

4. For a suitably defined sequence ε_n, with probability one

$$\limsup_{n\to\infty} \iint_T \left| \int_{\varepsilon_n}^\infty \ln x(\nu_n(dx,z) - \nu(dx,z)) \right| = 0, \tag{2.5.10}$$

$$\limsup_{n\to\infty} \left| \iint_T \int_0^{\varepsilon_n} \ln x\nu_n(dx,z)dsdt \right| = 0, \tag{2.5.11}$$

and for any fixed s,

$$\limsup_{n\to\infty} \left| \int_{(s,t)\in T} \int_0^{\varepsilon_n} \ln x\nu_n(dx,z)dt \right| = 0. \tag{2.5.12}$$

The details are omitted. Interested readers are referred to Bai and Silverstein (2006, 2009). We only give some comments and extensions in next section.

2.5.4 Comments and Extensions of Theorem 2.33

1. On the smoothness of the underlying distribution

The assumption that the real and imaginary parts of the entries of the matrix $\mathbf{X}_n$ have a bounded joint density is too restrictive because the circular law for a real Gaussian matrix does not follow from Theorem 2.33. In the sequel, we shall extend Theorem 2.33 to a more general case to cover the real Gaussian case, and, in general to random variables with bounded densities.

Theorem 2.34. *Assume that there are two directions such that the conditional density of the projection of the underlying random variable onto one direction given the projection onto the other direction is uniformly bounded and assume that the underlying distribution has zero mean and finite* $2+\eta$ *moment. Then the circular law holds.*

Suppose that the two directions are $(\cos(\theta_j), \sin(\theta_j))$, $j = 1, 2$. Then, the density condition is equivalent to

> *The conditional density of the linear combination* $\Re(x_{11})\cos(\theta_1) + \Im(x_{11})\sin(\theta_1) = \Re(e^{-i\theta_1}x_{11})$ *given* $\Re(x_{11})\cos(\theta_2) + \Im(x_{11})\sin(\theta_2) = \Re(e^{-i\theta_2}x_{11}) = \Im(ie^{-i\theta_2}x_{11})$ *is bounded.*

Consider the matrix $\mathbf{Y}_n = (y_{jk}) = e^{-i\theta_2+i\pi/2}\mathbf{X}_n$. The circular law $\frac{1}{\sqrt{n}}\mathbf{X}_n$ is obviously equivalent to the circular law for $\frac{1}{\sqrt{n}}\mathbf{Y}_n$. Then, the density condition for $\mathbf{Y}_n$ then becomes

> *The conditional density of* $\Re(y_{11})\sin(\theta_2 - \theta_1) + \Im(y_{11})\cos(\theta_2 - \theta_1)$ *given* $\Im(y_{11})$ *is bounded.*

This condition simply implies that $\sin(\theta_2 - \theta_1) \neq 0$. Thus, the density condition is further equivalent to

> *The conditional density of* $\Re(y_{11})$ *given* $\Im(y_{11})$ *is bounded.*

Therefore, we shall prove Theorem 2.34 under this latter condition.

2. Extension to the non-identical case

Reviewing the proofs of Theorems 2.33 and 2.34, one finds that the moment condition were used only in establishing convergence rate of $\nu_n(\cdot, z)$. To this end, we only need, for any $\delta > 0$ and some constant $\eta > 0$,

$$\frac{1}{n^2}\sum_{ij} \mathrm{E}|x_{ij}|^{2+\eta} I(|x_{ij}| \geq n^{\delta}) \to 0. \tag{2.5.13}$$

After the convergence rate is established, the proof of the circular law then reduces to showing (2.5.11) and (2.5.12). To guarantee this, we need only the following:

There are two directions such that the conditional density of the projection of each random variable x_{ij} *onto one direction given the projection onto the other direction is uniformly bounded.*

$$(2.5.14)$$

Therefore, we have the following theorem.

Theorem 2.35. *Assume that the entries of* $\mathbf{X}_n$ *are independent and have mean zero and variance* 1. *Also, we assume that conditions* (2.5.13) *and* (2.5.14) *are true. Then the circular law holds.*

3

Extreme Eigenvalues

In multivariate analysis, many statistics involved with a random matrix can be written as functions of integrals with respect to the ESD of the random matrix. When the LSD is known, one may want to apply the Helly-Bray Theorem to find approximate values of the statistics. However, the integrands are usually unbounded. For instance, the determinant of a random positive definite matrix can be written as an exponential function of the integral of $\log x$ with respect to the ESD where the integrand is unbounded both from below and above. Thus, one cannot use the LSD and Helly-Bray Theorem to find approximate values of the statistics. This would render the LSD useless. Fortunately, in most cases, the supports of the LSD's are compact intervals. Still, this does not mean that the Helly-Bray Theorem is applicable unless one can prove that the extreme eigenvalues of the random matrix remain in certain bounded intervals.

The investigation on limits of extreme eigenvalues is not only important in making the LSD useful when applying the Helly Bray Theorem, but also for its own practical interests. In Signal Processing, Pattern Recognition, Edge Detection and many other areas, the support of the LSD of the population covariance matrices consists of several disjoint pieces. It is important to know whether or not the LSD of the sample covariance matrices is also separated into the same number of disjoint pieces, under what conditions this is true, and whether or not there are eigenvalues falling into the spacings outside the support of the LSD of the sample covariance matrices.

The first work in this direction is due to Geman (1980). He proved that the largest eigenvalue of a sample covariance matrix tends to b $(= \sigma^2(1 + \sqrt{y})^2)$ when $p/n \to y \in (0, \infty)$ under a restriction on the growth rate of the moments of the underlying distribution. This work was generalized by Yin, Bai and Krishnaiah (1988) under the assumption of the existence of the fourth moment of the underlying distribution. In Bai, Silverstein and Yin (1988), it is further proved that if the fourth moment of the underlying distribution is infinite, then with probability one, the limsup of the largest eigenvalue of a sample covariance matrix is infinity. Combining the two results, we have in fact established the necessary and sufficient conditions for the existence of the limit of the largest eigenvalue of a large dimensional sample covariance matrix. In Bai and Yin (1988b), the necessary and sufficient conditions for the a.s. convergence of the extreme eigenvalues of a large Wigner matrix were found. The most difficult problem in this direction concerns the limit of the smallest eigenvalue of a large sample covariance matrix. In Yin, Bai and Krishnaiah (1983), it is proved that the lower limit of the smallest eigenvalue of a Wishart matrix has a positive lower bound if $p/n \to y \in (0, 1/2)$. Silverstein (1984) extended this work to allow $y \in (0,1)$. Further, Silverstein (1985) showed that the smallest eigenvalue of a standard Wishart matrix almost surely tends to $a\ (= (1-\sqrt{y})^2)$ if $p/n \to y \in (0,1)$. The most current result is due to Bai and Yin (1993) in which it is proved that the smallest (non-zero) eigenvalue of a large dimensional sample covariance matrix tends to $a = \sigma^2(1-\sqrt{y})^2$ when $p/n \to y \in (0, \infty)$ under the existence of the fourth moment of the underlying distribution.

In Bai and Silverstein (1998), it is shown that in any closed interval outside the support of the LSD of a sequence of large dimensional sample covariance matrices (when the population covariance matrix is not a multiple of the identity), with probability one, there are no eigenvalue for all large n. This work will be introduced in Section 3.3. This work leads to the exact separation of spectrum of large sample covariance matrices.

3.1 Wigner Matrix

In this section, we introduce some results related to Wigner matrix. The following theorem is a generalization of Bai and Yin (1988b) where the real case is considered.

Theorem 3.1. *Suppose that the diagonal elements of the Wigner matrix $\sqrt{n}\mathbf{W}_n = (\sqrt{n}w_{ij}) = (x_{ij})$ are iid real random variables and the elements above the diagonal are iid complex random variables and all these variables are independent. Then, the largest eigenvalue of $\mathbf{W}$ tends to $c > 0$ with probability one if and only if the following five conditions are true.*

$$\begin{array}{ll} \text{(i)} & \mathrm{E}((x_{11}^+)^2) < \infty, \\ \text{(ii)} & \mathrm{E}(x_{12}) \ \textit{is real and} \ \leq 0, \\ \text{(iii)} & \mathrm{E}(|x_{12} - \mathrm{E}(x_{12})|^2) = \sigma^2, \\ \text{(iv)} & \mathrm{E}(|x_{12}^4|) < \infty, \\ \text{(v)} & c = 2\sigma, \end{array} \tag{3.1.1}$$

where $x^+ = \max(x, 0)$.

By the symmetry of the largest and smallest eigenvalues of a Wigner matrix, one can easily derive the necessary and sufficient conditions for the existence of the limit of smallest eigenvalues of a Wigner matrix. Combining these conditions, we obtain the following theorem.

Theorem 3.2. *Suppose that the diagonal elements of the Wigner matrix* $\mathbf{W}_n$ *are iid real random variables and the elements above the diagonal are iid complex random variables and all these variables are independent. Then, the largest eigenvalue of* $\mathbf{W}$ *tends to* c_1 *and the smallest eigenvalue tends to* c_2 *with probability one if and only if the following five conditions are true.*

$$\begin{array}{ll} \text{(i)} & \mathrm{E}(x_{11}^2) < \infty, \\ \text{(ii)} & \mathrm{E}(x_{12}) = 0, \\ \text{(iii)} & \mathrm{E}(|x_{12}|^2) = \sigma^2, \\ \text{(iv)} & \mathrm{E}(|x_{12}^4|) < \infty, \\ \text{(v)} & c_1 = 2\sigma \ \ \textit{and} \ \ c_2 = -2\sigma. \end{array} \tag{3.1.2}$$

From the proof of Theorem 3.1, it is easy to see the following weak convergence theorem on the extreme eigenvalue of a large Wigner matrix.

Theorem 3.3. *Suppose that the diagonal elements of the Wigner matrix* $\sqrt{n}\mathbf{W}_n = (x_{ij})$ *are iid real random variables and the elements above the diagonal are iid complex random variables and all these variables are independent. Then, the largest eigenvalue of* $\mathbf{W}$ *tends to* $c > 0$ *in probability if and only if the following five conditions are true.*

$$\begin{array}{ll} \text{(i)} & \mathrm{P}(x_{11}^+ > \sqrt{n}) = o(n^{-1}), \\ \text{(ii)} & \mathrm{E}(x_{12}) \ \ \textit{is real and} \ \ \le 0, \\ \text{(iii)} & \mathrm{E}(|x_{12} - \mathrm{E}(x_{12})|^2) = \sigma^2, \\ \text{(iv)} & \mathrm{P}(|x_{12}| > \sqrt{n}) = o(n^{-2}), \\ \text{(v)} & c = 2\sigma. \end{array} \tag{3.1.3}$$

The proof of Theorem 3.1 consists of truncation, centralization and estimation of the trace of a high moment of $\mathbf{W}_n$. After the truncation and centralization, we may assume that the following conditions are true:

- $x_{ii} = 0$;
- $\mathrm{E}(x_{ij}) = 0$, $\sigma_n^2 = \mathrm{E}(|x_{ij}|^2) \le 1$, for $i \ne j$;
- $|x_{ij}| \le \delta_n\sqrt{n}$, for $i \ne j$;
- $\mathrm{E}|x_{ij}^\ell| \le b(\delta_n\sqrt{n})^{\ell-3}$, for some constant $b > 0$ and all $i \ne j$, $\ell \ge 3$.

To complete the proof of Theorem 3.1, we will prove, for any even integer k and real number $\eta > 2$,

$$\mathrm{P}(\lambda_{\max}(\mathbf{W}_n) \ge \eta) \le \mathrm{P}(\mathrm{tr}[(\mathbf{W}_n)^k] \ge \eta^k) \le \eta^{-k}\mathrm{E}(\mathrm{tr}(\mathbf{W}_n)^k) \tag{3.1.4}$$

under the above four assumptions.

That means, as a by-product, we also have proved the following theorem, which gives an estimation of the tail probability of the largest eigenvalue of $\mathbf{W}_n$. The estimate of the tail probability is very useful in many other limiting problems about $\mathbf{W}_n$.

Theorem 3.4. *Suppose the entries of* $\mathbf{W}$ *depend on* n *and satisfy*

$$\mathrm{E}(x_{jk}) = 0,\ \mathrm{E}(|x_{jk}^2|) \le \sigma^2,\ \mathrm{E}(|x_{jk}^\ell|) \le b(\delta_n\sqrt{n})^{\ell-3},\ (\ell \ge 3) \tag{3.1.5}$$

for some $b > 0$, *then for fixed* $\varepsilon > 0$ *and* $x > 0$,

$$\mathrm{P}(\lambda_{\max}(\mathbf{W}) \le 2\sigma + \varepsilon + x) = o(n^{-\ell}(2\sigma + \varepsilon + x)^{-2}), \tag{3.1.6}$$

which implies

$$\limsup \lambda_{\max}(\mathbf{W}) \le 2\sigma, a.s..$$

The proofs of the theorems can be found in Bai and Silverstein (2006).

3.2 Sample Covariance Matrix

We first introduce the following theorem.

Theorem 3.5. *Suppose that* $\{x_{jk},\ j,k = 1,2,\cdots\}$ *is a double array of iid random variables with mean zero and variance* σ^2 *and finite 4-th moment. Let* $\mathbf{X}_n = (x_{jk},\ j \le p, k \le n)$ *and* $\mathbf{S}_n = \frac{1}{n}\mathbf{X}\mathbf{X}^*$. *Then the largest eigenvalue of* $\mathbf{S}_n$ *tends to* $\sigma^2(1+\sqrt{y})^2$ *almost surely.*

If the 4-th moment of the underlying distribution is not finite, then with probability one, the limsup of the largest eigenvalue of $\mathbf{S}_n$ *is infinity.*

The real case of the first conclusion is due to Yin, Bai and Krishnaiah (1988) and the real case of the second conclusion is proved in Bai, Silverstein and Yin (1988). The proof of this theorem is almost the same as that of Theorem 3.1, the proof for the real case can be found in the referred papers. Thus the details are omitted and left as an exercise to the reader. Here, for our future use, we remark that, the proof of the above theorem can be extended to the following.

Theorem 3.6. *Suppose that the entries of the matrix* $\mathbf{X}_n = (x_{jkn},\ j \le p, k \le n)$ *are independent (not necessarily identically distributed) and satisfy*

1. $\mathrm{E}(x_{jkn}) = 0$,
2. $|x_{jkn}| \le \sqrt{n}\delta_n$,
3. $\max_{j,k} |\mathrm{E}|X_{jkn}|^2 - \sigma^2| \to 0$ *as* $n \to \infty$,
4. $\mathrm{E}|x_{jkn}|^\ell \le b(\sqrt{n}\delta_n)^{\ell-3}$ *for all* $\ell \ge 3$,

where $\delta_n \to 0$ *and* $b > 0$. *Let* $\mathbf{S}_n = \frac{1}{n}\mathbf{X}_n\mathbf{X}_n^*$. *Then, for any* $x > \varepsilon > 0$ *and integers* $j,k \ge 2$, *we have*

$$\mathrm{P}(\lambda_{\max}(\mathbf{S}_n) \ge \sigma^2(1+\sqrt{y})^2 + x) \le Cn^{-k}(\sigma^2(1+\sqrt{y})^2 + x - \varepsilon)^{-k}$$

for some constant $C > 0$.

In this section, we shall present a generalization to a result of Bai and Yin (1993). Assume that $\mathbf{X}_n$ is a $p \times n$ complex matrix and $\mathbf{S}_n = \frac{1}{n}\mathbf{X}_n\mathbf{X}_n^*$.

Theorem 3.7. *Assume that the entries of $\{x_{ij}\}$ is a double array of iid. complex random variables with mean zero, variance σ^2 and finite 4-th moment. Let $\mathbf{X}_n = (x_{ij};\, i \le p, j \le n)$ be the $p \times n$ matrix of the upper-left corner of the double array. If $p/n \to y \in (0,1)$, then, with probability one, we have*

$$\begin{aligned} -2\sqrt{y}\sigma^2 &\le \liminf_{n\to\infty} \lambda_{\min}(\mathbf{S}_n - \sigma^2(1+y)\mathbf{I}_n) \\ &\le \limsup_{n\to\infty} \lambda_{max}(\mathbf{S}_n - \sigma^2(1+y)\mathbf{I}_n) \le 2\sqrt{y}\sigma^2. \end{aligned} \tag{3.2.1}$$

From Theorem 3.7, one immediately gets the following Theorem.

Theorem 3.8. *Under the assumptions of Theorem 3.7, we have*

$$\lim_{n\to\infty} \lambda_{\min}(\mathbf{S}_n) = \sigma^2(1-\sqrt{y})^2 \tag{3.2.2}$$

and

$$\lim_{n\to\infty} \lambda_{\max}(\mathbf{S}_n) = \sigma^2(1+\sqrt{y})^2. \tag{3.2.3}$$

Denote the eigenvalues of $\mathbf{S}_n$ by $\lambda_1 \le \lambda_2 \le \cdots \le \lambda_n$. Write $\lambda_{\max} = \lambda_n$ and

$$\lambda_{\min} = \begin{cases} \lambda_1, & \text{if } p \le n, \\ \lambda_{p-n+1}, & \text{if } p > n. \end{cases}$$

Using the above convention, Theorem 3.8 is true for all $y \in (0,\infty)$.

The proof of Theorem 3.7 needs a basic lemma which may have its own interest. Thus, we present it below.

Lemma 3.9. *Under the conditions of Theorem 3.7, we have*

$$\limsup_{n\to\infty} \|\mathbf{T}_n(\ell)\| \le (2\ell+1)(\ell+1)y^{(\ell-1)/2}\sigma^{2\ell} \quad a.s., \tag{3.2.4}$$

where

$$\mathbf{T}_n(\ell) = n^{-\ell}\left(\sum{}' x_{av_1}\overline{x}_{u_1v_1}x_{u_1v_2}\overline{x}_{u_2v_2}\cdots x_{u_{\ell-1}v_\ell}\overline{x}_{bv_\ell}\right),$$

the summation $\sum'$ runs over for $v_1,\cdots,v_\ell = 1,2,\cdots,n$ and $u_1,\cdots,u_{\ell-1} = 1,2,\cdots,p$ subject to the restriction

$$a \ne u_1,\ u_1 \ne u_2,\cdots,u_{\ell-1} \ne b \quad \text{and} \quad v_1 \ne v_2,\ v_2 \ne v_3,\cdots,v_{\ell-1} \ne v_\ell.$$

Remark 3.10. *It seems that the finiteness of the fourth moment is also necessary for the almost sure convergence of the smallest eigenvalue of the large dimensional sample covariance matrix. However, we have at this point no idea how to prove it.*

3.2.1 Spectral Radius

Let $\mathbf{X}$ be an $n \times n$ matrix of iid complex random variables with mean zero and variance σ^2. In Bai and Yin (1986), large systems of linear equations and linear differential equations are considered. There, the norm of the matrix $(\frac{1}{\sqrt{n}}\mathbf{X})^k$ plays an important role in the stability of the solutions to those systems. The following theorem is established.

Theorem 3.11. *If* $\mathrm{E}(|x_{11}^4|) < \infty$, *then*

$$\limsup_{n\to\infty} \|(\frac{1}{\sqrt{n}}\mathbf{X})^k\| \le (1+k)\sigma^k, \quad \text{a.s..} \tag{3.2.1}$$

The proof of this theorem relies on, after truncation and centralization, the estimation of $\mathrm{E}([\mathrm{tr}(\frac{1}{\sqrt{n}}\mathbf{X})^k(\frac{1}{\sqrt{n}}\mathbf{X}^*)^k]^\ell)$. The details are omitted. Here, we introduce an important consequence on the spectral radius of $\frac{1}{\sqrt{n}}\mathbf{X}$, which plays an important role in establishing the circular law (See Section 4). This was also independently proved by Geman (1986), under additional restrictions on the growth of moments of the underlying distribution.

Theorem 3.12. *If* $\mathrm{E}(|x_{11}^4|) < \infty$, *then*

$$\limsup_{n\to\infty} |\lambda_{\max}(\frac{1}{\sqrt{n}}\mathbf{X})| \le \sigma, \quad \text{a.s.} \tag{3.2.2}$$

Theorem 3.12 follows from the fact that for any k,

$$\begin{aligned}&\limsup_{n\to\infty} |\lambda_{\max}(\tfrac{1}{\sqrt{n}}\mathbf{X})| = \limsup_{n\to\infty} |\lambda_{\max}[(\tfrac{1}{\sqrt{n}}\mathbf{X})^k]|^{1/k}\\ &\le \limsup_{n\to\infty} \|(\tfrac{1}{\sqrt{n}}\mathbf{X})^k\|^{1/k} \le (1+k)^{1/k}\sigma \to \sigma,\end{aligned}$$

by making $k \to \infty$.

Remark 3.13. *Checking the proof of Theorem 3.8, one finds that after truncation and centralization, the conditions for guaranteeing (3.2.1) are* $|x_{jk}| \le \delta_n\sqrt{n}$, $\mathrm{E}(|x_{jk}^2|) \le \sigma^2$ *and* $\mathrm{E}(|x_{jk}^3|) \le b$, *for some* $b > 0$. *This is useful in extending the circular law to the case where the entries are not identically distributed.*

3.3 Spectrum Separation

Spectrum separation can be considered as an extension of the limits of extreme eigenvalues of sample covariance matrix. In this section, we consider the matrix $\mathbf{B}_n = \frac{1}{n}\mathbf{T}^{1/2}\mathbf{X}_n\mathbf{X}_n^*\mathbf{T}_n^{1/2}$ where $\mathbf{T}_n^{1/2}$ is a Hermitian square root of the Hermitian non-negative definite $p \times p$ matrix $\mathbf{T}_n$, and investigate its spectral properties in relation to the eigenvalues of $\mathbf{T}_n$. A relationship is expected to exist since $\mathbf{B}_n$ can be viewed as the sample covariance matrix of n samples of the random vector $\mathbf{T}_n^{1/2}\mathbf{x}_1$, which has $\mathbf{T}_n$ for its population covariance matrix. When n is significantly larger than p, the law of large numbers tells us that $\mathbf{B}_n$ will be close to $\mathbf{T}_n$ with high probability. Consider then an interval $J \subset \mathbb{R}^+$ which does not contain any eigenvalues of $\mathbf{T}_n$ for all large n. For small y (the limit of p/n) it is reasonable to expect an interval $[a, b]$ close to J which contains no eigenvalues of $\mathbf{B}_n$. Moreover, the number of eigenvalues of $\mathbf{B}_n$ on one side of $[a, b]$ should match up with those of $\mathbf{T}_n$ on the same side of J. Under the assumptions on the entries of $\mathbf{X}_n$ given in Theorem 3.8 with $\sigma^2 = 1$ this can be proven by using Fan Ky inequality.

For notational convenience, defining $\lambda_0^{\mathbf{A}} = \infty$, suppose $\lambda_{i_n}^{\mathbf{T}_n}$ and $\lambda_{i_n+1}^{\mathbf{T}_n}$ lie, respectively, to the right and left of J. From Fan Ky inequality about eigenvalues of a product of two non-negative definite matrices, we have (using the fact that the spectra of $\mathbf{B}_n$ and $(1/n)\mathbf{X}_n\mathbf{X}_n^*\mathbf{T}_n$ are identical)

$$\lambda_{i_n+1}^{\mathbf{B}_n} \leq \lambda_1^{(1/n)\mathbf{X}_n\mathbf{X}_n^*}\lambda_{i_n+1}^{\mathbf{T}_n} \quad \text{and} \quad \lambda_{i_n}^{\mathbf{B}_n} \geq \lambda_p^{(1/n)\mathbf{X}_n\mathbf{X}_n^*}\lambda_{i_n}^{\mathbf{T}_n}. \tag{3.3.1}$$

From Theorem 3.8 we can, with probability one, ensure that $\lambda_1^{(1/n)\mathbf{X}_n\mathbf{X}_n^*}$ and $\lambda_p^{(1/n)\mathbf{X}_n\mathbf{X}_n^*}$ are as close as we please to one by making y suitably small. Thus an interval $[a, b]$ does indeed exist which separates the eigenvalues of $\mathbf{B}_n$ in exactly the same way the eigenvalues of $\mathbf{T}_n$ are split by J. Moreover a, b can be made arbitrarily close to the endpoints of J.

Even though the splitting of the support of F, the a.s. LSD of $F^{\mathbf{B}_n}$, is a function of y, splitting may occur regardless of whether y is small or not. Our goal is to extend the above result on exact separation beginning with any interval $[a, b]$ of $\mathbb{R}^+$ outside the support of F. We present an example of its importance which was the motivating force behind the pursuit of this topic. It arises from the detection problem in array signal processing. An unknown number q of sources emit signals onto an array of p sensors in a noise filled environment ($q < p$). If the population covariance matrix $\mathbf{R}$ of the vector of random values recorded from the sensors were known, then the value q can be determined from it, due to the fact that the multiplicity of the smallest eigenvalue of $\mathbf{R}$, attributed to the noise, is $p - q$. The matrix $\mathbf{R}$ is approximated by a sample covariance matrix $\widehat{\mathbf{R}}$ which, with a sufficiently large sample, will have, with high probability, $p-q$ *noise* eigenvalues clustering near each other and to the left of the other eigenvalues. The problem is, for p and/or q sizable the number of samples needed for $\widehat{\mathbf{R}}$ to adequately approximate $\mathbf{R}$ would be prohibitively large. However, if for p large, the number n of samples were to be merely on the same order of magnitude as p, then, under certain conditions on the signals and noise propagation, it is shown in Silverstein and Combettes (1992) that $F^{\widehat{\mathbf{R}}}$ would, with high probability, be close to the nonrandom LSD F. Moreover, it can be shown that for y sufficiently small, the support of F will split into two parts, with mass $(p - q)/p$ on the left, q/p on the right. In Silverstein and Combettes (1992) extensive computer simulations were performed to demonstrate that, at the least, the proportion of sources to sensors can be reliably estimated. It came as a surprise to find that, not only were there no eigenvalues outside the support of F, (except those near the boundary of the support), but the *exact* number of eigenvalues appeared on intervals slightly larger than those within the support of F. Thus, the simulations demonstrate that, in order to detect the *number* of sources in the large dimensional case, it is not necessary for $\widehat{\mathbf{R}}$ to be close to $\mathbf{R}$; the number of samples only needs to be large enough so that the support of F splits.

It is of course crucial to be able to recognize and characterize intervals outside the support of F, and to establish a correspondence with intervals outside the support of H, the LSD of $F^{\mathbf{T}_n}$. This is achieved through the Stieltjes transforms, $s_F(z)$ and $\underline{s}(z) \equiv s_{\underline{F}}(z)$, of, respectively, F and $\underline{F}$, where the latter denotes the LSD of $\underline{\mathbf{B}}_n \equiv (1/n)\mathbf{X}_n^*\mathbf{T}_n\mathbf{X}_n$. From Theorem 2.14, it is conceivable and will be proven that for each $z \in \mathbb{C}^+$, $s = s_F(z)$ is a solution to the equation

$$s = \int \frac{1}{t(1-y-yzs)-z} dH(t),$$

which is unique in the set $\{s \in \mathbb{C} : -(1-y)/z + ys \in \mathbb{C}^+\}$. Since the spectra of $\mathbf{B}_n$ and $\underline{\mathbf{B}}_n$ differ by $|p-n|$ zero eigenvalues, it follows that

$$F^{\underline{\mathbf{B}}_n} = (1-(p/n))I_{[0,\infty)} + (p/n)F^{\mathbf{B}_n},$$

from which we get

$$s_{F^{\underline{\mathbf{B}}_n}}(z) = -\frac{(1-p/n)}{z} + (p/n)s_{F^{\mathbf{B}_n}}(z), \quad z \in \mathbb{C}^+, \tag{3.3.2}$$

$$\underline{F} = (1-y)I_{[0,\infty)} + yF,$$

and

$$s_{\underline{F}}(z) = -\frac{(1-y)}{z} + ys_F(z), \quad z \in \mathbb{C}^+.$$

It follows that

$$s_F = -z^{-1} \int \frac{1}{1+ts_{\underline{F}}} dH(t), \tag{3.3.3}$$

for each $z \in \mathbb{C}^+$, $\underline{s} = s_{\underline{F}}(z)$, is the unique solution in $\mathbb{C}^+$ to the equation

$$\underline{s} = -\left(z - y\int \frac{t\,dH(t)}{1+t\underline{s}}\right)^{-1}, \tag{3.3.4}$$

and $s_{\underline{F}}(z)$ has an inverse, explicitly given by

$$z(s) = z_{y,H}(s) \equiv -\frac{1}{s} + y\int \frac{t\,dH(t)}{1+ts}. \tag{3.3.5}$$

Let $F^{y,H}$ denote $\underline{F}$, in order to express the dependence of the LSD of $F^{\underline{\mathbf{B}}_n}$ on the limiting dimension-to-sample size ratio y and LSD H of the population matrix. Then $s = s_{F^{y,H}}(z)$ has inverse $z = z_{y,H}(s)$.

From (3.3.5) much of the analytic behavior of F can be derived, (see Silverstein and Choi (1995)). This includes the continuous dependence of F on y and H, the fact that F has a continuous density on $\mathbb{R}^+$, and, most importantly for our present needs, a way of understanding the support of F. On any closed interval outside the support of $F^{y,H}$, $s_{F^{y,H}}$ exists and is increasing. Therefore on the range of this interval its inverse exists and is also increasing. In Silverstein and Choi (1995) the converse is shown to be true, along with some other results. We summarize the relevant facts in the following

Lemma 3.14. [*Silverstein and Choi* (1995)]. *Let for any c.d.f. G S_G denote its support and S_G^c, the complement of its support. If $u \in S^c_{F^{y,H}}$, then $s = s_{F^{y,H}}(u)$ satisfies:*

(1) $s \in \mathbb{R}\backslash\{0\}$,
(2) $-s^{-1} \in S^c_H$,
and
(3) $\frac{d}{ds} z_{y,H}(s) > 0$.
Conversely, if s satisfies (1) - (3), then $u = z_{y,H}(s) \in S^c_{F^{y,H}}$.

Thus by plotting $z_{y,H}(s)$ for $s \in \mathbb{R}$, the range of values where it is increasing yields $S^c_{F^{y,H}}$ (see Fig. 3.1).

Of course the supports of F and $F^{y,H}$ are identical on $\mathbb{R}^+$. The density function of $F^{0.1,H}$ is given in Fig. 3.2.

As for whether F places any mass at 0, it is shown in Silverstein and Choi (1995) that

$$F^{y,H}(0) = \max(0, 1 - y[1 - H(0)])$$

which implies

$$F(0) = \begin{cases} H(0), & y[1 - H(0)] \le 1, \\ 1 - y^{-1}, & y[1 - H(0)] > 1. \end{cases} \tag{3.3.6}$$

It is appropriate at this time to state a lemma which lists all the ways intervals in $S^c_{F^{y,H}}$ can arise with respect to the graph of $z_{y,H}(s)$, $s \in \mathbb{R}$. It also states the dependence of these intervals on y.

Lemma 3.15. (*a*) *If (t_1, t_2) is contained in S^c_H with $t_1, t_2 \in \partial S_H$ and $t_1 > 0$, then there is a $y_0 > 0$ for which $y < y_0 \Rightarrow$ there are two values $s^1_y < s^2_y$ in $[-t_1^{-1}, -t_2^{-1}]$ for which $(z_{y,H}(s^1_y), z_{y,H}(s^2_y)) \subset S^c_{F^{y,H}}$, with endpoints lying in $\partial S_{F^{y,H}}$, and $z_{y,H}(s^1_y) > 0$. Moreover,*

$$z_{y,H}(s^i_y) \to t_i, \quad \text{as } y \to 0 \tag{3.3.7}$$

for $i = 1, 2$. The endpoints vary continuously with y, shrinking down to a point as $y \uparrow y_0$, while $z_{y,H}(s^2_y) - z_{y,H}(s^1_y)$ is monotone in y.

(In the graph of $z_{y,H}(s)$, s^1_y and s^2_y are the local minimizer and maximizer in the interval $(-1/t_1, -1/t_2)$ and $z_{y,H}(s^1_y)$ and $z_{y,H}(s^2_y)$ are the local minimum and maximum values. As an example, notice the minimizers and

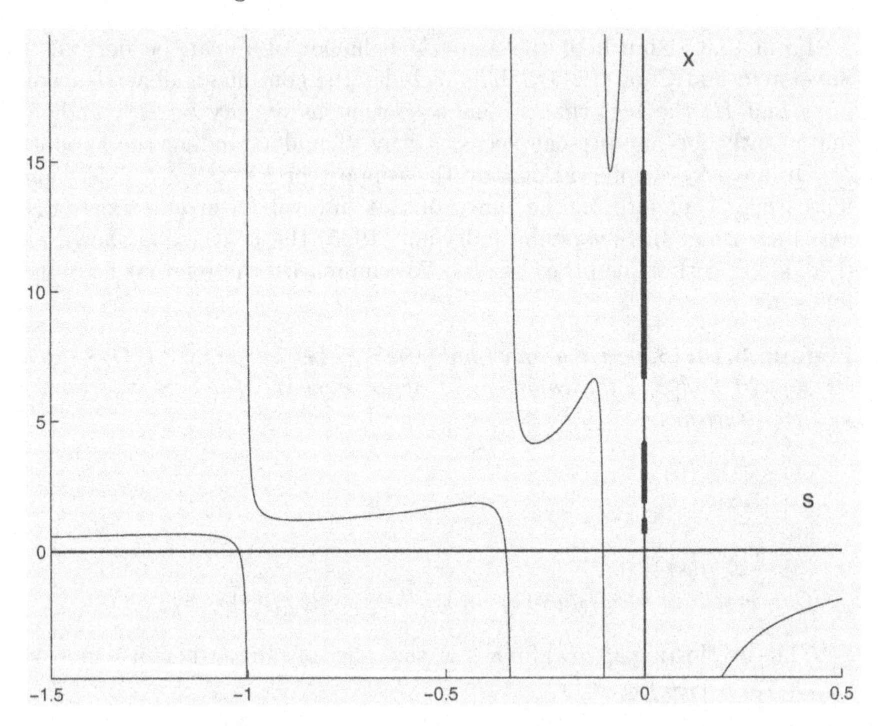

Fig. 3.1. The function $z_{0.1,H}(s)$ for a three-mass point H placing masses 0.2, 0.4 and 0.4 at three points $1, 3, 10$. The intervals of bold lines on the vertical axes are the support of $F^{0.1,H}$.

maximizers of the two curves in the middle of Figure 3.1)

(*b*) *If* $(t_3, \infty) \subset S_H^c$ *with* $0 < t_3 \in \partial S_H$, *then there exists* $s_y^3 \in [-1/t_3, 0)$ *such that* $z_{y,H}(s_y^3)$ *is the largest number in* $S_{F^{y,H}}$. *As* y *decreases from* ∞ *to* 0, (3.3.7) *holds for* $i = 3$ *with convergence monotone from* ∞ *to* t_3.

(The value s_y^3 is the right-most minimizer of the graph $z_{y,H}(s), s < 0$ and $z_{y,H}(s_y^3)$ is the largest local minimum value. See the curve immediately to the left of the vertical axis in Fig. 3.1)

(*c*) *If* $y[1 - H(0)] < 1$ *and* $(0, t_4) \subset S_H^c$ *with* $t_4 \in \partial S_H$, *then there exists* $s_y^4 \in (-\infty, -1/t_4]$ *such that* $z_{y,H}(s_y^4)$ *is the smallest positive number in* $S_{F^{y,H}}$, *and* (3.3.7) *holds with* $i = 4$, *the convergence being monotone from* 0 *as* y *decreases from* $[1 - H(0)]^{-1}$.

(The value s_y^4 is the left-most local maximizer and $z_{y,H}(s_y^4)$ is the smallest local maximum value, i.e., the smallest point of the support of $F^{y,H}$. See the left-most curve in Fig. 3.1).

(*d*) *If* $y[1 - H(0)] > 1$, *then, regardless of the existence of* $(0, t_4) \subset S_H^c$, *there exists* $s_y > 0$ *such that* $z_{y,H}(s_y) > 0$ *and is the smallest number in* $S_{F^{y,H}}$. *It decreases from* ∞ *to* 0 *as* y *decreases from* ∞ *to* $[1 - H(0)]^{-1}$.

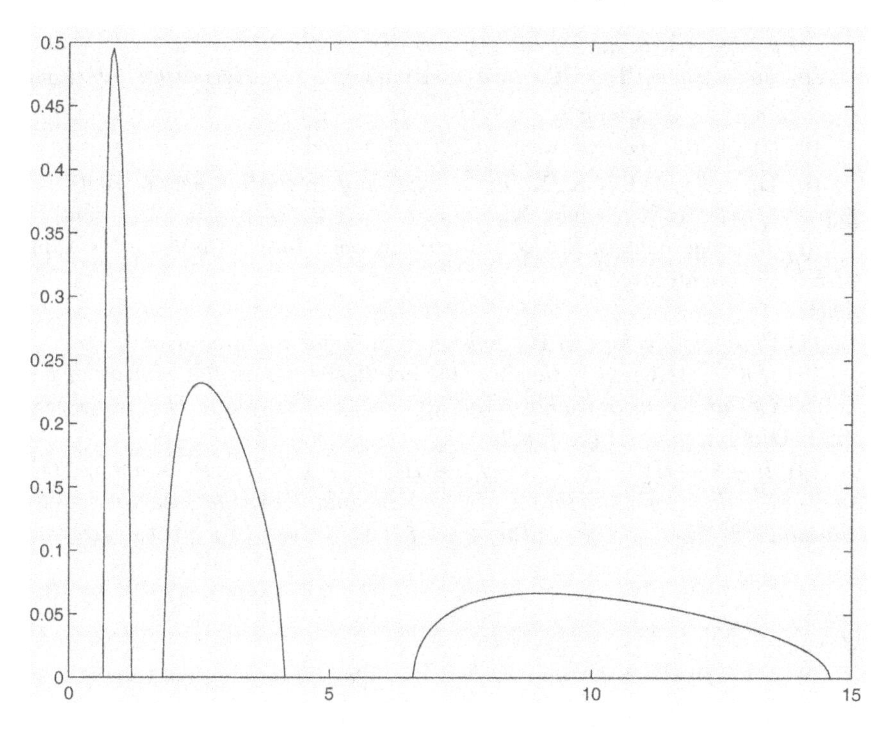

Fig. 3.2. The Density Function of $F^{0.1,H}$ with H defined in Figure 3.1

(In this case, the curve in Figure 3.1 should have a different shape. It will increase from $-\infty$ to the positive value $z_{y,H} > 0$ at s_y and then decrease to 0 as s increases from 0 to ∞.)

(e) *If* $H = I_{[0,\infty)}$, *that is,* H *places all mass at* 0, *then* $F = F^{y,I_{[0,\infty)}} = I_{[0,\infty)}$.

All intervals in $S^c_{F^{y,H}} \cap [0,\infty)$ *arise from one of the above. Moreover, disjoint intervals in* S^c_H *yield disjoint intervals in* $S^c_{F^{y,H}}$.

Thus, for interval $[a,b] \subset S^c_{F^{y,H}} \cap \mathbb{R}^+$, it is possible for $s_{F^{y,H}}(a)$ to be positive. This will occur only in case (d) of Lemma 3.15 when $b < z_{y,H}(s_y)$. For any other location of $[a,b]$ in $\mathbb{R}^+$ it follows from Lemma 3.14 that $s_{F^{y,H}}$ is negative, and

$$[-1/s_{F^{y,H}}(a), -1/s_{F^{y,H}}(b)] \tag{3.3.8}$$

is contained in S^c_H. This interval is the proper choice of J.

The main result can now be stated.

Theorem 3.16. *Assume:*

(a) *Assumptions in Theorem* 3.7 *hold:* x_{ij}, $i,j = 1,2,...$ *are i.i.d. random variables in* $\mathbb{C}$ *with* $\mathrm{E}x_{11} = 0$, $\mathrm{E}|x_{11}|^2 = 1$, *and* $\mathrm{E}|x_{11}|^4 < \infty$.

(b) $p = p(n)$ with $y_n = p/n \to y > 0$ as $p \to \infty$.

(c) For each n, $\mathbf{T} = \mathbf{T}_n$ is a nonrandom $p \times p$ Hermitian nonnegative definite matrix satisfying $H_n \equiv F^{\mathbf{T}_n} \xrightarrow{\mathcal{D}} H$, a c.d.f..

(d) $\|\mathbf{T}_n\|$, the spectral norm of $\mathbf{T}_n$ is bounded in n.

(e) $\mathbf{B}_n = (1/n)\mathbf{T}_n^{1/2}\mathbf{X}_n\mathbf{X}_n^*\mathbf{T}_n^{1/2}$, $\mathbf{T}_n^{1/2}$ any Hermitian square root of $\mathbf{T}_n$, $\underline{\mathbf{B}}_n = (1/n)\mathbf{X}_n^*\mathbf{T}_n\mathbf{X}_n$, where $\mathbf{X}_n = (x_{ij})$, $i = 1, 2, \cdots, p$, $j = 1, 2, \cdots, n$.

(f) Interval $[a, b]$ with $a > 0$ lies in an open interval outside the support of F^{y_n, H_n} for all large n.

Then:

(1) P(no eigenvalues of $\mathbf{B}_n$ appear in $[a, b]$ for all large n) = 1.

(2) If $y[1 - H(0)] > 1$, then x_0, the smallest value in the support of $F^{y,H}$, is positive, and with probability one $\lambda_n^{\mathbf{B}_n} \to x_0$ as $n \to \infty$. The number x_0 is the maximum value of the function $z_{y,H}(s)$ for $s \in \mathbb{R}^+$.

(3) If $y[1 - H(0)] \le 1$, or $y[1 - H(0)] > 1$ but $[a, b]$ is not contained in $(0, x_0)$, then by assumption (f) and Lemma 3.14, the interval (3.3.8) is contained in $S_{H_n}^c \cap \mathbb{R}^+$ for all n large. Let for these n, $i_n \ge 0$ be such that

$$\lambda_{i_n}^{\mathbf{T}_n} > -1/s_{F^{y,H}}(b) \qquad \text{and} \qquad \lambda_{i_n+1}^{\mathbf{T}_n} < -1/s_{F^{y,H}}(a). \tag{3.3.9}$$

Then

$$P(\lambda_{i_n}^{\mathbf{B}_n} > b \quad \text{and} \quad \lambda_{i_n+1}^{\mathbf{B}_n} < a \quad \text{for all large } n\ ) = 1.$$

Remark 3.17. *Conclusion (2) occurs when $n < p$ for large n in which case $\lambda_{n+1}^{\mathbf{B}_n} = 0$. Therefore exact separation should not be expected to occur for $[a, b] \subset [0, x_0]$. Regardless of their values, the $p - n$ smallest eigenvalues of $\mathbf{T}_n$ are essentially being converted by $\mathbf{B}_n$ to zero. It is worth noting that when $y[1 - H(0)] > 1$ and F and (consequently) H each has at least two non-connected members in their support in $\mathbb{R}^+$, the numbers of eigenvalues of $\mathbf{B}_n$ and $\mathbf{T}_n$ will match up in each respective member,* **except** *the left-most member. Thus the conversion to zero is affecting only this member.*

Remark 3.18. *The assumption of nonrandomness of $\mathbf{T}_n$ is made only for convenience. Using Fubini Theorem, Theorem 3.16 can easily be extended to random $\mathbf{T}_n$ (independent of x_{ij}) as long as the limit H is nonrandom and assumption (f) is true almost surely. At present it is unknown whether the boundedness of $\|\mathbf{T}_n\|$ can be relaxed.*

Conclusion (1) along with the results on the extreme eigenvalues of $(1/n)\mathbf{X}\mathbf{X}^*$ yield properties on the extreme eigenvalues of $\mathbf{B}_n$. Notice that the interval $[a, b]$ can also be unbounded, that is, $\limsup_n \|\mathbf{B}_n\|$ stays a.s. bounded (nonrandom bound). Also, when $p < n$ and $\lambda_p^{\mathbf{T}_n}$ is bounded away from 0 for all n, we can use

$$\lambda_p^{\mathbf{B}_n} \ge \lambda_p^{(1/n)\mathbf{X}\mathbf{X}^*} \lambda_p^{\mathbf{T}_n}$$

to conclude a nonrandom $b > 0$ exists for which a.s. $\lambda_p^{\mathbf{B}_n} > b$. Therefore we have the following

Corollary 3.19. *If $\|\mathbf{T}_n\|$ converges to the largest number in the support of H, then $\|\mathbf{B}_n\|$ converges a.s. to the largest number in the support of F. If the smallest eigenvalue of $\mathbf{T}_n$ converges to the smallest number in the support of H, and $y < 1$ then the smallest eigenvalue of $\mathbf{B}_n$ converges to the smallest number in the support of F.*

3.4 Tracy-Widom Law

3.4.1 TW Law for Wigner Matrix

In multivariate analysis, some statistics are defined by extreme eigenvalues of random matrices. That is, the limiting distribution of normalized extreme eigenvalues are of their special interest. In (1994), Tracy and Widom derived the limiting distribution of the largest eigenvalue of Wigner matrix when the entries is Gaussian distributed. The limiting law is named as the Tracy-Widom (TW) law in RMT. We shall introduce the TW law for Gaussian Winger matrix. Under the normality assumption, the density function of the ensemble is given

$$P(\mathbf{w})d\mathbf{w} = \exp(-\frac{\beta}{4} tr\mathbf{w}^*\mathbf{w})d\mathbf{w}$$

and the joint density of the eigenvalues is given by

$$p_{n\beta}(\lambda_1, \cdots, \lambda_n) = C_{n\beta} e^{-\frac{1}{2}\beta \sum \lambda_j^2} \prod_{j<k} |\lambda_j - \lambda_k|^\beta, \ -\infty < \lambda_1 < \cdots < \lambda_n < \infty,$$

where

$$\beta = \begin{cases} 1 & \text{for GOE,} \\ 2 & \text{for GUE,} \\ 4 & \text{for GSE,} \end{cases}$$

where GOE stands for *Gaussian Orthogonal Ensemble* for which all entries of the matrix are real normal random variables and whose distribution is invariant under real orthogonal similarity transformations; GUE stands for *Gaussian Unitary Ensemble* for which all entries of the matrix are complex normal random variables and whose distribution is invariant under complex Unitary similarity transformations; while GSE stands for *Gaussian Symplectic Ensemble* for which all entries of the matrix are normal quarterion random variables and whose distribution is invariant under symplectic transformations. Here, we need to give a note on quarterion Wigner matrix and symplectic transformations. We define 2×2 matrices

$$I_2 = \begin{pmatrix} 1 & 0 \\ 0 & 1 \end{pmatrix}, \ \mathbf{i} = \begin{pmatrix} i & 0 \\ 0 & -i \end{pmatrix}, \ \mathbf{j} = \begin{pmatrix} 0 & 1 \\ -1 & 0 \end{pmatrix} \ \text{and} \ \mathbf{k} = \begin{pmatrix} 0 & i \\ i & 0 \end{pmatrix}.$$

It is easy to verify that $\mathbf{i}^2 = \mathbf{j}^2 = \mathbf{k}^2 = \mathbf{ijk} = -I_2$. For any real numbers a_0, a_1, a_2, a_3, the 2×2 matrix of the linear combination

$$x = a_0 I_2 + a_1 \mathbf{i} + a_2 \mathbf{j} + a_3 \mathbf{k}$$

is called a quarterion. A GSE is a $2n \times 2n$ hermitian matrix $\mathbf{X} = (x_{ij})_{i,j=1}^n$ where x_{ij} is a quarterion with 4 coefficients being iid $N(0, 1/4)$ for $i \neq j$ and the diagonal elements of $x_{ii} = a_{0i} I_2$ with $a_{0i}, \overset{iid}{\sim} N(0,1)$.

It is well known that all eigenvalues of a GSE are real and have multiplicities 2 and thus it has two distinguished eigenvalues.

In Tracy and Widom (1994), the following theorem is proved.

Theorem 3.20. *Let λ_n denote the largest eigenvalue of an order n GOE, GUE or GSE, then*

$$n^{2/3}(\lambda_n - 2) \xrightarrow{\mathscr{D}} T_\beta$$

where T_β is a random variable whose distribution function F_β is given by

$$F_2(x) = \exp\left(- \int_x^\infty (t-x) q^2(t) dt \right)$$
$$F_1(x) = \exp\left(- \frac{1}{2} \int_x^\infty q(t) dt \right) [F_2(x)]^{1/2}$$
$$F_4(2^{-1/2}x) = \cosh\left(- \frac{1}{2} \int_x^\infty q(t) dt \right) [F_2(x)]^{1/2}$$

and $q(t)$ is the solution to the P_{II} differential equation

$$q'' = tq + 2q^3$$

satisfying the marginal condition

$$q(t) \sim Ai(t), \quad as\ t \to \infty$$

and Ai is the Airy function.

The description of Airy function is complicated and so are the TW distribution functions. For an intuitive understanding of the TW distributions, we present a graph of their densities, see Fig. 3.3.

3.4.2 TW Law for Sample Covariance Matrix

It is interesting that the normalized largest eigenvalue of standard Wishart matrix tends to the same TW law under the assumption of normality. The following result was established by Johnstone (2001).

We first consider the real case:

Theorem 3.21. *Suppose that λ_{max} denote the largest eigenvalue of a real Wishart matrix $W(n, I_p)$. Define*

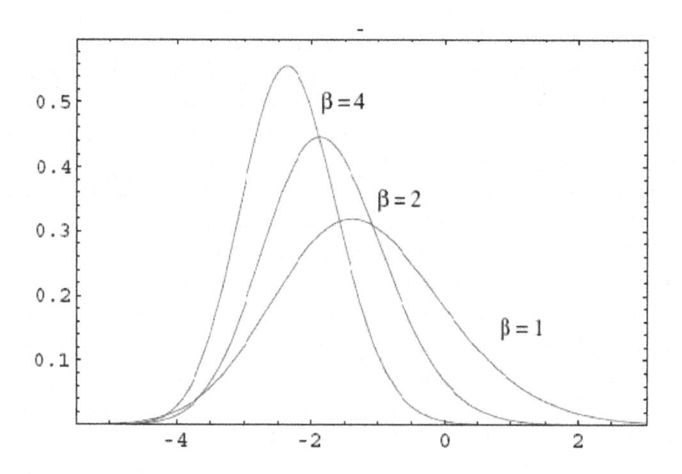

Fig. 3.3. The Density Function of F_β for $\beta = 1, 2, 4$.

$$\mu_{n,p} = (\sqrt{n-1} + \sqrt{p})^2$$

$$\sigma_{n,p} = (\sqrt{n-1} + \sqrt{p})\Big(\frac{1}{\sqrt{n-1}} + \frac{1}{\sqrt{p}}\Big)^{1/3}.$$

Then

$$\frac{\lambda_{max} - \mu_{n,p}}{\sigma_{n,p}} \xrightarrow{\mathscr{D}} W_1 \sim F_1.$$

where F_1 is the TW distribution with $\beta = 1$.

The complex Wishart case is due to Johansson (2000).

Theorem 3.22. *Suppose that λ_{max} denote the largest eigenvalue of a complex Wishart matrix $W(n, I_p)$. Define*

$$\mu_{n,p} = (\sqrt{n} + \sqrt{p})^2$$

$$\sigma_{n,p} = (\sqrt{n} + \sqrt{p})\Big(\frac{1}{\sqrt{n}} + \frac{1}{\sqrt{p}}\Big)^{1/3}.$$

Then

$$\frac{\lambda_{max} - \mu_{n,p}}{\sigma_{n,p}} \xrightarrow{\mathscr{D}} W_2 \sim F_2.$$

where F_2 is the TW distribution with $\beta = 2$.

4

Central Limit Theorems of Linear Spectral Statistics

4.1 Motivation and Strategy

As mentioned in the introduction, many important statistics in multivariate analysis can be written as functionals of ESD of some random matrices. The strong consistency of the ESD to LSD is not enough for more efficient statistical inferences, such as test of hypotheses, confidence regions, etc. In this chapter, we shall introduce some results on deeper properties of the convergence of ESD of large dimensional random matrices.

Let F_n be the ESD of random matrix which has an LSD F. We shall call

$$\hat{\theta} = \int f(x)dF_n(x) = \frac{1}{n}\sum_{k=1}^{n} f(\lambda_m)$$

a linear spectral statistic (LSS), associated with the given random matrix, which can be considered as an estimator of $\theta = \int f(x)dF(x)$. To test hypotheses about θ, it is necessary to know the limiting distribution of

$$G_n(f) = \alpha_n(\hat{\theta} - \theta) = \int f(x)dX_n(x),$$

where $X_n(x) = \alpha_n(F_n(x) - F(x))$ and $\alpha_n \to \infty$ is a suitably chosen normalizer such that $G_n(f)$ tends to a non-degenerate distribution.

Ideally, if for some choice of α_n, $X_n(x)$ tends to a limiting process $X(x)$ in the C space or D space equipped with the Skorokhod metric, then the limiting distribution of all LSS can be derived. Unfortunately, there is evidence indicating that $X_n(x)$ cannot tend to a limiting process in any metric space.

An example is given in Diaconis and Evans (2001), in which it is shown that if F_n is the empirical distribution function of the angles of eigenvalues of a Haar matrix, then for $0 \le \alpha < \beta < 2\pi$, the finite dimensional distributions of

$$\frac{\pi n}{\sqrt{\log n}}(F_n(\beta) - F_n(\alpha) - \mathrm{E}[F_n(\beta) - F_n(\alpha)])$$

converge weakly to $Z_{\alpha,\beta}$, jointly normal variables, standardized, with covariances

$$\operatorname{cov}(Z_{\alpha,\beta}, Z_{\alpha',\beta'}) = \begin{cases} 0.5, & \text{if } \alpha = \alpha' \text{ and } \beta \neq \beta', \\ 0.5, & \text{if } \alpha \neq \alpha' \text{ and } \beta = \beta', \\ -0.5, & \text{if } \alpha = \beta' \text{ or } \beta = \alpha', \\ 0, & \text{otherwise.} \end{cases}$$

This covariance structure cannot arise from a probability space on which $Z_{0,x}$ is defined as a stochastic process with measurable paths in $D[a,b]$ for any $0 < a < b < 2\pi$. Indeed, if so, then with probability one, for x decreasing to a, $Z_{0,x} - Z_{0,a}$ would converge to zero, which implies its variance would approach 0. But its variance remains at one. Furthermore, this result also shows that with any choice of α_n, $X_n(x)$ cannot tend to a non-trivial process in any metric space.

Therefore, we have to withdraw our attempts of looking for the limiting process of $X_n(x)$. Instead, we shall consider the convergence of $G_n(f)$ with $\alpha_n = n$. Earliest work dates back to Jonsson (1982) in which he proved the CLT for the centralized sum of the r-th power of eigenvalues of a normalized Wishart matrix. Similar work for the Wigner matrix was obtained in Sinai and Soshnikov (1998). Later, Johansson (1998) proved the CLT of linear spectral statistics of the Wigner matrix under density assumptions.

Because X_n tends to a weak limit implies the convergence of $G_n(f)$ for all continuous and bounded f, Diaconis and Evans' example shows that the convergence of $G_n(f)$ cannot be true for all f, at least for indicator functions. Thus in this chapter, we shall confine ourselves to the convergence of $G_n(f)$ to a normal variable when f is analytic in a region containing the support of F, for Wigner matrices and sample covariance matrices.

Our strategy will be as follows: Choose a contour $\mathcal{C}$ which encloses the support of F_n and F. Then, by the Cauchy integral formula, we have

$$f(x) = \frac{1}{2\pi i} \oint_{\mathcal{C}} \frac{f(z)}{z - x} dz. \tag{4.1.1}$$

By this formula, we can rewrite $G_n(f)$ as

$$G_n(f) = -\frac{1}{2\pi i} \oint_{\mathcal{C}} f(z)[n(s_n(z) - s(z)]dz, \tag{4.1.2}$$

where s_n and s are Stieltjes transforms of F_n and F respectively. So, the problem to find the limit distribution of $G_n(f)$ reduces to finding the limiting process of $M_n(z) = n(s_n(z) - s(z))$.

Before concluding this section, we present a lemma on estimation of moments of quadratic forms which is useful for the proofs of the CLT of LSS of both Wigner matrices and sample covariance matrices.

Lemma 4.1. *Suppose that x_i, $i = 1, \cdots, n$, are independent, with* $\mathrm{E}x_i = 0$, $\mathrm{E}|x_i|^2 = 1$, $\sup E|x_i|^4 = \nu < \infty$ *and* $|x_i| \leq \eta\sqrt{n}$ *with* $\eta > 0$. *Assume that* $\mathbf{A}$

is a complex matrix. Then for any given $2 \le p \le b\log(n\nu^{-1}\eta^4)$ *and* $b > 1$, *we have*

$$\mathrm{E}|\boldsymbol{\alpha}^*\mathbf{A}\boldsymbol{\alpha} - \mathrm{tr}(\mathbf{A})|^p \le \nu n^p(n\eta^4)^{-1}(40b^2\|\mathbf{A}\|\eta^2)^p,$$

where $\boldsymbol{\alpha} = (x_1, \cdots, x_n)^T$.

The proof of this lemma refers to Bai and Silverstein (2006, 2009).

4.2 CLT of LSS for Wigner Matrix

In this section, we shall consider the case where F_n is the ESD of the normalized Wigner matrix $\mathbf{W}_n$. More precisely, let $\mu(f)$ denote the integral of a function f with respect to a signed measure μ. Let $\mathcal{U}$ be an open set of the complex plane which contains the interval $[-2, 2]$, i.e. the support of semicircular law F.

For ease of statements, the set of complex Wigner matrices will called the *Complex Wigner Ensemble* (CWE) and the set of real Wigner matrices is called the *Real Wigner Ensemble* (RWE). In both cases, the entries are not necessarily identically distributed. If in addition the entries are Gaussian (with $\sigma^2 = 1$ and 2 for the CWE and RWE respectively), the above ensembles are the classical Gaussian unitary ensemble (GUE) and the Gaussian orthogonal ensemble (GOE) of random matrices.

Next define $\mathcal{A}$ to be the set of analytic functions $f : \mathcal{U} \mapsto \mathbb{C}$. We then consider the empirical process $G_n := \{G_n(f)\}$ indexed by $\mathcal{A}$, i.e.,

$$G_n(f) := n\int_{-\infty}^{\infty} f(x)[F_n - F](dx), \quad f \in \mathcal{A}. \tag{4.2.1}$$

To study the weak limit of G_n, we need conditions on the moments of the entries x_{ij} of the Wigner matrices $\sqrt{n}\mathbf{W}_n$. Note that the distributions of entries x_{ij} are allowed to depend on n, but the dependence on n is suppressed. Let

[M1] For all i, $\mathrm{E}|x_{ii}|^2 = \sigma^2 > 0$ and for all $i < j$, $\mathrm{E}|x_{ij}|^2 = 1$ and for CWE $\mathrm{E}x_{ij}^2 = 0$.

[M2] (homogeneity of 4-th moments) $M = \mathrm{E}|x_{ij}|^4$ for $i \ne j$;

[M3] (uniform tails) for any $\eta > 0$, as $n \to \infty$,

$$\frac{1}{\eta^4 n^2}\sum_{i,j}\mathrm{E}[|x_{ij}|^4 I(|x_{ij}| \ge \eta\sqrt{n})] = o(1).$$

Note that condition [M3] implies the existence of a sequence $\eta_n \downarrow 0$ such that

$$(\eta_n\sqrt{n})^{-4}\sum_{i,j}\mathrm{E}[|x_{ij}|^4 I(|x_{ij}| \ge \eta_n\sqrt{n})] = o(1). \tag{4.2.2}$$

Note that $\eta_n \to 0$ may be assumed to be as slow as desired. For definiteness, we assume that $\eta_n > 1/\log n$.

The main task of this section is to introduce the finite dimensional convergence of the empirical process G_n to a Gaussian process. That is, for any k elements $f_1, \cdots, f_m$ of $\mathcal{A}$, the vector $(G_n(f_1), \cdots, G_n(f_m))$ converges weakly to a p-dimensional Gaussian distribution.

Let $\{T_m\}$ be the family of Tchebychev polynomials and define for $f \in \mathcal{A}$ and any integer $\ell \geq 0$,

$$\begin{aligned}\tau_\ell(f) &= \frac{1}{2\pi}\int_{-\pi}^{\pi} f(2\cos(\theta))e^{i\ell\theta}d\theta \\ &= \frac{1}{2\pi}\int_{-\pi}^{\pi} f(2\cos(\theta))\cos(\ell\theta)d\theta \\ &= \frac{1}{\pi}\int_{-1}^{1} f(2t)T_\ell(t)\frac{1}{\sqrt{1-t^2}}dt. \end{aligned} \tag{4.2.3}$$

In order to give a unified statement for both ensembles, we introduce the parameter κ with values 1 and 2 for the complex and real Wigner ensemble respectively. Moreover set $\beta = \mathrm{E}(|x_{12}|^2 - 1)^2 - \kappa$. In particular, for the G.U.E. we have $\kappa = \sigma^2 = 1$ and for the G.O.E. we have $\kappa = \sigma^2 = 2$, and in both cases $\beta = 0$.

We quote the following theorem which was established in Bai and Yao (2005).

Theorem 4.2. *Under conditions [M1]—[M3], the spectral empirical process $G_n = (G_n(f))$ indexed by the set of analytic functions $\mathcal{A}$ converges weakly in finite dimension to a Gaussian process $G := \{G(f) : f \in \mathcal{A}\}$ with mean function $\mathrm{E}[G(f)]$ given by*

$$\frac{\kappa-1}{4}\{f(2)+f(-2)\} - \frac{\kappa-1}{2}\tau_0(f) + (\sigma^2-\kappa)\tau_2(f) + \beta\tau_4(f), \tag{4.2.4}$$

and the covariance function $c(f,g) := \mathrm{E}[\{G(f)-\mathrm{E}G(f)\}\{G(g)-\mathrm{E}G(g)\}]$ given by

$$\sigma^2\tau_1(f)\tau_1(g) + 2(\beta+1)\tau_2(f)\tau_2(g) + \kappa\sum_{\ell=3}^{\infty}\ell\tau_\ell(f)\tau_\ell(g) \tag{4.2.5}$$

$$= \frac{1}{4\pi^2}\int_{-2}^{2}\int_{-2}^{2} f'(t)g'(s)V(t,s)dtds, \tag{4.2.6}$$

where

$$\begin{aligned}V(t,s) &= \left(\sigma^2 - \kappa + \frac{1}{2}\beta ts\right)\sqrt{(4-t^2)(4-s^2)} \\ &\quad + \kappa\log\left(\frac{4-ts+\sqrt{(4-t^2)(4-s^2)}}{4-ts-\sqrt{(4-t^2)(4-s^2)}}\right). \end{aligned}\tag{4.2.7}$$

Note that our definition implies that the variance of $G(f)$ equals $c(f, \bar{f})$. Let $\delta_a(dt)$ be the Dirac measure at a point a. The mean function can also be written as

$$\mathrm{E}[G(f)] = \int_{\mathbb{R}} f(2t)d\nu(t), \tag{4.2.8}$$

with signed measure

$$\begin{aligned} d\nu(t) &= \frac{\kappa - 1}{4}[\delta_1(dt) + \delta_{-1}(dt)] \\ &+ \frac{1}{\pi}\left[-\frac{\kappa - 1}{2} + (\sigma^2 - \kappa)T_2(t) + \beta T_4(t)\right]\frac{1}{\sqrt{1-t^2}}I([-1,1])(t)dt. \end{aligned} \tag{4.2.9}$$

In the cases of G.U.E. and G.O.E., the covariance reduces to the third term in (4.2.5). The mean $\mathrm{E}[G(f)]$ is always zero for the G.U.E., since in this case $\sigma^2 = \kappa = 1$ and $\beta = 0$. As for the G.O.E., since $\beta = 0$ and $\sigma^2 = \kappa = 2$, we have

$$\mathrm{E}[G(f)] = \frac{1}{4}\{f(2) + f(-2)\} - \frac{1}{2}\tau_0(f).$$

Therefore the limit process is not necessarily centered.

Example 4.3. *Consider the case where $\mathcal{A} = \{f(x,t)\}$ and the stochastic process*

$$Z_n(t) = \sum_{k=1}^{n} f(\lambda_m, t) - n\int_{-2}^{2} f(x,t)F(dx).$$

If both f and $\partial f(x,t)/\partial t$ are analytic in x over a region containing $[-2,2]$, it follows easily from Theorem 4.2 that $Z_n(t)$ converges to a Gaussian process. Its finite-dimensional convergence is exactly the same as in Theorem 4.2, while its tightness can be obtained as a simple consequence of the same theorem.

4.2.1 Outlines of the Proof

Let $\mathcal{C}$ be the contour made by the boundary of the rectangle with vertices $(\pm a \pm iv_0)$ where $a > 2$ and $1 \geq v_0 > 0$. We can always assume that the constants $a - 2$ and v_0 are sufficiently small so that $\mathcal{C} \subset \mathcal{U}$.

Then, as mentioned in Section 4.1,

$$G_n(f) = -\frac{1}{2\pi i}\oint_{\mathcal{C}} f(z)n[s_n(z) - s(z)]dz, \tag{4.2.10}$$

where s_n and s are Stieltjes transforms of $\mathbf{W}_n$ and the semicircular law, respectively. The reader is reminded that the above equality may not be correct when some eigenvalues of $\mathbf{W}_n$ run outside the contour. A corrected version of (4.2.10) should be

$$G_n(f)I(B_n^c) = -\frac{1}{2\pi i}I(B_n^c)\oint_{\mathcal{C}} f(z)n[s_n(z) - s(z)]dz,$$

where $B_n = \{|\lambda_{ext}(\mathbf{W}_n)| \geq 1 + a/2\}$ and λ_{ext} denotes the smallest or largest eigenvalue of the matrix $\mathbf{W}_n$. But this difference will not be a matter because, after truncation and renormalization, for any $a > 2$ and $t > 0$,

$$\mathrm{P}(B_n) = o(n^{-t}). \tag{4.2.11}$$

This representation reduces our problem to showing that the process $M_n := (M_n(z))$ indexed by $z \notin [-2, 2]$ where

$$M_n(z) = n[s_n(z) - s(z)], \tag{4.2.12}$$

converges to a Gaussian process $M(z)$, $z \notin [-2, 2]$. We will show this conclusion by the following theorem.

Throughout this section, we set $\mathbb{C}_0 = \{z = u + iv \ : \ |v| \geq v_0\}$.

Theorem 4.4. *Under conditions* [M1]—[M3], *the process* $\{M_n(z);\ \mathbb{C}_0\}$ *converges weakly to a Gaussian process* $\{M(z);\ \mathbb{C}_0\}$ *with the mean and covariance functions given in Lemma* 4.5 *and Lemma* 4.6.

Since the mean and covariance functions of $M(z)$ are independent of v_0, the process $\{M(z);\ \mathbb{C}_0\}$ in Theorem 4.4 can be taken as a restriction of a process $\{M(z)\}$ defined on the whole complex plane, except the real axes. Further, by noticing the symmetry, $M(\bar{z}) = \overline{M(z)}$, and the continuity of the mean and covariance functions of $M(z)$ on the real axes except for $z \in [-2, 2]$, we may extend the process to $\{M(z);\ \Re z \notin [-2, 2]\}$.

Split the contour $\mathcal{C}$ as the union $\mathcal{C}_u + \mathcal{C}_l + \mathcal{C}_r + \mathcal{C}_0$, where $\mathcal{C}_l = \{z = -a + iy,\ \zeta_n n^{-1} < |y| \leq v_1\}$, $\mathcal{C}_r = \{z = a + iy,\ \zeta_n n^{-1} < |y| \leq v_1\}$ and $\mathcal{C}_0 = \{z = \pm a + iy,\ |y| \leq n^{-1}\zeta_n\}$, where $\zeta_n \to 0$ is a slowly varying sequence of positive constants. By Theorem 4.4 we get the following weak convergence

$$\int_{\mathcal{C}_u} M_n(z)dz \Rightarrow \int_{\mathcal{C}_u} M(z)dz.$$

To prove Theorem 4.2, we only need to show that for $j = l, r, 0$,

$$\lim_{v_1 \downarrow 0} \limsup_{n \to \infty} \mathrm{E}\left|\int_{\mathcal{C}_j} M_n(z)I(B_n^c)dz\right|^2 = 0, \tag{4.2.13}$$

and

$$\lim_{v_1 \downarrow 0} \mathrm{E}\left|\int_{\mathcal{C}_j} M(z)dz\right|^2 = 0. \tag{4.2.14}$$

Estimate (4.2.14) can be verified directly by the mean and variance functions of $M(z)$. The proof of (4.2.13) for the case $j = 0$ and the proof of (4.2.13) for $j = l$ and r will be technical and simple and thus omitted.

Step 1. Truncation and Renormalization

Choose $\eta_n > 1/\log n$ according to (4.2.2), we first truncate the variables as $\widehat{x}_{ij} = x_{ij}I(|x_{ij}| \le \eta_n\sqrt{n})$. We need to further normalize them by setting $\widetilde{x}_{ij} = (\widehat{x}_{ij} - \mathrm{E}\widehat{x}_{ij})/\sigma_{ij}$ for $i \ne j$ and $\widetilde{x}_{ii} = \sigma(\widehat{x}_{ii} - \mathrm{E}\widehat{x}_{ii})/\hat{\sigma}_{ii}$ where σ_{ij} are the standard deviation of $\widehat{x}_{ij}$.

Let $\widehat{F}_n$ and $\widetilde{F}_n$ be the ESD of the random matrices $(\frac{1}{\sqrt{n}}\widehat{x}_{ij})$ and $(\frac{1}{\sqrt{n}}\widetilde{x}_{ij})$, respectively. According to (4.2.1), we similarly define $\widehat{G}_n$ and $\widetilde{G}_n$. First observe that

$$\mathrm{P}(G_n \ne \widehat{G}_n) \le \mathrm{P}(F_n \ne \widehat{F}_n) = o(1). \tag{4.2.15}$$

Indeed,

$$\begin{aligned}\mathrm{P}(F_n \ne \widehat{F}_n) &\le \mathrm{P}\{\text{ for some } i,j,\ \widehat{x}_{ij} \ne x_{ij}\}\\ &\le \sum_{i,j}\mathrm{P}\left\{|x_{ij}| \ge \eta_n\sqrt{n}\right\}\\ &\le (\eta_n\sqrt{n})^{-4}\sum_{i,j}\mathrm{E}[|x_{ij}|^4 I(|x_{ij}| \ge \eta_n\sqrt{n})] = o(1).\end{aligned}$$

Secondly, as f is analytic, by conditions [M2] and [M3] we have

$$\begin{aligned}&\mathrm{E}\left|\widetilde{G}_n(f) - \widehat{G}_n(f)\right|^2\\ &\le Cn\mathrm{E}\sum_{j=1}^{n}|\tilde{\lambda}_{nj} - \hat{\lambda}_{nj}|^2\\ &\le C\left[\sum_{ij}[(n\eta_n^2)^{-2} + 2(n\eta_n^2)^{-3}]\mathrm{E}^2|x_{ij}^4|I(|x_{ij}| \ge \sqrt{n}\eta_n)\right]\\ &= o_p(1),\end{aligned}$$

where $\tilde{\lambda}_{nj}$ and $\hat{\lambda}_{nj}$ are the jth largest eigenvalues of the Wigner matrices $n^{-1/2}(\tilde{x}_{ij})$ and $n^{-1/2}(\hat{x}_{ij})$ respectively. Therefore the weak limit of the variables $(G_n(f))$ is not affected if the original variables x_{ij} are replaced by the normalized truncated variables $\widetilde{x}_{ij}$.

From the normalization, the variables $\widetilde{x}_{ij}$'s all have mean 0 and the same absolute second moments as the original variables. However the 4th moments of the off-diagonal elements are no longer homogenous and for the CWE, $\mathrm{E}x_{ij}^2$ is no longer 0. However, this does not matter because $\left|\sum_{i\ne j}[\mathrm{E}|x_{ij}^4| - \mathrm{E}|\widetilde{x}_{ij}|^4]\right| = o(n^{-2})$ and $\max_{i<j}|\mathrm{E}\widetilde{x}_{ij}^2| = O(1/n)$.

We now assume that the above conditions hold and we still use x_{ij} to denote the truncated and normalized variables $\widetilde{x}_{ij}$'s.

Step 2. Mean Function of M_n

For $z \in \mathbb{C}_0$, we have

$$\mathrm{E}M_n(z) = n[\mathrm{E}s_n(z) - s(z)].$$

Lemma 4.5. *The mean function* $\mathrm{E}M_n(z)$ *uniformly tends to*

$$b(z) = [1 + s'(z)]s(z)^3\Big[\sigma^2 - 1 + (\kappa - 1)s'(z)\beta s^2(z)\Big]$$

for $z \in \mathbb{C}_0$ *and for both ensembles CWE and RWE.*

Proof. We show that

$$\mathrm{E}M_n(z) - b(z) \to 0,$$

uniformly in $z \in \mathcal{C}_n = \mathcal{C}_u + \mathcal{C}_l + \mathcal{C}_r$.

By (2.1.14), we have

$$n\delta_n(z) = \sum_{k=1}^{n} \mathrm{E}\bigg(\frac{\varepsilon_k}{z + \mathrm{E}s_n(z) - \varepsilon_k}\bigg),$$

and

$$\mathrm{E}s_n(z) = \frac{-z + \mathrm{sgn}(\Im(z))\sqrt{z^2 - 4 + 4\delta_n(z)}}{2}, \tag{4.2.16}$$

where $\varepsilon_k = \frac{1}{\sqrt{n}}x_{kk} + \mathrm{E}s_n(z) - \frac{1}{n}\boldsymbol{\alpha}_k^*(\mathbf{W}_k - z\mathbf{I}_{n-1})^{-1}\boldsymbol{\alpha}_k$.

Using the identity for any integer p,

$$\frac{1}{u - \varepsilon} = \frac{1}{u}\left[1 + \frac{\varepsilon}{u} + \cdots + \frac{\varepsilon^p}{u^p} + \frac{\varepsilon^{p+1}}{u^p(u - \varepsilon)}\right],$$

we get

$$\begin{aligned} n\delta_n(z) &= \sum_{k=1}^{n} \frac{\mathrm{E}\varepsilon_k}{z + \mathrm{E}s_n(z)} + \sum_{k=1}^{n} \frac{\mathrm{E}\varepsilon_k^2}{(z + \mathrm{E}s_n(z))^2} \\ &\quad + \sum_{k=1}^{n} \frac{\mathrm{E}\varepsilon_k^3}{(z + \mathrm{E}s_n(z))^2(z + \mathrm{E}s_n(z) - \varepsilon_k)} \\ &= S_1 + S_2 + S_3. \end{aligned}$$

First we indicate that $S_3 = o(1)$ and the details are omitted sine it is routine and simple.

Next, we find the limit of $\mathrm{E}\varepsilon_k$. We have

$$\begin{aligned} n\mathrm{E}\varepsilon_k &= [\mathrm{E}\mathrm{tr}\mathbf{D}^{-1} - \mathrm{E}\mathrm{tr}\mathbf{D}_n^{-1}] \\ &= -\mathrm{E}\bigg(\frac{1 + n^{-1}\boldsymbol{\alpha}_k^*\mathbf{D}_k^{-2}\boldsymbol{\alpha}_k}{z + \mathrm{E}s_n(z) - \varepsilon_k}\bigg) \to s(z)(1 + s'(z)). \end{aligned}$$

This lead to $S_1 \to -s^2(1 + s')$.

Now, let us find the approximation of $\mathrm{E}\varepsilon_k^2$. By the previous estimation for $\mathrm{E}\varepsilon_k$, we have

$$\sum_{k=1}^{n} \mathrm{E}\varepsilon_k^2 = \sum_{k=1}^{n} \mathrm{E}(\varepsilon_k - \mathrm{E}\varepsilon_k)^2 + O(n^{-1}),$$

where $O(n^{-1})$ is uniform in $z \in \mathcal{C}_n$.

Furthermore by the definition of ε_k, we have

$$\varepsilon_k - \mathrm{E}\varepsilon_k = \frac{1}{\sqrt{n}} x_{kk} - n^{-1}[\boldsymbol{\alpha}_k^* \mathbf{D}_k^{-1} \alpha_k - \mathrm{E}\mathbf{D}_k^{-1}]$$
$$= \frac{1}{\sqrt{n}} x_{kk} - n^{-1}[\boldsymbol{\alpha}_k^* \mathbf{D}_k^{-1} \boldsymbol{\alpha}_k - \mathrm{tr}\mathbf{D}_k^{-1}] + n^{-1}[\mathrm{tr}\mathbf{D}_k^{-1} - \mathrm{Etr}\mathbf{D}_k^{-1}].$$

Therefore

$$\mathrm{E}[\varepsilon_k - \mathrm{E}\varepsilon_k]^2 = \frac{\sigma^2}{n} + \frac{1}{n^2}\mathrm{E}[\boldsymbol{\alpha}_k^* \mathbf{D}_k^{-1} \alpha_k - \mathrm{tr}\mathbf{D}_k^{-1}]^2 + \frac{1}{n^2}\mathrm{E}[\mathrm{tr}\mathbf{D}_k^{-1} - \mathrm{Etr}\mathbf{D}_k^{-1}]^2. \tag{4.2.17}$$

We indicate that (detailed proof omitted)

$$n^{-2}\mathrm{E}\left[\mathrm{tr}\mathbf{D}_k^{-1} - \mathrm{Etr}\mathbf{D}_k^{-1}\right]^2 = o(n^{-3/2}).$$

By simple calculation, for matrices $\mathbf{A} = (a_{ij})$ and $\mathbf{B} = (b_{ij})$, we have the identity

$$\mathrm{E}(\boldsymbol{\alpha}_k^* \mathbf{A} \boldsymbol{\alpha}_k - \mathrm{tr}\mathbf{A})(\boldsymbol{\alpha}_k^* \mathbf{B} \boldsymbol{\alpha}_k - \mathrm{tr}\mathbf{B})$$
$$= \mathrm{tr}\mathbf{AB} + \sum_{i,j} a_{ij} b_{ji} \mathrm{E}x_{ik}^2 \mathrm{E}\bar{x}_{jk}^2 + \sum_{i=1}^{n} a_{ii} b_{ii} (\mathrm{E}|x_{ik}|^4 - 2 - |\mathrm{E}x_{ik}^2|^2).$$

Combining this identity and the assumption that $\mathrm{E}x_{ij}^2 = o(1)$ for the CWE and $\mathrm{E}x_{ij}^2 = 1$ for the RWE, we have

$$n^{-1}\mathrm{E}[\boldsymbol{\alpha}_k^* \mathbf{D}_k^{-1} \boldsymbol{\alpha}_k - \mathrm{tr}\mathbf{D}_k^{-1}]^2 = \frac{\kappa}{n}\mathrm{E}\left[\mathrm{tr}\mathbf{D}_k^{-2}\right] + \frac{\beta}{n}\mathrm{E}\left[\sum_i d_{ii}^2\right] + o(1), \tag{4.2.18}$$

where d_{ii} are the diagonal entries of the matrix $\mathbf{D}_k^{-1}$.

It can be proven that

$$\max_{i,k} \mathrm{E}|d_{ii} - s(z)|^2 \to 0.$$

Hence,

$$n^{-1}\mathrm{E}[\boldsymbol{\alpha}_k^* \mathbf{D}_k^{-1} \boldsymbol{\alpha}_k - \mathrm{tr}\mathbf{D}_k^{-1}]^2 \to \kappa s'(z) + \beta s^2(z).$$

Consequently, we obtain

$$S_2 \to s^2(z)(\sigma^2 + \kappa s'(z) + \beta s(z)^2)).$$

Summing the above, we have reached

$$n\delta(z) \to s^2(z)\left(\sigma^2 - 1 + (\kappa - 1)s'(z) + \beta s^2(z)\right).$$

Finally, we obtain

$$\begin{aligned}
\mathrm{E}M_n(z) &= n(\mathrm{E}s_n(z) - s(z)) \\
&= \frac{n\mathrm{sgn}(\Im(z))}{2}\left(\sqrt{z^2 - 4 + 4\delta_n(z)} - \sqrt{z^2-4}\right) \\
&= \frac{2n\delta_n(z)\mathrm{sgn}(\Im(z))}{\sqrt{z^2 - 4 + 4\delta_n(z)} + \sqrt{z^2-4}} \\
&\to \frac{s^2(z)\left(\sigma^2 - 1 + (\kappa-1)s'(z) + \beta s^2(z)\right)}{\mathrm{sgn}(\Im(z))\sqrt{z^2-4}}.
\end{aligned}$$

The lemma is then proved from the above and the equality

$$\mathrm{sgn}(\Im(z))\sqrt{z^2-4} = -\frac{s(z)}{s'(z)}.$$

Step 3. Finite-Dimensional Convergence of $M_n - \mathrm{E}M_n$
We use the following martingale decomposition,

$$M_n(z) - \mathrm{E}M_n(z) = \sum_{k=1}^{n} \gamma_k,$$

where

$$\begin{aligned}
\gamma_k &= (\mathrm{E}_{k-1} - \mathrm{E}_k)\mathrm{tr}\mathbf{D}^{-1} \\
&= (\mathrm{E}_{k-1} - \mathrm{E}_k)(\mathrm{tr}\mathbf{D}^{-1} - \mathrm{tr}\mathbf{D}_k^{-1}) \\
&= (\mathrm{E}_{k-1} - \mathrm{E}_k)\frac{1 + \frac{1}{n}\boldsymbol{\alpha}_k^*(\mathbf{W}_k - z\mathbf{I}_{n-1})^{-2}\boldsymbol{\alpha}_k}{-z + \mathrm{E}s_n(z) + \varepsilon_k}.
\end{aligned} \tag{4.2.19}$$

We may prove that for any finite number of $z_1, \cdots, z_m$, $\{M_n(z_1) - \mathrm{E}M_n(z_1), \cdots, M_n(z_m) - \mathrm{E}M_n(z_m)\}$ tends to $2m$-dimensional normal with means 0. What we need is to derive the asymptotic variance-covariance.

Lemma 4.6. *Assume conditions* [M1]—[M3] *are satisfied. For any set of m points $\{z_s, s = 1, \cdots, m\}$ of $\mathbb{C}_0$, the random vector $\mathbf{Z}_n$ converge weakly to an m-dimensional zero-mean Gaussian distribution with covariance matrix given, with $s_j = s(z_j)$, by*

$$\begin{aligned}
\Gamma(z_i, z_j) &= \frac{\partial^2}{\partial z_i \partial z_j}\left[(\sigma^2 - \kappa)s_i s_j + \frac{1}{2}\beta(s_i s_j)^2 - \kappa \log(1 - s_i s_j)\right] \\
&= s_i' s_j' \left[\sigma^2 - \kappa + 2\beta s_i s_j + \frac{\kappa}{(1 - s_i s_j)^2}\right].
\end{aligned} \tag{4.2.20}$$

We omit the detailed proofs.

Step 4. Tightness of the Process $M_n(z) - \mathrm{E}M_n(z)$

It is enough to establish the following Hölder condition: for some positive constant K and $z_1, z_2 \in \mathbb{C}_0$,

$$\mathrm{E}|M_n(z_1) - M_n(z_2) - \mathrm{E}(M_n(z_1) - M_n(z_2))|^2 \le K|z_1 - z_2|^2. \qquad (4.2.21)$$

Recalling the martingale decomposition given in last step, we have

$$\begin{aligned}&\mathrm{E}|M_n(z_1) - M_n(z_2) - \mathrm{E}(M_n(z_1) - M_n(z_2))|^2\\ &= \sum_{k=1}^{n} \mathrm{E}|\gamma_k(z_1) - \gamma_k(z_2)|^2,\end{aligned}$$

where

$$\begin{aligned}\gamma_k(z) &= (\mathrm{E}_k - \mathrm{E}_{k-1})\sigma_k(z),\\ \sigma_k(z) &= \beta_k(z)(1 + \frac{1}{n}\gamma_k^* \mathbf{D}_k^{-2}\boldsymbol{\alpha}_k),\\ \beta_k(z) &= -\frac{1}{\frac{1}{\sqrt{n}}x_{kk} - z - \frac{1}{n}\boldsymbol{\alpha}_k \mathbf{D}_k^{-1}\boldsymbol{\alpha}_k}.\end{aligned}$$

Through routine arguments, (4.2.21) can be proven.

Step 5. Computation of the Mean and Covariance Function of $G(f)$

a. Mean Function

Let $\mathcal{C}$ be a contour as defined at the beginning of this section. By (4.2.10) and Lemma 4.5, we have

$$\begin{aligned}\mathrm{E}(G_n(f)) &= -\frac{1}{2\pi i}\oint_{\mathcal{C}} f(z)\mathrm{E}M_n(z)dz\\ \to \mathrm{E}(G(f)) &= -\frac{1}{2\pi i}\oint_{\mathcal{C}} f(z)\mathrm{E}M(z)dz\\ &= -\frac{1}{2\pi i}\oint_{\mathcal{C}} f(z)[1 + s'(z)]s^3(z)\Big[\sigma^2 - 1 + (\kappa - 1)s'(z) + \beta s^2(z)\Big]dz.\end{aligned}$$

Select $\rho < 1$ but so close to 1 that the contour

$$\mathcal{C}' = \{z = -(\rho e^{i\theta} + \rho^{-1}e^{-i\theta}) : 0 \le \theta < 2\pi\}$$

is completely contained in the analytic region of f. Note that when z runs a cycle along $\mathcal{C}'$ counterclockwise, s runs a cycle along the circle $|s| = \rho$ counterclockwise because $z = -(s + s^{-1})$[1]. By Cauchy's theorem, the above integral along $\mathcal{C}$ equals the integral along $\mathcal{C}'$. Thus, by changing variable z to s and noticing that $s' = s^2/(1 - s^2)$, we obtain

[1] The reason for choosing $|s| = \rho < 1$ is due to the fact that the mode of the Stieltjes transform of the semicircle law is less than 1.

$$
\begin{aligned}
&\mathrm{E}(G(f)) \\
&= -\frac{1}{2\pi i}\oint_{|s|=\rho} f(-s-s^{-1})s\Big[\sigma^2-1+(\kappa-1)\frac{s^2}{1-s^2}+\beta s^2\Big]ds.
\end{aligned}
$$

By setting $s=-e^{i\theta}$ and then $t=\cos\theta$, using $T_k(\cos\theta)=\cos(k\theta)$,

$$
\begin{aligned}
&-\frac{1}{2\pi i}\oint_{|s|=1} f(-s-s^{-1})s\Big[\sigma^2-1+(\kappa-1)\frac{s^2}{1-s^2}+\beta s^2\Big]ds \\
&= -\frac{1}{\pi}\int_0^{\pi} f(2\cos\theta)\Big[(\sigma^2-1)\cos 2\theta \\
&\qquad -\frac{1}{2}(\kappa-1)(1+2\cos 2\theta)+\beta\cos 4\theta\Big]d\theta \\
&= -\frac{1}{2}(\kappa-1)\tau_0(f)+(\sigma^2-\kappa)\tau_2(f)+\beta\tau_4(f).
\end{aligned}
$$

Let us evaluate the difference

$$
\frac{1}{2\pi i}\left[\oint_{|s|=1}-\oint_{|s|=\rho}\right] f(-s-s^{-1})s\Big[\sigma^2-1+(\kappa-1)\frac{s^2}{1-s^2}+\beta s^2\Big]ds.
$$

Note that the integrand has two poles on the circle $|s|=1$ with residuals $-\frac{1}{2}f(\pm 2)$ at points $s=\mp 1$. By contour integration, we have

$$
\begin{aligned}
&\frac{1}{2\pi i}\left[\oint_{|s|=1}-\oint_{|s|=\rho}\right] f(-s-s^{-1})s\Big[\sigma^2-1+(\kappa-1)\frac{s^2}{1-s^2}+\beta s^2\Big]ds \\
&\qquad\qquad = \frac{\kappa-1}{4}(f(2)+f(-2)).
\end{aligned}
$$

Putting together these two results gives the formula (4.2.4) for $\mathrm{E}[G(f)]$.

b. Covariance Function

Let $\mathcal{C}_j$, $j=1,2$, be two disjoint contours with vertices $\pm(2+\varepsilon_j)\pm iv_j$. The positive values of ε_j and v_j are chosen sufficiently small so that the two contours are contained in $\mathcal{U}$. By (4.2.10) and Theorem 4.4, we have

$$
\begin{aligned}
&\mathrm{Cov}(G_n(f),G_n(g)) \\
&= -\frac{1}{4\pi^2}\oint_{\mathcal{C}_1}\oint_{\mathcal{C}_2} f(z_1)g(z_2)\mathrm{Cov}(M_n(z_1),M_n(z_2))dz_1dz_2 \\
&= -\frac{1}{4\pi^2}\oint_{\mathcal{C}_1}\oint_{\mathcal{C}_2} f(z_1)g(z_2)\Gamma_n(z_1,z_2)dz_1dz_2+o(1) \\
&\longrightarrow c(f,g) = -\frac{1}{4\pi^2}\oint_{\mathcal{C}_1}\oint_{\mathcal{C}_2} f(z_1)g(z_2)\Gamma(z_1,z_2)dz_1dz_2,
\end{aligned}
$$

where $\Gamma(z_1,z_2)$ is given in (4.2.20).

By the proof of Lemma 4.6, we have

$$\Gamma(z_1, z_2) = \frac{\partial^2}{\partial z_1 \partial z_2} s(z_1)s(z_2)\widetilde{\Gamma}(z_1, z_2).$$

Integrating by parts, we obtain

$$\begin{aligned} c(f,g) &= -\frac{1}{4\pi^2} \oint_{\mathcal{C}_1} \oint_{\mathcal{C}_2} f'(z_1)g'(z_2)s(z_1)s(z_2)\widetilde{\Gamma}(z_1, z_2)dz_1 dz_2 \\ &= -\frac{1}{4\pi^2} \oint_{\mathcal{C}_1} \oint_{\mathcal{C}_2} A(z_1, z_2)dz_1 dz_2, \end{aligned}$$

where

$$\begin{aligned} A(z_1, z_2) = f'(z_1)g'(z_2)\Big[s(z_1)s(z_2)(\sigma^2 - \kappa) + \frac{1}{2}\beta s^2(z_1)s^2(z_2) \\ -\kappa \log(1 - s(z_1)s(z_2))\Big]. \end{aligned}$$

Let $v_j \to 0$ first and then $\varepsilon_j \to 0$. It is easy to simplify the above to

$$\begin{aligned} c(f,g) &= -\frac{1}{4\pi^2} \oint_{\mathcal{C}_1'} \oint_{\mathcal{C}_2'} f(z_1)g(z_2)\Gamma(z_1, z_2)dz_1 dz_2 \\ &= -\frac{1}{4\pi^2} \oint_{|s_1|=\rho_1} \oint_{|s_2|=\rho_2} f(-s_1 - s_1^{-1})g(-s_2 - s_2^{-1}) \\ &\quad \times(\sigma^2 - \kappa + 2\beta s_1 s_2 + \frac{\kappa}{(1 - s_1 s_2)^2})ds_1 ds_2. \end{aligned}$$

By Cauchy's theorem, we may change $\rho_2 = 1$ without affecting the value of the integral. Rewriting $\rho_1 = \rho$, expanding the fraction as a Taylor series and then making variable changes $s_1 = -\rho e^{i\theta_1}$ and $s_2 = -e^{i\theta_2}$, we obtain

$$\begin{aligned} &c(f,g) \\ &= \frac{1}{4\pi^2} \int_{[-\pi,\pi]^2} f(\rho e^{i\theta_1} + \rho^{-1}e^{-i\theta_1})g(2\cos\theta_2)\Big[\sigma^2 \rho e^{i(\theta_1+\theta_2)} \\ &\quad +2(\beta+1)\rho^2 e^{i2(\theta_1+\theta_2)} + \kappa \sum_{k=3}^{\infty} k\rho^k e^{ik(\theta_1+\theta_2)}\Big] d\theta_1 d\theta_2 \\ &= \sigma^2 \rho \tau_1(f,\rho)\tau_1(g) + 2(\beta+1)\rho^2 \tau_2(f,\rho)\tau_2(g) \\ &\quad +\kappa \sum_{k=3}^{\infty} k\rho^k \tau_k(f,\rho)\tau_k(g), \end{aligned}$$

where $\tau_k(f,\rho) = \frac{1}{2\pi} \int_{-\pi}^{\pi} f(\rho e^{i\theta} + \rho^{-1}e^{-i\theta})e^{ik\theta}d\theta$. By integration by parts, for $k \geq 3$ we have

$$\tau_k(f,\rho) = \frac{\rho^{-1}}{k}\tau_{k-1}(f',\rho) - \frac{\rho}{k}\tau_{k+1}(f',\rho)$$
$$= \frac{\rho^2}{k(k+1)}\tau_{k+2}(f'',\rho) - \frac{2}{k^2-1}\tau_k(f'',\rho) + \frac{\rho^{-2}}{k(k-1)}\tau_{k-2}(f'',\rho).$$

Since f'' is uniformly bounded in $\mathcal{U}$, we have $|\tau_k(f,\rho)| \le K/k(k-1)$ uniformly for all ρ close to 1. Then (4.2.5) follows from the dominated convergence theorem and letting $\rho \to 1$ under the summation.

4.3 CLT of LSS for Sample Covariance Matrices

In this section, we shall consider the CLT for LSS associated with the general form of a sample covariance matrix considered in Chapter 3, that is

$$\mathbf{B}_n = \frac{1}{n}\mathbf{T}^{1/2}\mathbf{X}_n\mathbf{X}_n^*\mathbf{T}^{1/2}.$$

Some limiting theorems on the ESD and the spectrum separation of $\mathbf{B}_n$ have been discussed in Chapter 3. In this section, we shall consider more special properties of the LSS constructed by eigenvalues of $\mathbf{B}_n$.

It has been proven that, under certain conditions, with probability 1, the ESD of $\mathbf{B}_n$ tends to a limit $F^{y,H}$ whose Stieltjes transform is the unique solution to

$$s = \int \frac{1}{\lambda(1-y-yzs)-z}dH(\lambda)$$

in the set $\{s \in \mathbb{C}^+ : -\frac{1-y}{z} + ys \in \mathbb{C}^+\}$.

Define $\underline{\mathbf{B}}_n \equiv (1/n)\mathbf{X}_n^*\mathbf{T}_n\mathbf{X}_n$ and denote its LSD and limiting Stieltjes transform as $\underline{F}^{y,H}$ and $\underline{s} = \underline{s}(z)$. Then the equation takes on a simpler form when $F^{y,H}$ is replaced by

$$\underline{F}^{y,H} \equiv (1-y)I_{[0,\infty)} + yF^{y,H},$$

namely

$$\underline{s}(z) \equiv s_{\underline{F}^{y,H}}(z) = -\frac{1-y}{z} + ys(z)$$

which has an inverse formula

$$z = z(\underline{s}) = -\frac{1}{\underline{s}} + y\int \frac{t}{1+t\underline{s}}dH(t). \tag{4.3.1}$$

Now, let us consider the linear spectral statistics defined as

$$\mu_n(f) = \int f(x)dF^{\mathbf{B}_n}(x).$$

Because the convergence $y_n \to y$ and $H_n \to H$ may be very slow, the difference $p(\mu_n(f) - \int f(x)dF^{y,H}(x))$ may not have a limiting distribution. Hence, we have to consider the limiting distribution of the normalized difference $p(\mu_n(f) - \int f(x)dF^{y_n,H_n}(x))$. For notational purpose, write

$$X_n(f) = \int f(x)dG_n(x),$$

where $G_n(x) = p(F^{\mathbf{B}_n}(x) - F^{y_n,H_n}(x))$.

The main result is stated in the following theorem which was presented in Bai and Silverstein (2004).

Theorem 4.7. *Assume that the X-variables satisfy the condition*

$$\frac{1}{np}\sum_{ij} \mathrm{E}|x_{ij}^4|I(|x_{ij}| \geq \sqrt{n}\eta) \to 0, \tag{4.3.2}$$

for any fixed $\eta > 0$ and the following additional conditions hold.

(a) *For each n, $x_{ij} = x_{ij}^{(n)}$, $i \leq p$, $j \leq n$ are independent.* $\mathrm{E}x_{ij} = 0$, $\mathrm{E}|x_{ij}|^2 = 1$, $\max_{i,j,n} \mathrm{E}|x_{ij}|^4 < \infty$, $p/n \to y$;

(b) $\mathbf{T}_n$ *is $p \times p$ non-random Hermitian non-negative definite with spectral norm bounded in p, with $F^{\mathbf{T}_n} \xrightarrow{\mathcal{D}} H$, a proper c.d.f.*

Let $f_1, \cdots, f_k$ be functions analytic on an open region containing the interval

$$[\liminf_n \lambda_{\min}^{\mathbf{T}_n} I_{(0,1)}(y)(1-\sqrt{y})^2, \limsup_n \lambda_{\max}^{\mathbf{T}_n}(1+\sqrt{y})^2]. \tag{4.3.3}$$

Then

(1) *the random vector*

$$(X_n(f_1), \cdots, X_n(f_k)) \tag{4.3.4}$$

forms a tight sequence in n.

(2) *If x_{ij} and $\mathbf{T}_n$ are real and* $\mathrm{E}(x_{ij}^4) = 3$, *then* (4.3.4) *converges weakly to a Gaussian vector $(X_{f_1}, \cdots, X_{f_k})$, with means*

$$\mathrm{E}X_f = -\frac{1}{2\pi i}\int_{\mathcal{C}} f(z) \frac{y\int \frac{\underline{s}(z)^3 t^2 dH(t)}{(1+t\underline{s}(z))^3}}{\left(1 - y\int \frac{\underline{s}(z)^2 t^2 dH(t)}{(1+t\underline{s}(z))^2}\right)^2} dz \tag{4.3.5}$$

and covariance function

$$\begin{aligned}&\mathrm{Cov}(X_f, X_g)\\ &= -\frac{1}{\pi^2}\int_{\mathcal{C}_1}\int_{\mathcal{C}_2} \frac{f(z_1)g(z_2)}{(\underline{s}(z_1) - \underline{s}(z_2))^2}\underline{s}'(z_1)\underline{s}'(z_2)dz_1 dz_2\end{aligned} \tag{4.3.6}$$

$(f, g \in \{f_1, \cdots, f_k\})$. *The contours in* (4.3.5) *and* (4.3.6) (*two in* (4.3.6), *which may be assumed to be non-overlapping*) *are closed and are taken in the positive direction in the complex plane, each enclosing the support of* $F^{y,H}$.

(3) *If* x_{ij} *is complex with* $\mathrm{E}(x_{ij}^2) = 0$ *and* $\mathrm{E}(|x_{ij}|^4) = 2$, *then* (2) *also holds, except the means are zero and the covariance function is* $1/2$ *times the function given in* (4.3.6).

This theorem can be viewed as an extension of results obtained in Jonsson (1982) where the entries of $\mathbf{X}_n$ are Gaussian, $\mathbf{T}_n = \mathbf{I}$ and $f_k = x^k$. The proof of the theorem consists of the following steps:

Step 1. Truncation and Normalization

We begin the proof of Theorem 4.7 here with the replacement of the entries of $\mathbf{X}_n$ with truncated and centralized variables. By condition (4.3.2), we may select $\eta_n \downarrow 0$ and such that

$$\frac{1}{np\eta_n^4} \sum_{ij} \mathrm{E}|x_{ij}|^4 I(|x_{ij}| \geq \eta_n \sqrt{n}) \to 0. \tag{4.3.7}$$

The convergence rate of the constants η_n can be arbitrarily slow and hence we may assume that $\eta_n n^{1/5} \to \infty$. Let $\widehat{\mathbf{B}}_n = (1/n)\mathbf{T}^{1/2}\widehat{\mathbf{X}}_n\widehat{\mathbf{X}}_n^*\mathbf{T}^{1/2}$ with $\widehat{\mathbf{X}}_n$ $p \times n$ having (i,j)-th entry $\hat{x}_{ij} = x_{ij} I_{\{|x_{ij}| < \eta_n \sqrt{n}\}}$. It is routine to prove that the variable truncation doesn't change the limiting distribution.

Define $\widetilde{\mathbf{B}}_n = (1/n)\mathbf{T}^{1/2}\widetilde{\mathbf{X}}_n\widetilde{\mathbf{X}}_n^*\mathbf{T}^{1/2}$ with $\widetilde{\mathbf{X}}_n$ $p \times n$ having (i,j)-th entry $\tilde{x}_{ij} = (\widehat{x}_{ij} - \mathrm{E}\widehat{x}_{ij})/\sigma_{ij}$, where $\sigma_{ij}^2 = \mathrm{E}|\widehat{x}_{ij} - \mathrm{E}\widehat{x}_{ij}|^2$. Similar to the Wigner case, one can prove that the above defined renormalization of the entries of $\mathbf{B}_n$ does change the limiting distribution of the LSS.

Therefore, we only need to find the limiting distribution of $\{\int f_j(x) d\widetilde{G}_n(x), j = 1, \cdots, k\}$. Hence, in the sequel, we shall assume the underlying variables are truncated at $\eta_n \sqrt{n}$, centralized and renormalized. For simplicity, we shall suppress all sub- or superscripts on the variables and assume that $|x_{ij}| < \eta_n \sqrt{n}$, $\mathrm{E}x_{ij} = 0$, $\mathrm{E}|x_{ij}|^2 = 1$, $\mathrm{E}|x_{ij}|^4 < \infty$, and for assumption made in Part (2) of Theorem 4.7 $\mathrm{E}|x_{ij}|^4 = 3 + o(1)$, while for assumption in (3) $\mathrm{E}x_{ij}^2 = o(1/n)$ and $\mathrm{E}|x_{ij}|^4 = 2 + o(1)$.

After truncation, centralization and renormalization, with modifications in the proof of Theorem 3.7, for any $\mu_1 > \limsup \|\mathbf{T}_n\|(1 + \sqrt{y})^2$ and $0 < \mu_2 < \liminf_n \lambda_{\min}^{\mathbf{T}_n} I_{(0,1)}(y)(1 - \sqrt{y})^2$, we have

$$P(\|\mathbf{B}_n\| \geq \mu_1) = o(n^{-\ell}), \tag{4.3.8}$$

and

$$P(\lambda_{\min}^{\mathbf{B}_n} \leq \mu_2) = o(n^{-\ell}), \tag{4.3.9}$$

The main proof of Theorem 4.7 will be given in the next sections.

Remarks on Convergence of Stieltjes Transforms

After truncation and centralization, our proof of the main theorem relies on establishing limiting results on

$$M_n(z) = p[s_{F^{\mathbf{B}_n}}(z) - s_{F^{y_n,H_n}}(z)] = n[s_{F^{\underline{\mathbf{B}}_n}}(z) - s_{\underline{F}^{y_n,H_n}}(z)],$$

or more precisely, on $\widehat{M}_n(\cdot)$, a truncated version of $M_n(\cdot)$ when viewed as a random two-dimensional process defined on a contour $\mathcal{C}$ of the complex plane, described as follows. Let $v_0 > 0$ be arbitrary. Let x_r be any number greater than the right end-point of interval (4.3.3). Let x_l be any negative number if the left end-point of (4.3.3) is zero. Otherwise choose $x_l \in (0, \liminf_n \lambda_{\min}^{\mathbf{T}_n} I_{(0,1)}(y)(1-\sqrt{y})^2)$. Let

$$\mathcal{C}_u = \{x + iv_0 : x \in [x_l, x_r]\}.$$

Then define

$$\mathcal{C}^+ \equiv \{x_l + iv : v \in [0, v_0]\} \cup \mathcal{C}_u \cup \{x_r + iv : v \in [0, v_0]\},$$

and $\mathcal{C} = \mathcal{C}^+ \cup \overline{\mathcal{C}^+}$. Further, we define now the subsets $\mathcal{C}_n$ of $\mathcal{C}^+$ on which $M_n(\cdot)$ agrees with $\widehat{M}_n(\cdot)$. Choose sequence $\{\varepsilon_n\}$ decreasing to zero satisfying for some $\alpha \in (0,1)$,

$$\varepsilon_n \geq n^{-\alpha}. \tag{4.3.10}$$

Let

$$\mathcal{C}_l = \begin{cases} \{x_l + iv : v \in [n^{-1}\varepsilon_n, v_0]\}, & \text{if } x_l > 0, \\ \{x_l + iv : v \in [0, v_0]\}, & \text{if } x_l < 0, \end{cases}$$

and

$$C_r = \{x_r + iv : v \in [n^{-1}\varepsilon_n, v_0]\}.$$

Then $\mathcal{C}_n = \mathcal{C}_l \cup \mathcal{C}_u \cup \mathcal{C}_r$. The process $\widehat{M}_n(\cdot)$ can now be defined. For $z = x + iv$ we have:

$$\widehat{M}_n(z) = \begin{cases} M_n(z), & \text{for } z \in \mathcal{C}_n, \\ M_n(x_r + in^{-1}\varepsilon_n), & \text{for } x = x_r, v \in [0, n^{-1}\varepsilon_n], \\ M_n(x_l + in^{-1}\varepsilon_n), & \text{for } x = x_l, v \in [0, n^{-1}\varepsilon_n]. \end{cases} \tag{4.3.11}$$

$\widehat{M}_n(\cdot)$ is viewed as a random element in the metric space $C(\mathcal{C}^+, \mathbb{R}^2)$ of continuous functions from $\mathcal{C}^+$ to $\mathbb{R}^2$. All of Chapter 2 of Billingsley (1968) applies to continuous functions from a set such as $\mathcal{C}^+$ (homeomorphic to $[0,1]$) to finite dimensional Euclidean space, with $|\cdot|$ interpreted as Euclidean distance.

We first prove the following lemma.

Lemma 4.8. *Under conditions* (a) *and* (b) *of Theorem* 4.7 $\{\widehat{M}_n(\cdot)\}$ *forms a tight sequence on* $\mathcal{C}^+$. *Moreover, if assumptions in* (2) *or* (3) *of Theorem*

4.7 on x_{ij} hold, then $\widehat{M}_n(\cdot)$ converges weakly to a two-dimensional Gaussian process $M(\cdot)$ satisfying for $z \in \mathcal{C}^+$ under the assumptions in (2)

$$\mathrm{E}M(z) = \frac{y \int \frac{\underline{s}(z)^3 t^2 dH(t)}{(1+t\underline{s}(z))^3}}{\left(1 - y \int \frac{\underline{s}(z)^2 t^2 dH(t)}{(1+t\underline{s}(z))^2}\right)^2} \tag{4.3.12}$$

and for $z_1, z_2 \in \mathcal{C}$,

$$\begin{aligned}\mathrm{Cov}(M(z_1), M(z_2)) &\equiv \mathrm{E}[(M(z_1) - \mathrm{E}M(z_1))(M(z_2) - \mathrm{E}M(z_2)) \\ &= 2\left(\frac{\underline{s}'(z_1)\underline{s}'(z_2)}{(\underline{s}(z_1) - \underline{s}(z_2))^2} - \frac{1}{(z_1 - z_2)^2}\right),\end{aligned} \tag{4.3.13}$$

while under the assumptions in (3) $\mathrm{E}M(z) = 0$, *and the "covariance" function analogous to* (4.3.13) *is* 1/2 *the RHS of* (4.3.13).

We now show how Theorem 4.7 follows from the above lemma. We use the identity

$$\int f(x)dG(x) = -\frac{1}{2\pi i}\int_{\mathcal{C}} f(z)s_G(z)dz \tag{4.3.14}$$

valid for any c.d.f. G and f analytic on an open set containing the support of G. The complex integral on the right is over any positively oriented contour enclosing the support of G and on which f is analytic. Choose v_0, x_r, and x_l so that $f_1, \cdots, f_k$ are all analytic on and inside the resulting $\mathcal{C}$.

Due to the a.s. convergence of the extreme eigenvalues of $(1/n)\mathbf{X}_n\mathbf{X}_n^*$ and the bounds

$$\lambda_{\max}^{\mathbf{AB}} \le \lambda_{\max}^{\mathbf{A}}\lambda_{\max}^{\mathbf{B}}, \qquad \lambda_{\min}^{\mathbf{AB}} \ge \lambda_{\min}^{\mathbf{A}}\lambda_{\min}^{\mathbf{B}},$$

valid for $n \times n$ Hermitian non-negative definite $\mathbf{A}$ and $\mathbf{B}$, we have with probability one

$$\liminf_{n\to\infty} \min(x_r - \lambda_{\max}^{\mathbf{B}_n}, \lambda_{\min}^{\mathbf{B}_n} - x_l) > 0.$$

It also follows that the support of F^{y_n,H_n} is contained in

$$[\lambda_{\min}^{\mathbf{T}_n} I_{(0,1)}(y_n)(1-\sqrt{y}_n)^2, \lambda_{\max}^{\mathbf{T}_n}(1+\sqrt{y}_n)^2].$$

Therefore for any $f \in \{f_1, \cdots, f_k\}$, with probability one

$$\int f(x)dG_n(x) = -\frac{1}{2\pi i}\int_{\mathcal{C}} f(z)M_n(z)dz$$

for all n large, where the complex integral is over $\mathcal{C}$ and $M_n(z) = \overline{M_n(\bar{z})}$ for $z \in \overline{\mathcal{C}^+}$. Moreover, with probability one, for all n large,

$$\begin{aligned}&\left|\int_{\mathcal{C}} f(z)(M_n(z) - \widehat{M}_n(z))dz\right| \\ &\le 4K\varepsilon_n(|\max(\lambda_{\max}^{\mathbf{T}_n}(1+\sqrt{y}_n)^2, \lambda_{\max}^{\mathbf{B}_n}) - x_r|^{-1} \\ &\quad + |\min(\lambda_{\min}^{\mathbf{T}_n} I_{(0,1)}(y_n)(1-\sqrt{y}_n)^2, \lambda_{\min}^{\mathbf{B}_n}) - x_l|^{-1})\end{aligned}$$

which converges to zero as $n \to \infty$. Here K is a bound on f over $\mathcal{C}$.

Since

$$\widehat{M}_n(\cdot) \longrightarrow \left(-\frac{1}{2\pi i}\int f_1(z)\,\widehat{M}_n(z)dz, \cdots, -\frac{1}{2\pi i}\int f_k(z)\,\widehat{M}_n(z)dz\right)$$

is a continuous mapping of $C(\mathcal{C}, \mathbb{R}^2)$ into $\mathbb{R}^k$, it follows that the above vector and subsequently (4.3.4) form tight sequences. Letting $M(\cdot)$ denote the limit of any weakly converging subsequence of $\{\widehat{M}_n(\cdot)\}$ we have the weak limit of (4.3.4) equal in distribution to

$$\left(-\frac{1}{2\pi i}\int_{\mathcal{C}} f_1(z)\,M(z)dz, \cdots, -\frac{1}{2\pi i}\int_{\mathcal{C}} f_k(z)\,M(z)dz\right).$$

The fact that this vector, under the assumptions in (2) or (3), is multivariate Gaussian follows from the fact that Riemann sums corresponding to these integrals are multivariate Gaussian, and that weak limits of Gaussian vectors can only be Gaussian. The limiting expressions for the mean and covariance follow immediately.

Notice the assumptions in (2) and (3) require x_{ij} to have the same first, second and fourth moments of either a real or complex Gaussian variable, the latter having real and imaginary parts i.i.d. $N(0, 1/2)$. We will use the terms "RSE" and "CSE" to refer to the Real and Complex Sample covariance matrix with these moment conditions.

The reason why concrete results are at present only obtained for the assumptions in (2) and (3) is mainly due to the identity

$$\begin{aligned}
&\mathrm{E}(\mathbf{x}_t^*\mathbf{A}\mathbf{x}_t - \mathrm{tr}\mathbf{A})(\mathbf{x}_t^*\mathbf{B}\mathbf{x}_t - \mathrm{tr}\mathbf{B}) \\
&= \sum_{i=1}^{p}(\mathrm{E}|x_{it}|^4 - |\mathrm{E}x_{it}^2|^2 - 2)a_{ii}b_{ii} \\
&+\mathrm{tr}\mathbf{A}_x\mathbf{B}_x^T + \mathrm{tr}\mathbf{A}\mathbf{B}
\end{aligned} \tag{4.3.15}$$

valid for $p \times p$ $\mathbf{A} = (a_{ij})$ and $\mathbf{B} = (b_{ij})$, where $\mathbf{x}_t$ is the t-th column of $\mathbf{X}_n$, $\mathbf{A}_x = (\mathrm{E}x_{it}^2 a_{ij})$, and $\mathbf{B}_x = (\mathrm{E}x_{it}^2 b_{ij})$ (note t is fixed). This formula will be needed in several places in the proof of Lemma 4.8. The assumptions in (3) leave only the last term on the right, whereas those in (2) leave the last two, but in this case the matrix $\mathbf{B}$ will always be symmetric. This also accounts for the relation between the two covariance functions, and the difficulty in obtaining explicit results more generally. As will be seen in the proof, whenever (4.3.15) is used, little is known about the limiting behavior of $\sum a_{ii}b_{ii}$ even when we assume the underlying distributions are identical.

Simple substitution reveals

$$\text{RHS of (4.3.6)} = -\frac{1}{2\pi^2}\int_{\mathcal{C}_1}\int_{\mathcal{C}_2}\frac{f(z(s_1))g(z(s_2))}{(s_1 - s_2)^2}d(s_1)d(s_2). \tag{4.3.16}$$

However, the contours depend on the z_1, z_2 contours and cannot be arbitrarily chosen. It is also true that

$$\text{RHS of (4.3.6)} = \frac{1}{\pi^2}\iint f'(x)g'(y)\log\left|\frac{\underline{s}(x)-\overline{\underline{s}}(y)}{\underline{s}(x)-\underline{s}(y)}\right|dxdy$$
$$= \frac{1}{2\pi^2}\iint f'(x)g'(y)\log\left(1+4\frac{\underline{s}_i(x)\underline{s}_i(y)}{|\underline{s}(x)-\underline{s}(y)|^2}\right)dxdy \tag{4.3.17}$$

and

$$\mathrm{E}X_f = \frac{1}{2\pi}\int f'(x)\arg\left(1-y\int\frac{t^2\underline{s}^2(x)}{(1+t\underline{s}(x))^2}dH(t)\right)dx. \tag{4.3.18}$$

Here for $0 \neq x \in \mathbb{R}$,

$$\underline{s}(x) = \lim_{z\to x}\underline{s}(z), \qquad z \in \mathrm{C}^+, \tag{4.3.19}$$

known to exist and satisfying (4.3.1), and $\underline{s}_i(x) = \Im\,\underline{s}(x)$. The term

$$j(x) = \arg\left(1-y\int\frac{t^2\underline{s}^2(x)}{(1+t\underline{s}(x))^2}dH(t)\right)$$

in (4.3.18) is well defined for almost every x and takes values in $(-\pi/2,\pi/2)$.

The proof of (4.3.17) and (4.3.18) can be induced by showing

$$k(x,y) \equiv \log\left(1+4\frac{\underline{s}_i(x)\underline{s}_i(y)}{|\underline{s}(x)-\underline{s}(y)|^2}\right) \tag{4.3.20}$$

to be Lebesgue integrable on $\mathbb{R}^2$. It is interesting to note that the support of $k(x,y)$ matches the support of $f^{y,H}$ on $\mathbb{R}-\{0\}$:

$$k(x,y)=0 \Longleftrightarrow \min(f^{y,H}(x), f^{y,H}(y)) = 0.$$

We also have $f^{y,H}(x) = 0 \Longrightarrow j(x) = 0$.

It is noteworthy to mention here a consequence of (4.3.17), namely that if the assumptions in (2) or (3) of Theorem 4.7 were to hold, then G_n, considered as a random element in $D[0,\infty)$ (the space of functions on $[0,\infty)$ that are right-continuous with left-hand limits, together with the Skorohod metric) cannot form a tight sequence in $D[0,\infty)$. Indeed, under either assumptions, if $G(x)$ were a weak limit of a subsequence, then, because of Theorem 4.7, it is straightforward to conclude that for any x_0 in the interior of the support of F and positive ε,

$$\int_{x_0}^{x_0+\varepsilon} G(x)dx$$

would be Gaussian, and therefore so would

$$G(x_0) = \lim_{\varepsilon\to 0}\frac{1}{\varepsilon}\int_{x_0}^{x_0+\varepsilon} G(x)dx.$$

However, the variance would necessarily be

$$\lim_{\varepsilon\to 0}\frac{1}{2\pi^2}\frac{1}{\varepsilon^2}\int_{x_0}^{x_0+\varepsilon}\int_{x_0}^{x_0+\varepsilon}k(x,y)dxdy=\infty.$$

Still, under the assumptions in (2) or (3), a limit may exist for $\{G_n\}$ when G_n is viewed as a linear functional

$$f\longrightarrow\int f(x)dG_n(x),$$

that is, a limit expressed in terms of a measure in a space of generalized functions. The characterization of the limiting measure of course depends on the space, which in turn relies on the set of test functions, which for now is restricted to functions analytic on the support of F. Work in this area is currently being pursued.

We emphasize here the importance of studying $G_n(x)$ which essentially balances $F^{\mathbf{B}_n}(x)$ with F^{y_n,H_n}, but not $F^{y,H}$ nor $\mathrm{E}F^{\mathbf{B}_n}(x)$. The LSD $F^{y,H}$ cannot be used simply because the convergence of $y_n\to y$ and that of $H_n\to H$ can be arbitrarily slow. It should be viewed as a mathematical convenience because the result is expressed as a limit theorem. More importantly, from the point of view of statistical inference, H_n can be viewed as a description of the current population and y_n is the ratio of dimension to sample size for the current sample. Thus, the statistical problems should be characterized by F^{y_n,H_n}, instead of $F^{y,H}$ or $\mathrm{E}F^{\mathbf{B}_n}(x)$.

Step 2. Convergence of Finite Dimensional Distributions

Write for $z\in\mathcal{C}_n$, $M_n(z)=M_n^1(z)+M_n^2(z)$ where

$$M_n^1(z)=p[s_{F^{\mathbf{B}_n}}(z)-\mathrm{E}s_{F^{\mathbf{B}_n}}(z)]$$

and

$$M_n^2(z)=p[\mathrm{E}s_{F^{\mathbf{B}_n}}(z)-s_{F^{y_n,H_n}}(z)].$$

In this section we will show for any positive integer m, the sum

$$\sum_{i=1}^m\alpha_iM_n^1(z_i)\qquad(\Im z_i\neq 0)$$

whenever it is real, is tight, and, under the assumptions in (2) or (3) of Theorem 4.7, will converge in distribution to a Gaussian random variable. Formula (4.3.13) will also be derived.

Let $\mathrm{E}_0(\cdot)$ denote expectation and $\mathrm{E}_j(\cdot)$ denote conditional expectation with respect to the σ-field generated by $\mathbf{x}_1,\cdots,\mathbf{x}_j$.

Using the martingale decomposition, we have

$$p[s_{F^{\mathbf{B}_n}}(z)-\mathrm{E}s_{F^{\mathbf{B}_n}}(z)]=\mathrm{tr}[\mathbf{D}^{-1}(z)-\mathrm{E}\mathbf{D}^{-1}(z)]$$
$$=-\sum_{j=1}^n(\mathrm{E}_j-\mathrm{E}_{j-1})\beta_j(z)\mathbf{r}_j^*\mathbf{D}_j^{-2}(z)\mathbf{r}_j,$$

where $\mathbf{r}_j = n^{-1/2}\mathbf{T}^{1/2}\mathbf{x}_j$, $\mathbf{D} = \mathbf{B}_n - z\mathbf{I}$, $\mathbf{D}_j = \mathbf{D} - \mathbf{r}_j\mathbf{r}_j^*$, and $\beta_j = (1 + \mathbf{r}_j^*\mathbf{D}_j^{-1}\mathbf{r}_j)^{-1}$.

By the same argument as Wigner case, one may prove the finite dimensional convergence while the asymptotic mean and covariances are as given in Lemma 4.8 and the details are omitted.

Step 3. Tightness of $M_n^1(z)$

The proof of the tightness of the sequence of random functions $\widehat{M}_n^1(z)$ for $z \in \mathcal{C}^+$ defined by (4.3.11) can be induced by verifying

(i) Tightness at any point in $[0, 1]$ ($\mathcal{C}^+$ for our case).

(ii)

$$\sup_{n; z_1, z_2 \in \mathcal{C}_n} \frac{\mathrm{E}|M_n^1(z_1) - M_n^1(z_2)|^2}{|z_1 - z_2|^2} \le K.$$

The details are omitted.

4.4 *F* Matrix

The CLT for LSS of large-dimensional sample covariance matrices opens a way for RMT to be applied to multivariate statistical analysis and many other areas, including large-dimensional data analysis, information theory, economics and finance, signal processing and wireless communications, *etc.* In multivariate statistical analysis, more statistics are defined by multivariate F-matrices than sample covariance matrices because the former is invariant under affine transformations. A multivariate F-matrix is defined as $\mathbf{F} = \mathbf{S}_1\mathbf{S}_2^{-1}$, where $\mathbf{S}_1$ and $\mathbf{S}_2$ are two sample covariance matrices constructed by two independent p-dimensional samples with sample sizes n_1 and n_2 respectively. Consider $\mathbf{S}_2^{-1}$ as $\mathbf{T}$ is the model $\mathbf{B}_n$ considered in last section, by considering the conditional distribution of the LSS defined by $\mathbf{F}$, the CLT of LSS for $\mathbf{F}$ can be obtained. However, when considering the conditional distribution, we have to choose the center in the normalization of the LSS for $\mathbf{F}$ as a random one which is not good for hypothesis tests in multivariate analysis. Extension of the results given in last section was obtained by Zheng (2007). We will gave a brief introduction of this result in this section.

A multivariate F-matrix is defined as a product of a sample covariance matrix and the inverse of another sample covariance matrix, independent of each other. More specifically, let $\{X_{ki} \in \mathbb{C}, i, k = 1, 2, \cdots\}$ and $\{Y_{ki} \in \mathbb{C}, i, k = 1, 2, \cdots\}$ are respectively double arrays of i.i.d. real or complex variables with mean zero and variance one. Write $\mathbf{X}_i = (X_{1i}, X_{2i}, \cdots, X_{pi})'$ and $\mathbf{Y}_i = (Y_{1i}, Y_{2i}, \cdots, Y_{pi})'$. The sample covariance matrices $S_1^{n_1}$ and $S_2^{n_2}$ from two populations are defined, respectively by

$$S_1^{n_1} = \left(\frac{1}{n_1}\sum_{i=1}^{n_1}\mathbf{X}_i\mathbf{X}_i^*\right)_{p\times p} \quad \text{and} \quad S_2^{n_2} = \left(\frac{1}{n_2}\sum_{i=1}^{n_2}\mathbf{Y}_i\mathbf{Y}_i^*\right)_{p\times p}. \tag{4.4.1}$$

Then, F-matrix is defined as

$$F = S_1^{n_1}(S_2^{n_2})^{-1} = \left(\frac{1}{n_1}\sum_{i=1}^{n_1}\mathbf{X}_i\mathbf{X}_i^*\right)\left(\frac{1}{n_2}\sum_{i=1}^{n_2}\mathbf{Y}_i\mathbf{Y}_i^*\right)^{-1} \tag{4.4.2}$$

where $n_2 > p$. A multivariate F-matrix is more fundamental and important in multivariate statistical analysis than sample covariance matrices since more testing statistics in multivariate statistical analysis are defined through F-matrix.

Suppose that $y_{n_1} = p/n_1 \to y_1 \in (0, +\infty)$ and $y_{n_2} = p/n_2 \to y_2 \in (0, 1)$. The LSD $F^{\{y_1,y_2\}}$ of multivariate F matrices in (4.4.2) has a density function

$$f^{\{y_1,y_2\}}(x) = \begin{cases} \dfrac{(1-y_2)\sqrt{(b-x)(x-a)}}{2\pi x(y_1+y_2x)}, & \text{when } a < x < b, \\ 0, & \text{otherwise} \end{cases} \tag{4.4.3}$$

where $a = \left(\dfrac{1-\sqrt{y_1+y_2-y_1y_2}}{1-y_2}\right)^2$, $b = \left(\dfrac{1+\sqrt{y_1+y_2-y_1y_2}}{1-y_2}\right)^2$ and a point mass $1 - 1/y_1$ at the origin if $y_1 > 1$.

The Stieltjes transform of the LSD $F^{\{y_1,y_2\}}$ is as follows

$$s_{\{y_1,y_2\}}(z) = \frac{1-y_1-z-zy_2-\sqrt{[(1-y_1)+z(1-y_2)]^2-4z}}{2z(y_1+zy_2)}$$

and

$$\underline{s}_{\{y_1,y_2\}}(z) = -\frac{y_1[z(1-y_2)+1-y_1]+2zy_2+y_1\sqrt{[(1-y_1)+z(1-y_2)]^2-4z}}{2z(y_1+zy_2)}$$

for any $z \in \mathbb{C}^+$ where $\underline{s}_{\{y_1,y_2\}}(z) = -\frac{1-y_1}{z} + y_1 \cdot s_{\{y_1,y_2\}}(z)$ (see P72 Bai and Silverstein (2006)).

In this section, we shall consider the limiting distribution of the normalized LSS of $\mathbf{F}$

$$\left(\int f_1(x)dG_n(x)dx, \cdots, \int f_m(x)dG_n(x)\right)$$

where $f_1, \cdots, f_m$ are functions analytic on an open region of the complex plane which covers the interval $[a, b]$ and

$$G_n(x) = p\left[F^{\mathbf{F}}(x) - F^{\{y_{n_1},y_{n_2}\}}(x)\right]$$

as $y_{n_1} = p/n_1 \to y_1 \in (0, +\infty)$ and $y_{n_2} = p/n_2 \to y_2 \in (0, 1)$.

4.4.1 Decomposition of $\mathbf{X}_{nf}$

We split $\mathbf{X}_{nf}$ as $\tilde{\mathbf{X}}_{nf} + \mathbf{X}_{nf}^{\dagger}$, where

$$\begin{aligned}
\tilde{X}_{nf} &= \left(\int f_1(x)d\tilde{G}_n(x)dx, \cdots, \int f_m(x)d\tilde{G}_n(x)\right)\\
\tilde{G}_n(x) &= p\left[F^{\mathbf{F}}(x) - F^{\{y_{n_1}, H_n\}}(x)\right]\\
X_{nf}^{\dagger} &= \left(\int f_1(x)dG_n^{\dagger}(x)dx, \cdots, \int f_m(x)dG_n^{\dagger}(x)\right)\\
G_n^{\dagger}(x) &= p\left[F^{\{y_{n_1}, H_n\}}(x) - F^{\{y_{n_1}, y_{n_2}\}}(x)\right].
\end{aligned}$$

We have the following theorem.

Theorem 4.9. *As $p/n_1 \to y_1 > 0$, $p/n_2 \to y_2 \in (0,1)$, given $\{\text{all } \mathbf{S}_2\}$, the conditional distribution of*

$$\left(\int f_1(x)d\tilde{G}_n(x), \cdots, \int f_m(x)d\tilde{G}_n(x)\right).$$

(i) is tight if $E|X_{11}^4| < \infty$;
(ii) converges weakly to a Gaussian vector $(\tilde{X}_{f_1}, \cdots, \tilde{X}_{f_m})$ with means

$$\mathrm{E}\tilde{X}_f = -\frac{1}{2\pi i}\oint_{\mathcal{C}_1} f(z)\frac{y_1\int \underline{s}(z)^3 t[t+\underline{s}(z)]^{-3}dF_2(t)}{\left\{1 - y_1\int \underline{s}(z)^2[t+\underline{s}(z)]^{-2}dF_2(t)\right\}^2}dz \tag{4.4.4}$$

and covariance function

$$\mathrm{Cov}(\tilde{X}_f, \tilde{X}_g) = -\frac{1}{2\pi^2}\oint_{\mathcal{C}_1}\oint_{\mathcal{C}_2}\frac{f(z_1)g(z_2)}{[\underline{s}(z_1) - \underline{s}(z_2)]^2}d\underline{s}(z_1)d\underline{s}(z_2), \tag{4.4.5}$$

if the variables $\mathbf{X}$'s are real and $\mathrm{E}(X_{11}^4) = 3$, $\mathrm{E}|Y_{11}|^4 < \infty$. The contours $\mathcal{C}_1$ and $\mathcal{C}_2$ in (4.4.4) and (4.4.5) are analogues of $\mathcal{C}$ subject to the restriction that the two in (4.4.5) do not intersect each other.

(iii) converges weakly to a normal vector with the means 0 and the covariance function is 1/2 the function given in (4.4.5), if $\mathbf{X}$'s are complex with $E\left(X_{11}^2\right) = 0$ and $E(|X_{11}^4|) = 2$ and $\mathrm{E}|Y_{11}|^4 < \infty$.

Here $H_n = F^{\mathbf{S}_2^{-1}}$ and $F_2 = F_{y_2}$ and $\underline{s}(z) = \underline{s}_{\{y_1, y_2\}}(z)$.

Theorem 4.9 is a consequence of Theorem 4.7.

Thus, to extend the CLT of LSS of sample covariance matrix to that of F-matrix, we need only consider the limit distribution of

$$\mathbf{X}_{nf}^{\dagger} = \left(\int f_1(x)dG_n^{\dagger}(x)dx, \cdots, \int f_m(x)dG_n^{\dagger}(x)\right).$$

4.4.2 Limiting Distribution of $\mathbf{X}_{nf}^{\dagger}$

Again, using Cauchy formula to the analytic functions

$$f(x) = \frac{1}{2\pi i}\oint_{\mathcal{C}} \frac{f(z)dz}{z-x},$$

we have

$$X_{nf}^{\dagger} = -\frac{p}{2\pi i}\oint_{\mathcal{C}} f(z)\Big(s_{\{y_{n_1},H_n\}}(z) - s_{n0}(z)\Big)dz$$

where $s_{\{y_{n_1},H_n\}}(z)$ and $s_{n0}(z)$ are the Stieltjes transforms of conditional LSD $F^{y_{n_1},H_n}$ and the LSD $F^{\{y_{n_1},y_{n_2}\}}$, respectively.

Let $\underline{s}_{\{y_{n_1},H_n\}}(z) = -\frac{1-y_{n_1}}{z} + y_{n_1}s_{\{y_{n_1},H_n\}}(z)$ and $\underline{s}_{\{y_{n_1},y_{n_2}\}}(z) = -\frac{1-y_{n_1}}{z} + y_{n_1}s_{\{y_{n_1},y_{n_2}\}}(z)$. In remainder of the section, for brevity, $s_{\{y_{n_1},y_{n_2}\}}(z)$ and $\underline{s}_{\{y_{n_1},y_{n_2}\}}(z)$ will be simply written as $s_{n0}(z)$ and $\underline{s}_{n0}(z)$, if no confusion.

Then, we have

$$X_{nf}^{\dagger} = -\frac{n_1}{2\pi i}\oint_{\mathcal{C}} f(z)\Big(\underline{s}_{\{y_{n_1},H_n\}}(z) - \underline{s}_{n0}(z)\Big)dz.$$

By (3.3.5), we have

$$\begin{aligned}
z &= -\frac{1}{\underline{s}_{\{y_{n_1},H_n\}}(z)} + y_{n_1}\int \frac{t}{1+t\underline{s}_{\{y_{n_1},H_n\}}(z)}dH_n(t) \\
&= -\frac{1}{\underline{s}_{\{y_{n_1},H_n\}}(z)} + y_{n_1}\int \frac{1}{t+\underline{s}_{\{y_{n_1},H_n\}}(z)}dF_{n_2}(t) \\
z &= -\frac{1}{\underline{s}_{n0}(z)} + y_{n_1}\int \frac{t}{1+t\underline{s}_{n0}(z)}dH_0(t) \\
&= -\frac{1}{\underline{s}_{n0}(z)} + y_{n_1}\int \frac{1}{t+\underline{s}_{n0}(z)}dF_{y_{n_2}}(t),
\end{aligned}$$

where F_{n_2} and $F_{y_{n_2}}$ are the ESD and LSD of the sample covariance matrix $\mathbf{S}_2$.

Making difference of the above equations, one gets

$$\begin{aligned}
n_1[\underline{s}_{\{y_{n_1},H_n\}}(z) - \underline{s}_{n0}(z)] &= \frac{p\underline{s}_{\{y_{n_1},H_n\}}(z)\underline{s}_{n0}(z)[s_{n_2}(-\underline{s}_{n0}(z)) - s_{y_{n_2}}(-\underline{s}_{n0}(z))]}{1-y_{n_1}\int \frac{\underline{s}_{\{y_{n_1},H_n\}}(z)\underline{s}_{n0}(z)}{(t+\underline{s}_{\{y_{n_1},H_n\}}(z))(t+\underline{s}_{n0}(z))}dF_{n_2}(t)} \\
&= \frac{n_2\underline{s}_{\{y_{n_1},H_n\}}(z)\underline{s}_{n0}(z)[\underline{s}_{n_2}(-\underline{s}_{n0}(z)) - \underline{s}_{y_{n_2}}(-\underline{s}_{n0}(z))]}{1-y_{n_1}\int \frac{\underline{s}_{\{y_{n_1},H_n\}}(z)\underline{s}_{n0}(z)}{(t+\underline{s}_{\{y_{n_1},H_n\}}(z))(t+\underline{s}_{n0}(z))}dF_{n_2}(t)} \\
&\sim \frac{n_2\underline{s}^2(z)[\underline{s}_{n_2}(-\underline{s}_{n0}(z)) - \underline{s}_{y_{n_2}}(-\underline{s}_{n0}(z))]}{1-y_1\int \frac{\underline{s}^2(z)}{(t+\underline{s}(z))^2}dF_{y_2}(t)}.
\end{aligned}$$

Then, we can prove the following theorem.

Theorem 4.10. *(i) if* $E|Y_{11}^4| < \infty$*, the random vector*

$$\mathbf{X}_{nf}^{\dagger} = \left(\int f_1(x)dG_n^{\dagger}(x), \cdots, \int f_m(x)dG_n^{\dagger}(x)\right) \tag{4.4.6}$$

forms a tight sequence;

(ii) if the variables $\mathbf{Y}$*'s are real and* $E(Y_{11}^4) = 3$*, then* $\mathbf{X}_{nf}^{\dagger}$ *converges weakly to a Gaussian vector* $\mathbf{X}_f^{\dagger} = (X_{f_1}^{\dagger}, \cdots, \tilde{X}_{f_m}^{\dagger})$ *with means*

$$\mathrm{E}X_f^{\dagger} = \frac{1}{2\pi i} \oint_{\mathcal{C}} f(z) \frac{y_2 \underline{s}_2(-\underline{s}(z))}{(1+\underline{s}_2(-\underline{s}(z)))[(1+s_2(-\underline{s}(z)))^2 - y_2 s_2^2(-\underline{s}(z))]} d\underline{s}_2(-\underline{s}(z)) \tag{4.4.7}$$

and covariance function

$$\mathrm{Cov}(X_f^{\dagger}, X_g^{\dagger}) = -\frac{1}{2\pi^2} \oint_{\mathcal{C}_1} \oint_{\mathcal{C}_2} f(z_1) g(z_2)$$
$$\left(\frac{\underline{s}_2'(-\underline{s}(z_1))\underline{s}_2'(-\underline{s}(z_2))}{(s_2(-\underline{s}(z_1)) - s_2(-\underline{s}(z_2)))^2} - \frac{1}{(\underline{s}(z_1) - \underline{s}(z_2))^2}\right) d\underline{s}(z_1) d\underline{s}(z_2). \tag{4.4.8}$$

The contours $\mathcal{C}$*,* $\mathcal{C}_1$ *and* $\mathcal{C}_2$ *in (4.4.7) and (4.4.8) are analogues of* $\mathcal{C}$ *subject to the restriction that the two in (4.4.8) do not intersect each other.*

(iii) if $\mathbf{Y}$*'s are complex with* $E\left(Y_{11}^2\right) = 0$ *and* $E(|Y_{11}^4|) = 2$*, then (ii) also holds, except the means are zero and the covariance function is 1/2 the function given in (4.4.8), where* H_n *and* H_0 *are the ESD and LSD of* $\mathbf{S}_2^{-1}$*, and* $\underline{s} = \underline{s}_{\{y_1, y_2\}}$*.*

The conclusion (i) can be proved by the same approach as Bai and Silverstein (2004). As for conclusions (ii) and (iii), making use of Lemma 4.8, for the real case,

$$n_2[\underline{s}_{n_2}(z) - \underline{s}_{y_{n_2}}(z)]$$

tends to a Gaussian stochastic process on any given contour $\mathcal{C}_2^{\dagger}$ which encloses the interval $[(1 - \sqrt{y_2})^2, (1 + \sqrt{y_2})^2]$ with means

$$\frac{y_2 \underline{s}_2^3(z)(1 + \underline{s}_2(z))}{\left((1 + \underline{s}_2(z))^2 - y_2 \underline{s}_2^2(z)\right)^2}$$

and covariances

$$2\left(\frac{\underline{s}_2'(z_1)\underline{s}_2'(z_2)}{(\underline{s}_2(z_1) - \underline{s}_2(z_2))^2} - \frac{1}{(z_1 - z_2)^2}\right), \tag{4.4.9}$$

where $\underline{s}_2(z)$ is the corresponding Stieltjes transform of the MP-law with the ratio parameter $y = y_2 = \lim p/n_2$. For the complex case, the above conclusion remain true except that the asymptotic mean is 0 and covariance function is half of the amount given in (4.4.9).

Therefore the real case,

$$n_2[\underline{s}_{n_2}(-\underline{s}_{n0}(z)) - \underline{s}_{y_{n_2}}(-\underline{s}_{n0}(z))]$$

tends to a Gaussian stochastic process $M^{\dagger}(z)$ on the contour $\mathcal{C}$ and covariances

$$\mathrm{E}M^{\dagger}(z) = \frac{y_2\underline{s}^2(z)\underline{s}_2^3(-\underline{s}(z))(1+\underline{s}_2(-\underline{s}(z)))}{(1-y_1\int\frac{\underline{s}^2(z)}{(t+\underline{s}(z))^2}dF_{y_2}(t))[(1+\underline{s}_2(-\underline{s}(z)))^2-y_2\underline{s}_2^2(-\underline{s}(z))]^2}$$

$$\mathrm{E}M^{\dagger}(z_1)M^{\dagger}(z_2) = 2\Big(\frac{\underline{s}_2'(-\underline{s}(z_1))\underline{s}_2'(-\underline{s}(z_2))}{(s_2(-\underline{s}(z_1))-s_2(-\underline{s}(z_2)))^2}-\frac{1}{(\underline{s}(z_1)-\underline{s}(z_2))^2}\Big)$$
$$\times\Big(\frac{\underline{s}^2(z_1)\underline{s}^2(z_2)}{(1-y_1\int\frac{\underline{s}^2(z_1)}{(t+\underline{s}(z_1))^2}dF_{y_2}(t))(1-y_1\int\frac{\underline{s}^2(z_2)}{(t+\underline{s}(z_2))^2}dF_{y_2}(t))}\Big). \quad (4.4.10)$$

For the complex case, the above conclusion remain true except that the asymptotic mean is 0 and covariance function is half of the amount given in (4.4.10).

Now, we simplify the expressions. By (4.3.1), the Stieltjes transform $\underline{s}$ satisfies

$$z = -\frac{1}{\underline{s}(z)}+\frac{y_1}{t+\underline{s}(z)}dF_2(t), \quad (4.4.11)$$

where F_2 is the LSD of $\mathbf{S}_2$ as $p/n_2\to y_2\in(0,1)$. Therefore, by differentiating both sides of (4.4.11), we obtain

$$\underline{s}'(z) := \frac{d\underline{s}(z)}{dz} = \frac{\underline{s}^2(z)}{1-\int\frac{y_1\underline{s}^2(z)}{(t+\underline{s}(z))^2}dF_2(t)}. \quad (4.4.12)$$

Similarly,

$$\underline{s}_2'(z) := \frac{d\underline{s}_2(z)}{dz} = \frac{\underline{s}_2^2(z)}{1-\frac{y_2\underline{s}_2^2(z)}{(1+\underline{s}_2(z))^2}}. \quad (4.4.13)$$

Hence,

$$\mathrm{E}M^{\dagger}(z) = -\frac{y_2\underline{s}_2(-\underline{s}(z))}{(1+\underline{s}_2(-\underline{s}(z)))[(1+s_2(-\underline{s}(z)))^2-y_2s_2^2(-\underline{s}(z))]}\frac{d}{dz}\underline{s}_2(-\underline{s}(z))$$
$$\mathrm{E}M^{\dagger}(z_1)M^{\dagger}(z_2) =$$
$$2\Big(\frac{\underline{s}_2'(-\underline{s}(z_1))\underline{s}_2'(-\underline{s}(z_2))}{(s_2(-\underline{s}(z_1))-s_2(-\underline{s}(z_2)))^2}-\frac{1}{(\underline{s}(z_1)-\underline{s}(z_2))^2}\Big)\underline{s}'(z_1)\underline{s}'(z_2). \quad (4.4.14)$$

4.4.3 The Limiting Distribution of $\mathbf{X}_{nf}$

It is easy to see that the limiting vectors $\tilde{\mathbf{X}}_f$ and $\mathbf{X}_f^{\dagger}$ are independent of each other. Therefore, the limiting distribution of $\mathbf{X}_{nf}$ is the same as $\tilde{\mathbf{X}}_f+\mathbf{X}_f^{\dagger}$. Before write out our conclusion, let's simplify the expressions of asymptotic means and covariances.

By (4.4.4) and (4.4.14), we have

$$\mathrm{E}(X_f)=\frac{1}{4\pi i}\oint_{\mathcal{C}}f(z)d\log\Big(1-\int\frac{y_1\underline{s}^2(z)}{[t+\underline{s}(z)]^2}dF_{y_2}(t)\Big)\Big(1-\frac{y_2\cdot[\underline{s}_{y_2}(-\underline{s}(z))]^2}{[1+\underline{s}_{y_2}(-\underline{s}(z))]^2}\Big).$$

After certain simplification, we have

$$\left(1-\int \frac{y_1\underline{s}^2(z)}{[t+\underline{s}(z)]^2}dF_{y_2}(t)\right)\left(1-\frac{y_2\cdot[\underline{s}_{y_2}(-\underline{s}(z))]^2}{[1+\underline{s}_{y_2}(-\underline{s}(z))]^2}\right)$$
$$=\frac{(1-y_2)s_0^2(z)+2s_0(z)+1-y_1}{(1+s_0(z))^2},$$

where $s_0(z)=\underline{s}_{y_2}(-\underline{s}(z))$ which satisfies the equation

$$z=-\frac{s_0(s_0+1-y_1)}{(1-y_2)s_0+1}. \tag{4.4.15}$$

Set $s_0=-\frac{1+hr\xi}{1-y_2}$, where $h^2=y_1+y_2-y_1y_2$ and $r>1$ but very close to 1, $|\xi|=1$. Then, by (4.4.15),

$$z=-\frac{(1+hr\xi)(1+hr^{-1}\bar{\xi})}{(1-y_2)^2}=\frac{1+h^2+h(r+r^{-1})\cos(\theta)+ih(r-r^{-1})\sin(\theta)}{(1-y_2)^2}.$$

This shows that when ξ runs a cycle along the unit circle anticlockwise, z runs a cycle anticlockwise and the cycle incloses the interval $[a,b]$, where $a,b=\frac{(1\mp h)^2}{(1-y_2)^2}$.

Also, we have

$$\frac{(1-y_2)s_0^2(z)+2s_0(z)+1-y_1}{(1+s_0(z))^2}=(1-y_2)\frac{r^2\xi^2-1}{(r\xi+h^{-1}y_2)^2}$$

Therefore, as $r\downarrow 1$,

$$\mathrm{E}X_f=\frac{1}{4\pi i}\oint_{|\xi|=1} f\left(\frac{1+h^2+2h\Re(\xi)}{(1-y_2)^2}\right)\left(\frac{1}{r\xi+1}+\frac{1}{r\xi-1}-\frac{2}{r\xi+h^{-1}y_2}\right)d\xi+o(1). \tag{4.4.16}$$

By (4.4.5) and (4.4.8), we obtain

$$\mathrm{cov}(X_f,X_g)=-\frac{1}{2\pi^2}\oint_{C_1}\oint_{C_2} f(z_1)g(z_2)\frac{1}{(s_0(z_1)-s_0(z_2))^2}ds_0(z_1)ds_0(z_2).$$

Making variable change $s_0(z_j)=-\frac{1+hr_j\xi_j}{1-y_2}$, where $r_2>r_1>1$. Similarly, one can prove that as $r_2\downarrow 1$,

$$\mathrm{Cov}(X_f,X_g)=-\frac{1}{2\pi^2}\oint\oint_{|\xi_1|=|\xi_2|=1} f\left(\frac{1+h^2+2h\Re(\xi_1)}{(1-y_2)^2}\right)$$
$$g\left(\frac{1+h^2+2h\Re(\xi_2)}{(1-y_2)^2}\right)\frac{d\xi_1 d\xi_2}{(r_1\xi_1-r_2\xi_2)^2}+o(1). \tag{4.4.17}$$

Hence, by (4.4.16) and (4.4.17), we have

Theorem 4.11. *Assume that the* $\mathbf{X}$*-variables satisfy the condition*

$$\frac{1}{n_1 p}\sum_{ij} E|X_{ij}^4| \cdot I\left(|X_{ij}| \geq \sqrt{n}\eta\right) \to 0,$$

for any fixed $\eta > 0$ *and the* $\mathbf{Y}$*-variables satisfy a similar condition. Also, we assume the following additional conditions hold.*

$(a)'$ *For each* n, X_{ij}, Y_{ij}, $i \leq p, j \leq n$ *are independent and* $EX_{ij} = EY_{ij} = 0$, $E|X_{ij}|^2 = E|Y_{ij}|^2 = 1$, $\sup E|X_{ij}|^4 < \infty$, $\sup E|Y_{ij}|^4 < \infty$, $y_{n_1} = p/n_1 \to y_1 \in (0, +\infty)$, $y_{n_2} = p/n_2 \to y_2 \in (0,1)$ *and*

$(b)'$ *Let* $f_1, \cdots, f_m$ *be functions analytic in an open region containing the interval*

$$\left[I_{(0,1)}(y_1) \cdot \frac{(1-\sqrt{y_1})^2}{(1+\sqrt{y_2})^2}, \quad \frac{(1+\sqrt{y_1})^2}{(1-\sqrt{y_2})^2}\right]$$

where $I_{(0,1)}(y_1)(1-\sqrt{y_1})^2/(1+\sqrt{y_2})^2$ *is the lower bound of the smallest eigenvalues and* $(1+\sqrt{y_1})^2/(1-\sqrt{y_2})^2$ *is an upper bound of the largest eigenvalues of* $S_1^{n_1}(S_2^{n_2})^{-1}$, *respectively. Then*

$(i)'$ *the random vector*

$$\left(\int f_1(x)dG_{n_1,n_2}(x), \cdots, \int f_m(x)dG_{n_1,n_2}(x)\right) \tag{4.4.18}$$

forms a tight sequence.

$(ii)'$ *If* X_{ij}, Y_{ij} *are real and* $\mathrm{E}(X_{ij}^4) = E(Y_{ij}^4) = 3$, *then* (4.4.18) *converges weakly to a Gaussian vector* $(X_{f_1}, \cdots, X_{f_m})$ *with means*

$$\mathrm{E}X_f = \frac{1}{4\pi i}\oint_{|\xi|=1} f\left(\frac{1+h^2+2h\Re(\xi)}{(1-y_2)^2}\right)\left(\frac{1}{r\xi+1}+\frac{1}{r\xi-1}-\frac{2}{r\xi+h^{-1}y_2}\right)d\xi+o(1) \tag{4.4.19}$$

and covariance function

$$\mathrm{Cov}(X_f, X_g) = -\frac{1}{2\pi^2}\oint\oint_{|\xi_1|=|\xi_2|=1} f\left(\frac{1+h^2+2h\Re(\xi_1)}{(1-y_2)^2}\right) g\left(\frac{1+h^2+2h\Re(\xi_2)}{(1-y_2)^2}\right)\frac{d\xi_1 d\xi_2}{(r_1\xi_1 - r_2\xi_2)^2} + o(1) \tag{4.4.20}$$

where $f, g \in \{f_1, \cdots, f_m\}$ *and* $r, r_2 \downarrow 1$.

$(iii)'$ *If* X_{ij}, Y_{ij} *are complex with* $E(X_{ij}^2) = E(Y_{ij}^2) = 0$ *and* $E|X_{ij}|^4 = E|Y_{ij}|^4 = 2$, *then* (ii) *also holds, except the means are zero and the covariance function is* $1/2$ *the function given in* (4.4.20).

Now, we consider three examples.

Example 1: If $f = \log(a + bx)$, $f' = \log(a' + b'x)$, $a, a', b, b' > 0$ and the F matrix is real, then

$$\mathrm{E}X_f = \frac{1}{2}\log\left(\frac{(c^2 - d^2)h^2}{(ch - y_2 d)^2}\right)$$
$$cov(X_f, X_{f'}) = 2\log(cc'/(cc' - dd')),$$

where $c > d > 0$, $c' > d' > 0$ satisfying $c^2 + d^2 = a(1 - y_2)^2 + b(1 + h^2)$, $c'^2 + d'^2 = a'(1 - y_2)^2 + b'(1 + h^2)$, $cd = bh$ and $c'd' = b'h$.

In fact, we have

$$\begin{aligned}
\mathrm{E}(X_f) &= \lim_{r\downarrow 1}\frac{1}{4\pi i}\oint_{|\xi|=1}\log(|c+d\xi|^2)\left(\frac{1}{r\xi+1}+\frac{1}{r\xi-1}-\frac{2}{r\xi+h^{-1}y_2}\right)d\xi \\
&= \lim_{r\downarrow 1}\frac{1}{4\pi i}\oint_{|\xi|=1}\log(|c+d\xi^{-1}|^2)\left(\frac{1}{r\xi^{-1}+1}+\frac{1}{r\xi^{-1}-1}-\frac{2}{r\xi^{-1}+h^{-1}y_2}\right)\xi^{-2}d\xi \\
&= \lim_{r\downarrow 1}\left\{\frac{1}{8\pi i}\oint_{|\xi|=1}\log(|c+d\xi|^2)\left(\frac{1}{r\xi+1}+\frac{1}{r\xi-1}-\frac{2}{r\xi+h^{-1}y_2}+\frac{1}{\xi(r+\xi)}\right.\right. \\
&\quad \left.\left.+\frac{1}{\xi(r-\xi)}-\frac{2}{\xi(r+h^{-1}y_2\xi)}\right)d\xi\right\} \\
&= \lim_{r\downarrow 1}\Re\left\{\frac{1}{8\pi i}\int_{|\xi|=1}\log((c+d\xi)^2)\left(\frac{1}{r\xi+1}+\frac{1}{r\xi-1}-\frac{2}{r\xi+h^{-1}y_2}+\frac{1}{\xi(r+\xi)}\right.\right. \\
&\quad \left.\left.+\frac{1}{\xi(r-\xi)}-\frac{2}{\xi(r+h^{-1}y_2\xi)}\right)d\xi\right\} \\
&= \frac{1}{4}\left(\log[(c^2-d^2)^2]-2\log[(c-y_2 dh^{-1})^2]\right).
\end{aligned}$$

Furthermore,

$$\begin{aligned}
&\mathrm{cov}(X_f, X_{f'}) \\
&= -\frac{1}{2\pi^2}\oint\oint_{|\xi_1|=|\xi_2|=1}\log(|c+d\xi_1|^2)\log(|c'+d'\xi_2|^2)\frac{d\xi_1 d\xi_2}{(r_1\xi_1 - r_2\xi_2)^2} \\
&= \frac{1}{2\pi^2}\oint_0^{2\pi}\oint_0^{2\pi}\log(|c+de^{i\theta_1}|^2)\log(|c'+d'e^{i\theta_2}|^2)\frac{e^{i\theta_1}e^{i\theta_2}d\theta_1 d\theta_2}{(r_1 e^{i\theta_1} - r_2 e^{i\theta_2})^2} \\
&= \frac{1}{4\pi^2}\oint_0^{2\pi}\oint_0^{2\pi}\log(|c+de^{i\theta_1}|^2)\log(|c'+d'e^{i\theta_2}|^2)\left[\frac{e^{i\theta_1}e^{i\theta_2}}{(r_1 e^{i\theta_1} - r_2 e^{i\theta_2})^2}\right. \\
&\quad \left.+\frac{e^{-i\theta_1}e^{-i\theta_2}}{(r_1 e^{-i\theta_1} - r_2 e^{-i\theta_2})^2}\right]d\theta_1 d\theta_2 \\
&= \frac{1}{4\pi^2}\oint_0^{2\pi}\log(|c'+d'e^{i\theta_2}|^2)\Re\left\{\oint_0^{2\pi}\log((c+de^{i\theta_1})^2)\left[\frac{e^{i\theta_1}e^{i\theta_2}}{(r_1 e^{i\theta_1} - r_2 e^{i\theta_2})^2}\right.\right. \\
&\quad \left.\left.+\frac{e^{-i\theta_1}e^{-i\theta_2}}{(r_1 e^{-i\theta_1} - r_2 e^{-i\theta_2})^2}\right]d\theta_1\right\}d\theta_2
\end{aligned}$$

$$= \frac{1}{4\pi^2} \oint_0^{2\pi} \log(|c' + d' e^{i\theta_2}|^2) \Re \left\{ \oint_{|\xi_1|=1} \log((c + d\xi_1)^2) \left[\frac{e^{i\theta_2}}{i(r_1\xi_1 - r_2 e^{i\theta_2})^2} \right.\right.$$

$$\left.\left. + \frac{\xi_1^{-2} e^{-i\theta_2}}{i(r_1\xi_1 - r_2 e^{-i\theta_2})^2} \right] d\xi_1 \right\} d\theta_2 \text{ (the first term is analytic)}$$

$$= \frac{1}{4\pi^2} \oint_0^{2\pi} \log(|c' + d' e^{i\theta_2}|^2) \Re \left\{ \oint_{|\xi_1|=1} \log((c + d\xi_1)^2) \frac{e^{i\theta_2}}{i(r_1 e^{i\theta_2} - r_2\xi_1)^2} \right] d\xi_1 \right\} d\theta_2$$

$$= \frac{1}{2\pi} \oint_0^{2\pi} \log(|c' + d' e^{i\theta_2}|^2) \Re \left[\frac{2d e^{i\theta_2}}{r_2(cr_2 + dr_1 e^{i\theta_2})} \right] d\theta_2$$

$$= \Re \left\{ \frac{1}{2\pi} \oint_0^{2\pi} \log((c' + d' e^{i\theta_2})^2) \left[\frac{d e^{i\theta_2}}{r_2(cr_2 + dr_1 e^{i\theta_2})} + \frac{d e^{-i\theta_2}}{r_2(cr_2 + dr_1 e^{-i\theta_2})} \right] d\theta_2 \right.$$

$$= \Re \left\{ \frac{1}{2\pi i} \oint_{|\xi_2|=1} \log((c' + d'\xi_2)^2) \left[\frac{d}{r_2(cr_2 + dr_1\xi_2)} + \frac{d}{r_2\xi_2(cr_2\xi_2 + dr_1)} \right] d\xi_2 \right.$$

$$= 2r_1^{-1} r_2^{-1} [\log(c') - \log(c' - r_1 dd'/r_2 c)] \to 2\log(cc'/(cc' - dd')).$$

Example 2: For any positive integers $k \geq r \geq 1$, we have

$$\begin{aligned} \mathrm{E}(X_{x^r}) &= \frac{1}{4\pi i} \oint_{|\xi|=1} \frac{(1 + h\xi)^r (1 + h\xi^{-1})^r}{|1 - y_2|^{2r}} \left(\frac{1}{\xi + 1} + \frac{1}{\xi - 1} - \frac{2}{\xi + h^{-1} y_2} \right) d\xi \\ &= \frac{1}{2(1 - y_2)^{2r}} \left[(1 - h)^{2r} + (1 + h)^{2r} - 2(1 - y_2)^r (1 - h^2/y_2)^r \right. \\ &\quad \left. - \sum_{\substack{i \leq r \\ j,k \geq 0 \\ i-j=2k+1}} \binom{r}{j} \binom{r}{i} h^{j+i} + 2 \sum_{\substack{i \leq r \\ j,k \geq 0 \\ i-j=k+1}} \binom{r}{j} \binom{r}{i} h^{j+i} \left(-\frac{h}{y_2} \right)^{k-1} \right] \end{aligned}$$

and

$$\begin{aligned} &\mathrm{Cov}(X_{x^r}, X_{x^k}) \\ &= -\frac{1}{2\pi^2 (1-y_2)^{2r+2k}} \lim_{1<r_1<r_2 \to 1} \oint\oint_{|\xi_1|=|\xi_2|=1} \frac{|1+h\xi_1|^{2r} |1+h\xi_2|^{2k}}{(r_1\xi_1 - r_2\xi_2)^2} d\xi_1 d\xi_2 \\ &= \frac{2 \cdot r! \cdot k!}{(y_2-1)^{2r+2k}} \sum_{j=0}^{r-1} (j+1) \cdot \left\{ \left(\sum_{l_3=j+1}^{[\frac{1+j+r}{2}]} \frac{(y_1+y_2-y_1y_2+1)^{r-2l_3} \cdot (y_1+y_2-y_1y_2)^{l_3}}{(-1-j+l_3)! \cdot (1+j+r-2l_3)! \cdot l_3!} \right) \right. \\ &\quad \left. \times \left(\sum_{l_3'=0}^{[\frac{k-j-1}{2}]} \frac{(y_1+y_2-y_1y_2+1)^{k-2l_3'} \cdot (y_1+y_2-y_1y_2)^{l_3'}}{(j+1+l_3')! \cdot (k-j-1-2l_3')! \cdot l_3'!} \right) \right\} \end{aligned}$$

where $[a]$ is the integer part of a, that is, the maximum integer less than or equal to a.

Example 3: If $f = e^x$, then

$$\mathrm{E}(X_f) = \frac{1}{2}\Bigg(e^{(1-h)^2/(1-y_2)^2} + e^{(1+h)^2/(1-y_2)^2} - 2e^{(1-y_2)(1-h^2/y_2)/(1-y_2)^2}$$

$$- e^{2/(1-y_2)^2} \sum_{\substack{j,k,l\geq 0 \\ j-k=2l+1}} \frac{h^{j+k}}{j!k!(1-y_2)^{2j+2k}} - 2e^{2/(1-y_2)^2}$$

$$\times \sum_{\substack{j,k,l\geq 0 \\ j-k=2l+1}} \frac{h^{j+k}}{j!k!(1-y_2)^{2j+2k}}\left(-\frac{h}{y_2}\right)^{l-1}\Bigg).$$

$$\mathrm{Var}(X_f) = -\frac{1}{2\pi^2}\oint_{|\xi_1|=1}\oint_{|\xi_2|=1} \frac{e^{|1+h\xi_1|^2/(1-y_2)^2} e^{|1+h\xi_2|^2/(1-y_2)^2}}{(\xi_1-\xi_2)^2} d\xi_1 d\xi_2$$

$$= \sum_{j,k=1}^{+\infty} \frac{1}{j!k!}\left(-\frac{1}{2\pi^2}\oint\oint \frac{|1+h\xi_1|^{2j}|1+h\xi_2|^{2k}}{(\xi_1-\xi_2)^2(1-y_2)^{2j+2k}} d\xi_1 d\xi_2\right).$$

5

Limiting Behavior of Eigenmatrix of Sample Covariance Matrix

Let $\mathbf{X}_n = (X_{ij})$ be an $p \times n$ matrix of i.i.d. complex random variables and let $\mathbf{T}_n$ be an $p \times p$ nonnegative definite Hermitian matrix with a square root $T_n^{\frac{1}{2}}$. In this chapter, we shall consider the matrix $\mathbf{A}_n = \frac{1}{n}\mathbf{T}_n^{1/2}\mathbf{X}_n\mathbf{X}_n^*\mathbf{T}_n^{1/2}$. If $\mathbf{T}_n$ is nonrandom, then $\mathbf{A}_n$ can be considered as a sample covariance matrix of a sample drawn from a population with distribution of $\mathbf{T}_n^{1/2}\mathbf{X}_{\cdot,1}$, where $\mathbf{X}_{\cdot,1} = (X_{11}, \cdots, X_{n1})'$. If $\mathbf{T}_n$ is an inverse of another sample covariance matrix, then the multivariate F matrix can be considered as a special case of the matrix $\mathbf{A}_n$.

The limiting properties of the eigenvalues of a sample covariance matrix $\mathbf{A}_n$ are intensively investigated in the literature. However, relatively less work was done about the limiting behavior of eigenvectors of $\mathbf{A}_n$. Some results on this aspect can be found in Silverstein (1984, 1989, 1990). The reason that more attention has been focused on the ESD of sample covariance matrices may be due to the origin of Random Matrix Theory (RMT) which is the Quantum Mechanics (QM), where the eigenvalues of large dimensional random matrices are used to describe energy levels of particles in a large system. Another reason is the difficulty to describe the limiting behavior of the eigenmatrix (the matrix of orthonormal eigenvectors) because the dimension is growing to infinity.

In applications of RMT to many other disciplines, such as statistics, wireless communications, *e.g.*, the CDMA (Code-Divition Multiple Access) systems, MIMO (Mulyiple input multiple output) system, finance and economics *etc.*, the role of eigenmatrix might be more important than the eigenvalues. For example, in signal processing, for signals received by linearly spaced sensors, the estimates of the directions of arrivals (DOA) are based on the noise eigenspace. In principal component analysis or factor analysis, the directions of the principal components are the eigenvectors corresponding to the largest eigenvalues.

To be motivated, consider the eigenmatrix of a Wishart matrix which has an Haar distribution, that is, the uniform distribution over the group of unitary matrices (or orthogonal matrices for real case). It is conceivable that the eigenmatrix of large sample covariance matrix should be "asymptotically Haar distributed", at least when the population covariance matrix is a multiple of identity. However, we are facing a problem on how to formulate the terminology "asymptotically Haar distributed" because their dimensions are increasing. In this paper, we shall adopt the method of Silverstein (1989, 1990). If $\mathbf{U}$ has a Haar measure over the orthogonal matrices, then, for any unit vector $\mathbf{x} \in \mathbb{R}^p$, $\mathbf{y} = \mathbf{U}\mathbf{x} = (y_1, \cdots, y_p)'$ has a uniform distribution over the unit sphere $\mathscr{S}_p = \{\mathbf{x} \in \mathbb{R}^p;\ \|\mathbf{x}\| = 1\}$. If $\mathbf{z} = (z_1, \cdots, z_p)' \sim N(0, I_p)$, then $\mathbf{y}$ has the same distribution as $\mathbf{z}/\|\mathbf{z}\|$.

Now, define a stochastic process $Y_p(t)$ in space $D(0,1)$ by

$$\begin{aligned} Y_p(t) &= \sqrt{\frac{p}{2}} \sum_{i=1}^{[pt]} (|y_i|^2 - \frac{1}{p}) \\ &\stackrel{d}{=} \sqrt{\frac{p}{2}} \frac{1}{\|\mathbf{z}\|^2} \sum_{i=1}^{[pt]} (|z_i|^2 - \frac{\|\mathbf{z}\|^2}{p}) \end{aligned} \tag{5.0.1}$$

where $[a]$ denotes the greatest integer $\leq a$. From the second equality, it is easy to see that $Y_n(t)$ converges to a Brownian Bridge $B(t)$ (BB), when p converges to infinity. Thus, we are interested in whether this property is reserved for general sample covariance matrices.

Let $U_n \Lambda_n U_n^*$ denote the spectral decomposition of $\mathbf{A}_n$, where $\Lambda_n = \mathrm{diag}(\lambda_1, \lambda_2, \cdots, \lambda_p)$, $U_n = (u_{ij})$ is a unitary matrix consisting of the orthonormal eigenvectors of $\mathbf{A}_n$. Assume that $\mathbf{x}_n \in \mathbb{C}^p$, $\|\mathbf{x}_n\| = 1$, is an arbitrary non-random unit vector and $\mathbf{y} = (y_1, y_2, \cdots, y_p)^* = U_n^* \mathbf{x}_n$. We define a stochastic process by the way of (5.0.1). If $\mathbf{U}_n$ is "asymptotically Haar distributed", then $\mathbf{y}$ should be "asymptotically uniformly distributed" over the unit sphere and $Y_n(t)$ should tend to a BB. Our main goal is to examine the limiting properties of the vector $\mathbf{y}$ through the stochastic process $Y_n(t)$.

5.1 Earlier Work by Silverstein

In Silverstein (1984), it is assumed that:
i) For all n, i, j, X_{ij} are iid. with

$$\mathrm{E}(X_{11}) = 0, \mathrm{E}X_{11}^2 = 1; \tag{5.1.1}$$

$$\mathrm{E}(X_{11}^4) = 3, \tag{5.1.2}$$

$$\text{all moments exist} \tag{5.1.3}$$

$$\mathrm{E}|X_{11}|^m \leq m^{\alpha m}, \text{ for all } m \geq 2 \text{ and some } \alpha > 0. \tag{5.1.4}$$

ii) $\mathbf{T}_n \equiv \mathbf{I}_p$;
iii) $\frac{p}{n} \to y > 0$ as $n \to \infty$.

Theorem 5.1. *Let $\{\mathbf{x}_n\}$, $\mathbf{x}_n \in \mathbb{R}^p$, be an arbitrary sequence of nonrandom unit vectors. Then, under assumptions (5.1.1), (5.1.2) and (5.1.3),*

$$\left\{\int_0^\infty x^i dY_p(F_n(t))\right\}_{i=1}^\infty \xrightarrow{\mathscr{D}} \left\{\int_0^\infty x^i dB(F_y(t))\right\}_{i=1}^\infty, \tag{5.1.5}$$

where F_n is the ESD of $\mathbf{A}_n$, F_y is the MP law and $B(t)$ is the standard Brownian Motion.

Moreover, if (5.1.2) does not hold, then there exists sequences $\mathbf{x}_n$ for which (5.1.5) (with $h(x) = x^i$) fails to converge in distribution.

By Taylor expansion and some tedious estimation, Silverstein also proved that

Theorem 5.2. *Let $\{f_i\}_{i=1}^\infty$ be functions defined on $[0, \infty)$, where for each $i, f_i(0) = 0$, and f_i is analytic at 0, with corresponding radius of convergence greater than $(1 + \sqrt{y})^2$. Let $\{\mathbf{x}_n\}$, $\mathbf{x}_n \in \mathbb{R}^p$, be an arbitrary sequence of non-random unit vectors. Then, under assumptions (5.1.1), (5.1.2) and (5.1.4),*

$$\left\{\int_0^\infty f_i(x) dY_p(F_n(t))\right\}_{i=1}^\infty \xrightarrow{\mathscr{D}} \left\{\int_0^\infty f_i(x) dB(F_y(t))\right\}_{i=1}^\infty, \tag{5.1.6}$$

as $n \to \infty$.

In (1989), he further proved

Theorem 5.3. *(a) We have*

$$\begin{aligned}
&\left\{\sqrt{p/2}(\mathbf{x}_n' \mathbf{A}_n^r \mathbf{x}_n - \frac{1}{n}\mathrm{tr}\mathbf{A}_n^r)\right\}_{r=1}^\infty \\
&= \left\{\int_0^\infty t^r dY_p((F_n(t)))\right\}_{r=0}^\infty \\
&\xrightarrow{\mathscr{D}} \left\{\int_a^b t^r dB(F_y(t))\right\}_{r=0}^\infty, \quad \text{as } n \to \infty
\end{aligned} \tag{5.1.7}$$

for every sequence $\{\mathbf{x}_n\}$, $\mathbf{x}_n \in \mathbb{R}^p$, $\|\mathbf{x}_n\| = 1$, if and only if

$$\mathrm{E}(X_{11}) = 0, \mathrm{E}|X_{11}^2| = 1, \mathrm{E}(|X_{11}^4| = 3. \tag{5.1.8}$$

(b) If $\int_0^\infty t dY_p(F_n(t))$ is to converge in distribution to any random variable for each of the $\{\mathbf{x}_n\}$, sequences $\{(1, 0, cdots, 0)^T\}$, $\{(1/\sqrt{p}, \cdots, 1/\sqrt{p})^T\}$, then necessarily $\mathrm{E}(|X_{11}^4| < \infty) < \infty$ and $\mathrm{E}(X_{11}) = 0$.

(c) If $\mathrm{E}(X_{11}^4) < \infty$ but $\mathrm{E}[(X_{ll} - \mathrm{E}(X_{11}))4]/(Var(X_{11}))2 \neq 3$, then there exist sequences $\{\mathbf{x}_n\}$ for which fails to converge in distribution.

Silverstein's work provided certain evidences for the conjecture that $Y_p(t)$ weakly tends to a Brownian Bridge but not yet prove the conjecture. Theorem 5.3 proved that the conjecture is generally not true, unless the 4th moment of the underlying distribution is also the same as normal distributions. Silverstein's 1990 work does prove that the conjecture is true for some special $\mathbf{x}_n$-sequences.

Theorem 5.4. *Let X_{11} be standardized, symmetric and $\mathrm{E}X_{11}^4 < \infty$ and $\mathbf{x}_n = (\pm 1/\sqrt{p}, \cdots, \pm 1/\sqrt{p})^T$. Then $Y_p(t)$ tends to the standardized Brownian Bridge.*

Remark 5.5. *It seems that there is a contradictory between Theorems 5.3 and 5.4. In fact, conclusion (a) of Theorem 5.3 says if for all sequences $\{\mathbf{x}_n\}$ the convergence (5.1.7) holds, then $\mathrm{E}|X_{11}^4| = 3$. Theorem 5.4 only requires some choices of $\{\mathbf{x}_n\}$.*

5.2 Further extension of Silverstein's Work

In this section, we return to the general form of sample covariance matrix with a general non-negative definite matrix $\mathbf{T}$. We define the vector $\mathbf{y}_n$ as before and define a new empirical distribution function, which is simply denoted as VESD, based on the components of $\mathbf{y}_n$ as

$$F_1^{\mathbf{A}_n}(x) = \sum_{i=1}^{p} |y_i|^2 I(\lambda_i \leq x). \tag{5.2.1}$$

Recall that the ESD of $\mathbf{A}_n$ is

$$F^{\mathbf{A}_n}(x) = \frac{1}{p} \sum_{i=1}^{p} I(\lambda_i \leq x).$$

Then, it follows that

$$Y_p(x) = \sqrt{\frac{p}{2}} (F_1^{\mathbf{A}_n}(x) - F^{\mathbf{A}_n}(x)).$$

The investigation of $Y_p(t)$ is then converted to one concerning the difference $F_1^{\mathbf{A}_n}(x) - F^{\mathbf{A}_n}(x)$ of the two empirical distributions.

It is obvious that $F_1^{\mathbf{A}_n}(x)$ is a random probability distribution function and its Stieltjes transform is given by

$$m_{F_1^{\mathbf{A}_n}}(z) = \mathbf{x}_n^*(\mathbf{A}_n - z\mathbf{I})^{-1}\mathbf{x}_n. \tag{5.2.2}$$

As it has been seen, the difference between $F_1^{\mathbf{A}_n}(x)$ and $F^{\mathbf{A}_n}(x)$ is only in their different weights at the eigenvalues of $\mathbf{A}_n$. However, it will be proven that although these two empirical distributions have different weights they have the identical limit, which is included in Theorem 5.6 below.

To investigate the convergence of $Y_p(x)$, we consider its linear functional which is defined as

$$\hat{X}_n(g) = \int g(x) dY_p(x),$$

where g is a bounded continuous function. It turns out that

$$\hat{X}_n(g) = \sqrt{\frac{p}{2}}\left[\sum_{j=1}^{p} |y_j^2| g(\lambda_j) - \frac{1}{p}\sum_{j=1}^{p} g(\lambda_j)\right]$$
$$= \sqrt{\frac{p}{2}}\left[\int g(x) dF_1^{\mathbf{A}_n}(x) - \int g(x) dF^{\mathbf{A}_n}(x)\right].$$

Proving the convergence of $\hat{X}_n(g)$ under general conditions is difficult. Following the idea of the CLT of LSS, described in last chapter, we shall prove the CLT for those g which are analytic over the support of limiting spectral distribution of $\mathbf{A}_n$. To this end, let

$$G_n(x) = \sqrt{p}(F_1^{\mathbf{A}_n}(x) - F^{y_n,H_n}(x))$$

where $y_n = \frac{p}{n}$ and $F^{y_n,H_n}(x)$ denotes the limiting distribution by substituting y_n for y and H_n for H in $F^{y,H}$.
The main results of this section are formulated in the following three theorems.

Theorem 5.6. *Suppose:*
(1) For each n $X_{ij} = X_{ij}^n, i, j = 1, 2, \cdots$, are i.i.d. complex random variables with $EX_{11} = 0, E|X_{11}|^2 = 1$ and $E|X_{11}|^4 < \infty$.
(2) $\mathbf{x}_n \in \mathbb{C}_1^n = \{\mathbf{x} \in \mathbb{C}^n, \| \mathbf{x} \| = 1\}$ and $\lim_{n\to\infty} \frac{p}{n} = y \in (0, \infty)$.
(3) $\mathbf{T}_n$ is non-random Hermitian non-negative definite with its spectral norm bounded in n, with $H_n = F^{\mathbf{T}_n} \xrightarrow{\mathscr{D}} H$, a proper distribution function and with $\mathbf{x}_n^*(\mathbf{T}_n - z\mathbf{I})^{-1}\mathbf{x}_n \to s_{F^H}(z)$, where $s_{F^H}(z)$ denotes the Stieltjes transform of $H(t)$, then, it holds that

$$F_1^{\mathbf{A}_n}(x) \to F^{y,H}(x) \quad a.s.$$

Remark 5.7. *The condition $\mathbf{x}_n^*(T_n - zI)^{-1}\mathbf{x}_n \to s_{F^H}(z)$ is critical for our Theorem 5.6 as well as for our main theorems give later. At first, we indicate that if $\mathbf{T}_n = b\mathbf{I}$ for some positive constant b or more generally $\lambda_{\max}(\mathbf{T}_n) - \lambda_{\min}(\mathbf{T}_n) \to 0$, the condition $\mathbf{x}_n^*(\mathbf{T}_n - z\mathbf{I})^{-1}\mathbf{x}_n \to s_{F^H}(z)$ holds uniformly for all $\mathbf{x}_n \in \mathbb{C}_1^n$.*

We also indicate that this condition doesn't require that T_n has to be a multiple of an identity. As an application of this remark, one sees that the eigenmatrix of a sample covariance matrix transform $\mathbf{x}_n$ to a unit vector whose entries absolute values are near to $1/\sqrt{p}$. Consequently, the condition $\mathbf{x}_n^*(T_n - zI)^{-1}\mathbf{x}_n \to m_{F^H}(z)$ holds when $\mathbf{T}_n$ is the inverse of another sample covariance matrix which is independent of $\mathbf{X}_n$. Therefore, the multivariate F matrix satisfies the assumptions of Theorem 5.6.

In general, the condition may not hold for all $\mathbf{x}_n \in \mathbb{C}_1^p$. However, there always exist some $\mathbf{x}_n \in \mathbb{C}_1^p$ such that this condition holds, say $\mathbf{x}_n = (\mathbf{u}_1 + \cdots + \mathbf{u}_p)/\sqrt{p}$, where $\mathbf{u}_1, \cdots, \mathbf{u}_p$ are the orthonormal eigenvectors of spectral decomposition of $\mathbf{T}_n$.

Applying Theorem 5.6, we get some interesting results:

Corollary 5.8. *Let $(\mathbf{A}_n^m)_{ii}, m = 1, 2, \cdots,$ denote the i-th diagonal elements of matrices $\mathbf{A}_n^m$. Under the conditions of Theorem 5.6 for $\mathbf{x}_n = \mathbf{e}_{ni}$, then for any fixed m,*

$$\lim_{n\to\infty} \left| (A_n^m)_{ii} - \int x^m dF^{c,H}(x) \right| \to 0, \quad a.s.,$$

where $\mathbf{e}_{ni}$ is the p-vector with i-th element 1 and others 0.

Remark 5.9. *If $\mathbf{T}_n = b\mathbf{I}$ for some positive constant b or more generally $\lambda_{\max}(\mathbf{T}_n) - \lambda_{\min}(\mathbf{T}_n) \to 0$, there is a better result, i.e.,*

$$\lim_{n\to\infty} \max_i \left| (\mathbf{A}_n^m)_{ii} - \int x^m dF^{y,H}(x) \right| \to 0, \quad a.s.. \tag{5.2.3}$$

(The proof of this corollary follows easily from the uniform convergence of the condition (3) of Theorem 5.6 for all $\mathbf{x}_n \in \mathbb{C}_1^p$, with careful checking the proof of Theorem 5.6.)

More generally, we have

Corollary 5.10. *If $f(x)$ is a bounded function and the assumptions of Theorem 5.6 are satisfied, then*

$$\sum_{j=1}^p |y_j^2| f(\lambda_j) - \frac{1}{p}\sum_{j=1}^p f(\lambda_j) \to 0 \quad a.s..$$

Remark 5.11. *The proof of the above Corollaries are immediate. Applying Corollary 5.10, Theorem 5.6 of Eldar and Chen (2003) can be extended to a more general case without difficulty.*

Theorem 5.12. *In addition to the conditions of Theorem 5.6, we further assume that*
(4) $g_1, \cdots, g_k$ are defined and analytic on an open region $\mathcal{D}$ of the complex plane, which contains the real interval

$$[\liminf_n \lambda_{\min}^{\mathbf{T}_n} I_{(0,1)}(y)(1-\sqrt{y})^2, \limsup_n \lambda_{\max}^{\mathbf{T}_n}(1+\sqrt{y})^2]. \tag{5.2.4}$$

(5)

$$\sup_z \sqrt{n} \left| \mathbf{x}_n^*(\underline{m}_{F^{y_n,H_n}}(z)\mathbf{T}_n + I)^{-1}\mathbf{x}_n - \int \frac{1}{\underline{s}_{F^{y_n,H_n}}(z)t+1} dH_n(t) \right| \to 0,$$

as $n \to \infty$.
Then the following conclusions hold.
(a) The random vectors

$$\left(\int g_1(x)dG_n(x), \cdots, \int g_k(x)dG_n(x) \right) \tag{5.2.5}$$

form a tight sequence.

(b) If X_{11} and $\mathbf{T}_n$ are real and $EX_{11}^4 = 3$, the above random vector converges weakly to a Gaussian vector $X_{g_1}, \cdots, X_{g_k}$, with zero means and covariance function

$$Cov(X_{g_1}, X_{g_2}) \tag{5.2.6}$$
$$= -\frac{1}{2\pi^2} \int_{\mathcal{C}_1} \int_{\mathcal{C}_2} g_1(z_1) g_2(z_2) \frac{(z_2 \underline{s}(z_2) - z_1 \underline{s}(z_1))^2}{y^2 z_1 z_2 (z_2 - z_1)(\underline{s}(z_2) - \underline{s}(z_1))} dz_1 dz_2.$$

The contours $\mathcal{C}_1$ and $\mathcal{C}_2$ in the above equality are disjoint, both contained in the analytic region for the functions $(g_1, \cdots, g_k)$ and both enclosing the support of F^{y_n, H_n} for all large p.

(c) If X_{11} is complex, with $EX_{11}^2 = 0$ and $E|X_{11}|^4 = 2$, then the conclusions (a) and (b) still hold, but the covariance function reduces to half of the quantity given in (5.2.6).

Remark 5.13. *If $\mathbf{T}_n = b\mathbf{I}$ for some positive constant b or more generally $\sqrt{p}(\lambda_{\max}(\mathbf{T}_n) - \lambda_{\min}(\mathbf{T}_n)) \to 0$, then the condition (5) holds uniformly for all $\mathbf{x}_n \in \mathbb{C}_1^n$.*

Remark 5.14. *Indeed, we also establish the central limit theorem for $\hat{X}_n(g)$ according to Theorem 1.1 of Bai and Silverstein (2004) and Theorem 5.12. Besides the Theorem 5.12 which holds for more general function $g(x)$, the following theorem reveals more similarities between the process $Y_n(t)$ and the BB.*

Theorem 5.15. *Besides the assumptions of Theorem 5.12, $H(x)$ satisfies*

$$\int \frac{dH(t)}{(1 + t\underline{s}(z_1))(1 + t\underline{s}(z_2))} - \int \frac{dH(t)}{(1 + t\underline{s}(z_1))} \int \frac{dH(t)}{(1 + t\underline{s}(z_2))} = 0, \tag{5.2.7}$$

then all results of Theorem 4.4 remain true. Moreover, formula (5.2.6) can be simplified to

$$Cov(X_{g_1}, X_{g_2}) \tag{5.2.8}$$
$$= \frac{2}{y} \left(\int g_1(x) g_2(x) dF^{y,H}(x) - \int g_1(x_1) dF^{y,H}(x_1) \int g_2(x_2) dF^{y,H}(x_2) \right).$$

Remark 5.16. *Obviously, (5.2.7) holds when $\mathbf{T}_n = b\mathbf{I}$. Actually, (5.2.7) holds if and only if $H(x)$ is a degenerate distribution. To see it, one only need to choose z_2 to be the complex conjugate of z_1.*

Remark 5.17. *Theorem 5.15 extends the Theorem of Silverstein (1989). First, one sees that the r-th moment of $F_1^{\mathbf{A}_n}(x)$ is $\mathbf{x}_n^* \mathbf{A}_n^r \mathbf{x}_n$, which is a special case with $g(x) = x^r$. Then applying Theorem 5.15 with $\mathbf{T}_n = b\mathbf{I}$ and combining with the Theorem 1.1 of Bai and Silverstein (2004), one can get* *the sufficient part of (a) in the Theorem of Silverstein (1989). Actually, for* $\mathbf{T}_n = \mathbf{I}$, *formula (5.2.6) can be simplified to*

$$Cov(X_{g_1}, X_{g_2}) = \frac{2}{y}\left(\int g_1(x)g_2(x)dF_y(x) - \int g_1(x_1)dF_y(x_1)\int g_2(x_2)dF_y(x_2)\right) \tag{5.2.9}$$

where $F_y(x)$ *is a special case of* $F^{y,H}(x)$ *as* $\mathbf{T}_n = \mathbf{I}$.

Similar to the proof of Theorem 4.7, the proof of Theorem 5.12 needs an intermediate Lemma 5.18 stated below about the limiting property of the Stieltjes transform

$$M_n(z) = \sqrt{n}(m_{F_1^{A_n}}(z) - m_{F^{c_n,H_n}}(z)),$$

restricted on a contour $\mathcal{C}$ in a complex plane, described as follows. Let u_r be a number which is larger than the right end-point of interval (5.2.4) and u_l be a negative number if the left end-point of interval (5.2.4) is zero, otherwise let $u_l \in (0, \liminf_n \lambda_{\min}^{\mathbf{T}_n} I_{(0,1)}(y)(1-\sqrt{y})^2)$. Let $v_0 > 0$ be arbitrary. Define

$$\mathcal{C}_u = \{u + iv_0 : u \in [u_l, u_r]\}.$$

Then, the contour

$$\mathcal{C} = \mathcal{C}_u \cup \{u_l + iv : v \in [0, v_0]\} \cup \{u_r + iv : v \in [0, v_0]\}$$
$$\cup \{\text{ their symmetric parts below the real axis }\}.$$

Under the conditions of Theorem 5.12, for later use, we may select the contour $\mathcal{C}$ in the region on which the functions g's are analytic.

As in the proof of Theorem 4.7, due to technical difficulty, we will consider $M_n^*(z)$, a truncated version of $M_n(z)$. Choose a sequence of positive numbers $\{\delta_n\}$ such that for $0 < \rho < 1$

$$\delta_n \downarrow 0, \quad \delta_n \geq n^{-\rho}. \tag{5.2.10}$$

Write

$$\mathcal{C}_l = \begin{cases} \{u_l + iv : v \in [n^{-1}\delta_n, v_0]\} \text{ if } & u_l > 0, \\ \{u_l + iv : v \in [0, v_0]\} \quad \text{if} & u_l < 0 \end{cases}$$

and

$$\mathcal{C}_r = \{u_r + iv : v \in [n^{-1}\delta_n, v_0]\}.$$

Let $\mathcal{C}_0 = \mathcal{C}_u \cup \mathcal{C}_l \cup \mathcal{C}_r$. Now, for $z = u + iv$, we can define the process

$$M_n^*(z) = \begin{cases} M_n(z) \\ \quad \text{if} \quad z \in \mathcal{C}_0 \cup \bar{\mathcal{C}}_0 \\ \frac{nv+\delta_n}{2\delta_n} M_n(u_r + in^{-1}\delta_n) + \frac{\delta_n - nv}{2\delta_n} M_n(u_r - in^{-1}\delta_n) \\ \quad \text{if} \quad u = u_r, v \in [-n^{-1}\delta_n, n^{-1}\delta_n] \\ \frac{nv+\delta_n}{2\delta_n} M_n(u_l + in^{-1}\delta_n) + \frac{\delta_n - nv}{2\delta_n} M_n(u_l - in^{-1}\delta_n) \\ \quad \text{if} \quad u = u_l > 0, v \in [-n^{-1}\delta_n, n^{-1}\delta_n]. \end{cases}$$

$M_n^*(z)$ can be viewed as a random element in the metric space $C(\mathcal{C}, \mathbb{R}^2)$ of continuous functions from $\mathcal{C}$ to $\mathbb{R}^2$. We shall prove the following Lemma.

Lemma 5.18. *Under the assumptions of Theorem 5.6, (4) and (5) of Theorem 5.12, $M_n^*(z)$ forms a tight sequence on $\mathcal{C}$. Furthermore, when the conditions in (b) and (c) of Theorem 5.12 on X_{11} hold, for $z \in \mathcal{C}$, $M_n^*(z)$ converges to a Gaussian process $M(\cdot)$ with zero mean and for $z_1, z_2 \in \mathcal{C}$, under the assumptions in (b),*

$$Cov(M(z_1), M(z_2)) = \frac{2(z_2\underline{s}(z_2) - z_1\underline{s}(z_1))^2}{y^2 z_1 z_2 (z_2 - z_1)(\underline{s}(z_2) - \underline{s}(z_1))} \tag{5.2.11}$$

while under the assumptions in (c), covariance function similar to (5.2.11) is the half of the value of (5.2.11).

For details of the proof, the reader refers to Bai, Miao and Pan (2007).

5.3 Projecting the Eigenmatrix to a d-dimensional Space

The approach introduced in last two sections is main to project the eigenmatrix of the sample covariance matrix onto an arbitrary unit vector. Due to the need of application to Markowitz optimal portfolio principle, we need to limiting properties of projections of the eigenmatrix of large sample covariance matrix onto more two two vectors. Before introducing main results, we first formulate the problem. For simplicity, we assume the matrix $\mathbf{T}_n = \mathbf{I}_p$.

Suppose that $\{x_{jk}, j, k = 1, 2, \cdots\}$ is a set of double array complex random variables which are independent and identically distributed (iid) with mean zero and variance 1. Let $\mathbf{x}_j = (x_{1j}, \cdots, x_{pj})^T$ and $\mathbf{X} = (\mathbf{x}_1, \cdots, \mathbf{x}_n)$, we define

$$\mathbf{S}_n = \frac{1}{n}\sum_{k=1}^{n} \mathbf{x}_k \mathbf{x}_k^* = \frac{1}{n}\mathbf{X}\mathbf{X}^* \tag{5.3.1}$$

where $\mathbf{x}_k^*$ and $\mathbf{X}^*$ are the transposes of the complex conjugates of $\mathbf{x}_k$ and $\mathbf{X}$ respectively.

Recently, applications of the RMT become more and more important in modern statistics and other areas. For example, incorporating the RMT into the theory of the optimal mean-variance portfolio (Markowitz 1952, 1959), Bai, Liu and Wong (2006) proved that when the number of assets is large, the conventional return estimate[1] for the optimal portfolio return obtained by substituting the sample mean and sample covariance matrix into the parameters in the theoretic optimal portfolio return induces serious errors which have to be corrected before the return estimates could be implemented and become practically applicable.

As mentioned in earlier sections, the universal belief in the RMT is the conjecture that the eigenmatrix, of $\mathbf{S}_n$, is asymptotically Haar-distributed.

[1] They call it plug-in return estimate.

This conjecture has been hanged on for long but yet could not be solved due to the difficulty in describing the "asymptotically Haar-distributed" property when the dimension p increases to infinity.

Denote the spectral decomposition of $\mathbf{S}_n$ by $\mathbf{U}_n^* \Lambda \mathbf{U}_n$. If $\mathbf{x}_{ij}$ is normally distributed, $\mathbf{U}_n$ has a Haar measure on the orthogonal matrices and is independent of the eigenvalues in Λ.

For any general large sample covariance, that is, without the normality assumption, for any given unit vector $\mathbf{x}_n \in \mathbb{C}_1^p$, we constructed a process $Y_p(t)$ and tried to examine the behavior of the $Y_p(t)$ process.

In this section, we investigate the asymptotics of eigenmatrix of any general large sample covariance matrix $\mathbf{S}_n$ when $\mathbf{x}_n$ runs over a subset of the p-dimensional unit sphere in which $\mathbb{C}_1^p = \{\mathbf{x}_n : \ \|\mathbf{x}_n\| = 1, \ \mathbf{x}_n \in \mathbb{C}^p\}$.

Intuitively, if $\mathbf{U}_n$ is Haar-distributed, then, for any pair of unit p-vectors $\mathbf{x}$ and $\mathbf{y}$ satisfying $\mathbf{x} \perp \mathbf{y}$, $(\mathbf{U}_n\mathbf{x}, \mathbf{U}_n\mathbf{y})$ has the same joint distribution as

$$(\mathbf{z}_1 \quad \mathbf{z}_2) \begin{pmatrix} \mathbf{z}_1'\mathbf{z}_1 & \mathbf{z}_1'\mathbf{z}_2 \\ \mathbf{z}_2'\mathbf{z}_1 & \mathbf{z}_2'\mathbf{z}_2 \end{pmatrix}^{-1/2}, \tag{5.3.2}$$

where $\mathbf{z}_1$ and $\mathbf{z}_2$ are two independent p-vectors, each contains p iid standard normal variables. Note that as n tends to infinity, we have

$$\frac{1}{p} \begin{pmatrix} \mathbf{z}_1'\mathbf{z}_1 & \mathbf{z}_1'\mathbf{z}_2 \\ \mathbf{z}_2'\mathbf{z}_1 & \mathbf{z}_2'\mathbf{z}_2 \end{pmatrix} \to \mathbf{I}_2. \tag{5.3.3}$$

Thus, any functionals defined by these two random vectors should be asymptotically independent of each other. We shall extend this consideration to describe $\mathbf{U}_n$ to be asymptotically Haar-distributed as follows.

More generally, consider $\mathbf{x}$ and $\mathbf{y}$ to be two p-vectors with an angle θ. Then, we may find two orthonormal vectors $\boldsymbol{\alpha}_1$ and $\boldsymbol{\alpha}_2$ such that

$$\mathbf{x} = \|\mathbf{x}\|\boldsymbol{\alpha}_1, \quad \mathbf{y} = \|\mathbf{y}\|(\boldsymbol{\alpha}_1 \cos\theta + \boldsymbol{\alpha}_2 \sin\theta).$$

By (5.3.2) and (5.3.3), we have

$$\mathbf{U}_n\mathbf{x} \sim p^{-1/2}\|\mathbf{x}\|\mathbf{z}_1, \quad \mathbf{U}_n\mathbf{y} \sim p^{-1/2}\|\mathbf{y}\|(\mathbf{z}_1 \cos\theta + \mathbf{z}_2 \sin\theta). \tag{5.3.4}$$

Let $\sigma > 0$ be a positive constant. Now, consider the following three quantities

$$\mathbf{x}^*(\mathbf{S}_n + \sigma\mathbf{I})^{-1}\mathbf{x}, \quad \mathbf{x}^*(\mathbf{S}_n + \sigma\mathbf{I})^{-1}\mathbf{y}, \quad \mathbf{y}^*(\mathbf{S}_n + \sigma\mathbf{I})^{-1}\mathbf{y}. \tag{5.3.5}$$

We hypothesize that if $\mathbf{U}_n$ is asymptotically Haar-distributed and is asymptotically independent of Λ, then the above three quantities should be asymptotically equivalent to

$$p^{-1}\|\mathbf{x}\|^2\mathbf{z}_1^*(\Lambda+\sigma\mathbf{I})^{-1}\mathbf{z}_1,$$
$$p^{-1}\|\mathbf{x}\|\|\mathbf{y}\|\mathbf{z}_1^*(\Lambda+\sigma\mathbf{I})^{-1}(\mathbf{z}_1\cos\theta+\mathbf{z}_2\sin\theta),$$
$$p^{-1}\|\mathbf{y}\|^2(\cos\theta\mathbf{z}_1+\sin\theta\mathbf{z}_2)^*(\Lambda+\sigma\mathbf{I})^{-1}(\mathbf{z}_1\cos\theta+\mathbf{z}_2\sin\theta). \quad (5.3.6)$$

We shall investigate stochastic processes defined by such functionals in this section. By using the Stieltjes transform of sample covariance matrix, we have

$$p^{-1}\mathbf{z}_1^*(\Lambda+\sigma\mathbf{I})^{-1}\mathbf{z}_1 \to m(\sigma) = -\frac{1+\sigma-y-\sqrt{(1+y+\sigma)^2-4y}}{2y\sigma}, \quad a.s.,$$

where $m(\sigma)$ is a solution of the quadratic equation

$$m(1+\sigma-y+y\sigma m)-1=0. \quad (5.3.7)$$

Readers may refer to Bai (1999) for more details. By the same argument, we conclude that

$$p^{-1}(\cos\theta\mathbf{z}_1+\sin\theta\mathbf{z}_2)^*(\Lambda+\sigma\mathbf{I})^{-1}(\mathbf{z}_1\cos\theta+\mathbf{z}_2\sin\theta)\to m(\sigma), \quad a.s..$$

In addition, applying the results in Section 5.2, it is straightforward to show that, for the complex case,

$$p^{-1/2}[\mathbf{z}_1^*(\Lambda+\sigma I)^{-1}\mathbf{z}_1-pm_n(\sigma)]\to N(0,W), \quad (5.3.8)$$

while, for the real case, the limiting variance is $2W$, where

$$m_n(\sigma) = -\frac{1+\sigma-y_n-\sqrt{(1+y_n+\sigma)^2-4y}}{2y_n\sigma}$$
$$y_n = p/n$$
$$W = W(\sigma_1,\sigma_2) = \frac{m(\sigma_1)m(\sigma_2)}{(1+\sigma_1+\sigma_1 y m(\sigma_1))(1+\sigma_2+\sigma_2 y m(\sigma_2))-y}.$$

Here, the definitions of "real case" and "complex case" are given in Theorem 5.6 as stated below. Moreover, by the same argument, one could obtain a similar result for

$$p^{-1/2}[\cos\theta\mathbf{z}_1+\sin\theta\mathbf{z}_2)^*(\Lambda+\sigma\mathbf{I})^{-1}(\mathbf{z}_1\cos\theta+\mathbf{z}_2\sin\theta)-pm_n(\sigma)]. \quad (5.3.9)$$

In this section we shall further normalize the second term in (5.3.6) and, thereafter, derive the CLT for the joint distribution of the three terms in (5.3.6) after normalization. In addition, more notably, in this paper we will establish some limiting behaviors of the processes defined by these normalized quantities as discussed in the next section.

5.3.1 Main Results

Let $\mathbb{S}=\mathbb{S}_p$ be a subset of the unit p-sphere $\mathbb{C}_1^p$ indexed by an m-dimensional hyper-cube $T=[0,2\pi]^m$. For example, for any m arbitrarily chosen orthogonal unit p-vectors $\mathbf{x}_1,\cdots,\mathbf{x}_{m+1}\in\mathbb{C}_1^p$, we define

$$\mathbb{S} = \left\{ \mathbf{x}(\mathbf{t}) = \mathbf{x}_1 \cos t_1 + \mathbf{x}_2 \sin t_1 \cos t_2 + \cdots + \mathbf{x}_{m+1} \sin t_1 \cdots \sin t_m, \mathbf{t} \in T \right\}. \tag{5.3.10}$$

If $\mathbb{S}$ is chosen in (5.3.10), then the inner product $\mathbf{x}_n(\mathbf{t}_1)^*\mathbf{x}(\mathbf{t}_2)$ is a function of $\mathbf{t}_1$ and $\mathbf{t}_2$ only (*i.e.* independent of n). Also, the norm of the difference (we will call it the norm difference) $\|\mathbf{x}_n(\mathbf{t}_1) - \mathbf{x}_n(\mathbf{t}_2)\|$ satisfies the Lipschitz condition. Thus, if the time index set is chosen arbitrarily, we could assume that the angle, $\vartheta(\mathbf{t}_1, \mathbf{t}_2)$, between $\mathbf{x}(\mathbf{t}_1)$ and $\mathbf{x}(\mathbf{t}_2)$ tends to a function of $\mathbf{t}_1$ and $\mathbf{t}_2$ with their norm difference satisfies the Lipschitz condition.

Thereafter, we define a stochastic process $\mathbf{Y}_n(\mathbf{u}, \sigma)$ mapping from the time index set $T \times T \times I$ to $\mathbb{S}$ with $I = [\sigma_{10}, \sigma_{20}]$ $(0 < \sigma_{10} < \sigma_{20})$ as follows:

$$Y_n(\mathbf{u}, \sigma) = \sqrt{p}\Big(\mathbf{x}(\mathbf{t}_1)^*(\mathbf{S}_n + \sigma\mathbf{I})^{-1}\mathbf{x}(\mathbf{t}_2) - \mathbf{x}(\mathbf{t}_1)^*\mathbf{x}(\mathbf{t}_2)m_n(\sigma)\Big),$$

where $(\mathbf{u}, \sigma) = (\mathbf{t}_1, \mathbf{t}_2, \sigma) \in T \times T \times I$. The following theorem is then obtained:

Theorem 5.19. *Assume that the entries of* $\mathbf{X}$ *are iid with mean 0, variance 1 and finite* 4th *moment. Also, if the variables are complex, we further assume* $\mathrm{E}X_{11}^2 = 0$ *and* $\mathrm{E}|X_{11}|^4 = 2$*, this case is referred as the* **complex case**. *If the variables are real, we assume* $\mathrm{E}X_{11}^4 = 3$ *and refer it as the* **real case**. *Then, the process* $Y_n(\mathbf{t}_1, \mathbf{t}_2, \sigma)$ *converges weakly to a Gaussian process* $Y(\mathbf{t}_1, \mathbf{t}_2, \sigma)$ *with mean zero and variance-covariance function* $\mathrm{E}Y(\mathbf{t}_1, \mathbf{t}_2, \sigma_1)Y(\mathbf{t}_3, \mathbf{t}_4, \sigma_2)$ *satisfying:*

$$\mathrm{E}Y(\mathbf{t}_1, \mathbf{t}_2, \sigma_1)Y(\mathbf{t}_3, \mathbf{t}_4, \sigma_2) = \vartheta(\mathbf{t}_1, \mathbf{t}_4)\vartheta(\mathbf{t}_3, \mathbf{t}_2)W(\sigma_1, \sigma_2)$$

for the complex case and

$$\mathrm{E}Y(\mathbf{t}_1, \mathbf{t}_2, \sigma_1)Y(\mathbf{t}_3, \mathbf{t}_4, \sigma_2) = (\vartheta(\mathbf{t}_1, \mathbf{t}_4)\vartheta(\mathbf{t}_3, \mathbf{t}_2) + \vartheta'(\mathbf{t}_1, \mathbf{t}_3)\overline{\vartheta'}(\mathbf{t}_4, \mathbf{t}_2))W(\sigma_1, \sigma_2)$$

for the real case where

$$W(\sigma_1, \sigma_2) = \frac{ym(\sigma_1)m(\sigma_2)}{(1 + \sigma_1 + \sigma_1 ym(\sigma_1))(1 + \sigma_2 + \sigma_2 ym(\sigma_2)) - y}$$

and

$$\vartheta(\mathbf{t}, \mathbf{s}) = \lim \mathbf{x}_n^*(\mathbf{t})\mathbf{x}_n(\mathbf{s}),$$
$$\vartheta'(\mathbf{t}, \mathbf{s}) = \lim \mathbf{x}_n^*(\mathbf{t})\bar{\mathbf{x}}_n(\mathbf{s}).$$

We note that by the results given Section 5.2, we conclude

$$\sqrt{p}\Big[\mathbf{x}(\mathbf{t}_1)^*(\mathbf{S}_n + \sigma\mathbf{I})^{-1}\mathbf{x}(\mathbf{t}_1) - m_n(\sigma)\Big] \to N(0, W)$$

for the complex case while the asymptotic variance is $2W$ for the real case.

More generally, if $\mathbf{x}$ and $\mathbf{y}$ are two orthonormal vectors, applying Theorem 5.19, we obtain the limiting distribution of the three quantities in (5.3.5) with normalization that

$$\sqrt{p}\begin{pmatrix} \mathbf{x}^*(\mathbf{S}_n+\sigma\mathbf{I})^{-1}\mathbf{x}-m_n(\sigma) \\ \mathbf{x}^*(\mathbf{S}_n+\sigma\mathbf{I})^{-1}\mathbf{y} \\ \mathbf{y}^*(\mathbf{S}_n+\sigma\mathbf{I})^{-1}\mathbf{y}-m_n(\sigma) \end{pmatrix} \longrightarrow N\left(\begin{pmatrix} 0 \\ 0 \\ 0 \end{pmatrix}, \begin{pmatrix} W & 0 & 0 \\ 0 & W & 0 \\ 0 & 0 & W \end{pmatrix}\right) \tag{5.3.11}$$

for the complex case while the asymptotic covariance matrix is

$$\begin{pmatrix} 2W & 0 & 0 \\ 0 & W & 0 \\ 0 & 0 & 2W \end{pmatrix}$$

for the real case.

Remark 5.20. *This theorem shows that the three quantities in* (5.3.5) *are asymptotically independent of one another. Thus, it provides a stronger reasoning to support the conjecture that* $\mathbf{U}_n$ *is asymptotically Haar-distributed than those established in previous literature.*

In many situations, we are interested to extend the process $Y_n(\mathbf{u},\sigma)$ to a region $T\times T\times D$ where D is a compact subset of the complex plane and is disjoint with the interval $[a,b]$, the support of the MPL. Actually, we can define a complex measure by putting complex mass $\bar{x}_j(\mathbf{t}_1)y_j(\mathbf{t}_2)$ at λ_j, the jth eigenvalue of S_n. Then, the Stieltjes transform of this complex measure is

$$s_n(z)=\mathbf{x}^*(\mathbf{S}_n-z\mathbf{I})^{-1}\mathbf{y},$$

where $z=\mu+i\nu$ with $\nu\neq 0$. If $\mathbf{x}^*\mathbf{y}$ is a constant (or have a limit, we still denote it as $\mathbf{x}^*\mathbf{y}$ for simplicity), it follows from Theorem 5.19 that

$$\mathbf{x}^*(\mathbf{S}_n-z\mathbf{I})^{-1}\mathbf{y}\to\mathbf{x}^*\mathbf{y}s(z),$$

where $s(z)$ is the Stieltjes transform of MPL, *i.e.*

$$s(z)=\frac{1-z-y+\sqrt{(1-z+y)^2-4y}}{2yz}$$

in which by convention, the square root $\sqrt{z}$ takes the one with imaginary part to have the same sign as the imaginary part of z. By definition, $m(\sigma)=s(-\sigma)=\lim_{\nu\downarrow 0}s(-\sigma+i\nu)$. In calculating the limit, we follow the conventional sign of the square root of a complex number that the real part of $\sqrt{(-\sigma+iv-1-y)^2-4y}$ should have the opposite sign of v and thus

$$m(\sigma)=-\frac{1+\sigma-y-\sqrt{(1+y+\sigma)^2-4y}}{2y\sigma}.$$

Now, we are ready to extend the process $Y_n(\mathbf{u}, \sigma)$ to

$$Y_n(\mathbf{u}, z) = \sqrt{p}(\mathbf{x}^*(\mathbf{t}_1)(S_n - zI)^{-1}\mathbf{x}(\mathbf{t}_2) - \mathbf{x}^*(\mathbf{t}_1)\mathbf{x}(\mathbf{t}_2)s(z, y_n)$$

where $s(z, y_n)$ is the Stieltjes transform of the LSD of S_n with y replaced by y_n. Then, we obtain the following theorem:

Theorem 5.21. *Under the conditions of Theorem 5.19, the process* $Y_n(\mathbf{u}, z)$ *tends to a* **multivariate** *Gaussian process* $Y(\mathbf{u}, z)$ *with mean 0 and covariance function* $\mathrm{E}(Y(\mathbf{u}_1, z_1)Y(\mathbf{u}_2, z_2))$ *satisfying*

$$\mathrm{E}(Y(\mathbf{u}_1, z_1)Y(\mathbf{u}_2, z_2)) = \vartheta(\mathbf{t}_1, \mathbf{t}_4)\vartheta(\mathbf{t}_3, \mathbf{t}_2)W(\mathbf{z}_1, z_2) \tag{5.3.12}$$

for the complex case and

$$\mathrm{E}(Y(\mathbf{u}, z_1)Y(\mathbf{u}, z_2)) = (\vartheta(\mathbf{t}_1, \mathbf{t}_4)\vartheta(\mathbf{t}_3, \mathbf{t}_2) + \vartheta'(\mathbf{t}_1, \mathbf{t}_3)\overline{\vartheta'}(\mathbf{t}_4, \mathbf{t}_2))W(\mathbf{z}_1, z_2) \tag{5.3.13}$$

for the real case where $\mathbf{u}_1 = (\mathbf{x}(\mathbf{t}_1), \mathbf{x}(\mathbf{t}_2))$ *and* $\mathbf{u}_2 = (\mathbf{x}(\mathbf{t}_3), \mathbf{x}(\mathbf{t}_4))$ *and*

$$W(z_1, z_2) = \frac{ys(z_1)s(z_2)}{(1 + z_1 + z_1ys(z_1))(1 + z_2 + z_2ys(z_2)) - y}.$$

Suppose that $f(x)$ is a function analytic on a region containing the interval $[a, b]$. Then, we may construct a linear spectral statistic with respect to complex measure as defined earlier; that is,

$$\sum_{j=1}^{p} f(\lambda_j)\bar{x}_j(\mathbf{t}_1)y_j(\mathbf{t}_2).$$

We then consider the normalized quantity

$$X_n(f) = \sqrt{p}\left(\sum_{j=1}^{p} f(\lambda_j)\bar{x}_j(\mathbf{t}_1)y_j(\mathbf{t}_2) - \mathbf{x}^*(\mathbf{t}_1)\mathbf{y}(\mathbf{t}_2)\int f(x)dF_{y_n}(x)\right), \tag{5.3.14}$$

where F_y is the standardized MP law. By applying Cauchy formula

$$f(x) = \frac{1}{2\pi i}\int_{\mathcal{C}} \frac{f(z)}{z - x}dz,$$

we obtain

$$X_n(f, \mathbf{u}) = -\frac{\sqrt{p}}{2\pi i}\left(\int_{\mathcal{C}}[\mathbf{x}^*(\mathbf{t}_1)(\mathbf{S}_n - z\mathbf{I})^{-1}\mathbf{y}(\mathbf{t}_2) - \mathbf{x}^*(\mathbf{t}_1)\mathbf{y}(\mathbf{t}_2)s_n(z)]f(z)dz\right), \tag{5.3.15}$$

where $\mathbf{u} = (\mathbf{t}_1, \mathbf{t}_2)$ and

$$s_n(z) = \frac{1 - z - y_n + \sqrt{(1 - z + y_n)^2 - 4y_n}}{2y_nz}.$$

Thereafter, we obtain the following corollaries:

Corollary 5.22. *Under the conditions of Theorem 5.19, for any k functions $f_1, \cdots, f_k$ analytic in a region containing the interval $[a, b]$, the k-dimensional process*

$$(X_n(f_1, \mathbf{u}_1), \cdots, X_n(f_k, \mathbf{u}_k))$$

tends to the k-dimensional stochastic **multivariate** Gaussian processes with mean zero and covariance functions satisfying

$$\mathrm{E}(X(f, \mathbf{u})X(g, \mathbf{v})) = -\frac{\theta}{4\pi^2}\int_{c_1}\int_{c_2} W(z_1, z_2)f(z_1)g(z_2)dz_1dz_2$$

where $\theta = \vartheta(\mathbf{t}_1, \mathbf{t}_4)\vartheta(\mathbf{t}_3, \mathbf{t}_2)$ for complex case and $= \vartheta(\mathbf{t}_1, \mathbf{t}_4)\vartheta(\mathbf{t}_3, \mathbf{t}_2) + \vartheta'(\mathbf{t}_1, \mathbf{t}_3)\overline{\vartheta'}(\mathbf{t}_4, \mathbf{t}_2)$ for real case.

Corollary 5.23. *The covariance function in Corollary* 5.8 *can also be written as*

$$\mathrm{E}(X(f, \mathbf{u})X(g, \mathbf{v})) = \theta\Big(\int_a^b f(x)g(x)dF_y(x) - \int_a^b f(x)dF_y(x)\int_a^b g(x)dF_y(x)\Big)$$

where θ has been defined in Corollary 5.8,

5.3.2 Sketch of Proof of Theorem 5.19

To prove Theorem 5.19, we first present the following lemmas.

Lemma 5.24. *Under the conditions of Theorem 5.19, for any $\mathbf{x}_n, \mathbf{y}_n \in \mathrm{C}_1^p$, we have*

$$\sqrt{n}(\mathbf{x}_n^*\mathrm{E}(S_n + \sigma I)^{-1}\mathbf{y}_n - \mathbf{x}_n^*\mathbf{y}_n m_n(\sigma)) \to 0.$$

Proof. When $\mathbf{y}_n = \mathbf{x}_n$, Lemma 5.24 in our paper reduces to the conclusion (5.5) $\to 0$ of Bai, Miao and Pan (2007). Thus, to complete the proof, one could simply keep $\mathbf{x}_n^*$ unchanged and substitute $\mathbf{x}_n$ by $\mathbf{y}_n = (\mathbf{x}_n^*\mathbf{y}_n)\mathbf{x}_n + \mathbf{z}_n$ in the proof of the above conclusion. Thereafter, the proof of this lemma follows.

Lemma 5.25. *Under the conditions of Theorem 5.19, for any matrix M_j bounded in norm and independent of $\mathbf{s}_j$,*

$$\max_j \Big|\frac{1}{n}(\mathbf{s}_j^* M_j \mathbf{s}_j - \mathrm{tr}M_j)\Big| \xrightarrow{a.s.} 0. \tag{5.3.16}$$

The proof of this lemma could be easily obtained from applying the truncation technique and invoking Lemma 2.7 of Bai and Silverstein (1998).

To prove Theorem 5.19, by Lemma 5.24, it is sufficient to show that $Y_n(\mathbf{u}, \sigma) - \mathrm{E}Y_n(\mathbf{u}, \sigma)$ tends to the limit process $Y(\mathbf{u}, \sigma)$. To this end, we shall first prove the finite dimensional convergence in Step 1 and, thereafter, prove the tightness in Step 2.

Step 1. Finite Dimensional Convergence

Under the assumption of finite fourth moment, we may truncate the random variables X_{ij} at $\varepsilon_n n^{1/4}$ where $\varepsilon_n \to 0$ and then renormalize them. Thus, it is reasonable to impose an additional assumption that $|X_{ij}| \le \varepsilon_n \sqrt{n}$.

Suppose $\mathbf{r}_j$ denote the jth column of $\frac{1}{\sqrt{n}}\mathbf{X}_n$. We let $\mathbf{A}(\sigma) = \mathbf{S}_n + \sigma\mathbf{I}$, $\mathbf{A}_j(\sigma) = \mathbf{A}(\sigma) - \mathbf{r}_j\mathbf{r}_j^*$ and let $\mathbf{x}_n$, $\mathbf{y}_n$ be any two vectors in $\mathbb{C}_1^p$. Thereafter, we define

$$\begin{aligned}\xi_j(\sigma) &= \mathbf{r}_j^*\mathbf{A}_j^{-1}(\sigma)\mathbf{r}_j - \frac{1}{n}\mathrm{tr}\mathbf{A}_j^{-1}(\sigma),\\ \gamma_j &= \mathbf{r}_j^*\mathbf{A}_j^{-1}\mathbf{y}_n\mathbf{x}_n^*\mathbf{A}_j^{-1}(\sigma)\mathbf{r}_j - \frac{1}{n}\mathbf{x}_n^*\mathbf{A}_j^{-1}(\sigma)\mathbf{A}_j^{-1}(\sigma)\mathbf{y}_n\end{aligned}$$

and

$$\beta_j(\sigma) = \frac{1}{1+\mathbf{r}_j^*\mathbf{A}_j^{-1}(\sigma)\mathbf{r}_j}, b_j(\sigma) = \frac{1}{1+n^{-1}\mathrm{tr}\mathbf{A}_j^{-1}(\sigma)}, \bar{b} = \frac{1}{1+n^{-1}\mathrm{E}tr\mathbf{A}^{-1}(\sigma)}.$$

In addition, we define the σ-field $\mathcal{F}_j = \sigma(\mathbf{r}_1, \cdots, \mathbf{r}_j)$ and let $\mathrm{E}_j(.)$ denote the conditional expectation when $\mathcal{F}_j$ is given. The term E_0 stands for the unconditional expectation.

Using the martingale decomposition, we have

$$\begin{aligned}\mathbf{A}^{-1}(\sigma) - \mathrm{E}\mathbf{A}^{-1}(\sigma) &= \sum_{j=1}^{n}(\mathrm{E}_j - \mathrm{E}_{j-1})[\mathbf{A}^{-1}(\sigma) - \mathbf{A}_k^{-1}(\sigma)]\\ &= \sum_{j=1}^{n}(\mathrm{E}_j - \mathrm{E}_{j-1})\beta_j\mathbf{A}_j^{-1}(\sigma)\mathbf{r}_j\mathbf{r}_j^*\mathbf{A}_j^{-1}(\sigma). \quad (5.3.17)\end{aligned}$$

Then,

$$\begin{aligned}Y_n(\mathbf{u}, \sigma) &= \sqrt{p}\sum_{j=1}^{n}(\mathrm{E}_j - \mathrm{E}_{j-1})[\mathbf{x}(\mathbf{t}_1)^*\mathbf{A}^{-1}(\sigma) - \mathbf{A}_k^{-1}(\sigma)\mathbf{x}(\mathbf{t}_2)]\\ &= \sqrt{p}\sum_{j=1}^{n}(\mathrm{E}_j - \mathrm{E}_{j-1})\beta_j\mathbf{x}(\mathbf{t}_1)^*\mathbf{A}_j^{-1}(\sigma)\mathbf{r}_j\mathbf{r}_j^*\mathbf{A}_j^{-1}(\sigma)\mathbf{x}(\mathbf{t}_2).\end{aligned}$$

Consider the K-dimensional distribution of $\{Y_n(\mathbf{u}_1, \sigma_1), \cdots, Y_n(\mathbf{u}_K, \sigma_K)\}$, where $(\mathbf{u}_i, \sigma_i) = (\mathbf{t}_{i1}, \mathbf{t}_{i2}, \sigma_i) \in T \times T \times I$. Invoking Theorem 35.12 of Billingsley (1995), for any constant a_i, $i = 1, \cdots, K$, we have

$$\sum_{i=1}^{K} a_i(Y_n(\mathbf{u}_i, \sigma_i) - \mathrm{E}Y_n(\mathbf{u}_i, \sigma_i)) \Rightarrow N(0, \boldsymbol{\alpha}'\Sigma\boldsymbol{\alpha}),$$

where

$$\boldsymbol{\alpha} = (a_1, \cdots, a_K)',$$
$$\Sigma_{ij} = \mathrm{E}Y(\mathbf{t}_{i1}, \mathbf{t}_{i2}, \sigma_i)Y(\mathbf{t}_{j1}, \mathbf{t}_{j2}, \sigma_j) = \vartheta(\mathbf{t}_{i1}, \mathbf{t}_{j2})\vartheta(\mathbf{t}_{j1}, \mathbf{t}_{i2})W(\sigma_i, \sigma_j)$$

for the complex case and

$$\Sigma_{ij} = (\vartheta(\mathbf{t}_{i1}, \mathbf{t}_{j2})\vartheta(\mathbf{t}_{j1}, \mathbf{t}_{i2}) + \vartheta'(\mathbf{t}_{i1}, \mathbf{t}_{j1})\bar{\vartheta}'(\mathbf{t}_{j2}, \mathbf{t}_{i2}))W(\sigma_i, \sigma_j)$$

for the real case.

For this end, we will verify the Liapounov condition and calculate the asymptotic covariance matrix Σ in next subsections.

It is routine to verify **the Liapounov condition** and hence the finite dimensional convergence reduces to finding the asymptotic covariances of $Y_n(\sigma)$. The details are omitted.

Before deriving the asymptotic covariances, we need **simplify $Y_n(\mathbf{u}) - \mathrm{E}Y_n(\mathbf{u})$ in equivalent forms.**

For any $\mathbf{x}_n, \mathbf{y}_n \in \mathbb{C}_1^p$, from (5.3.17), we have

$$\begin{aligned}
&\mathbf{x}_n^*\mathbf{A}^{-1}(\sigma)\mathbf{y}_n - \mathrm{E}\mathbf{x}_n^*\mathbf{A}^{-1}(\sigma)\mathbf{y}_n \\
&= \sum_{j=1}^n \Big(\bar{b}\mathrm{E}_j\gamma_j + \mathrm{E}_j(\beta_j - \bar{b})\gamma_j \\
&\quad + (\mathrm{E}_j - \mathrm{E}_{j-1})b_j(\sigma)\beta_j(\sigma)\xi_j(\sigma)\mathbf{r}_j^*\mathbf{A}_j^{-1}(\sigma)\mathbf{y}_n\mathbf{x}_n^*\mathbf{A}_j^{-1}(\sigma)\mathbf{r}_j\Big). \quad (5.3.18)
\end{aligned}$$

For the third term on the right hand side of (5.3.18), we have

$$\begin{aligned}
&\mathrm{E}\left|\sqrt{p}\sum_{j=1}^n(\mathrm{E}_j - \mathrm{E}_{j-1})b_j(\sigma)\beta_j(\sigma)\xi_j(\sigma)\mathbf{r}_j^*\mathbf{A}_j^{-1}(\sigma)\mathbf{y}_n\mathbf{x}_n^*\mathbf{A}_j^{-1}(\sigma)\mathbf{r}_j\right|^2 \\
&= p\sum_{j=1}^n \mathrm{E}\left|(\mathrm{E}_j - \mathrm{E}_{j-1})b_j(\sigma)\beta_j(\sigma)\xi_j(\sigma)\mathbf{r}_j^*\mathbf{A}_j^{-1}(\sigma)\mathbf{y}_n\mathbf{x}_n^*\mathbf{A}_j^{-1}(\sigma)\mathbf{r}_j\right|^2 \\
&\le p\sum_{j=1}^n\left(\mathrm{E}|\xi_j(\sigma)|^4\mathrm{E}|\mathbf{r}_j^*\mathbf{A}_j^{-1}(\sigma)\mathbf{y}_n\mathbf{x}_n^*\mathbf{A}_j^{-1}(\sigma)\mathbf{r}_j|^4\right)^{1/2} = o(n^{-1/2}).
\end{aligned}$$

For the second term on the right hand side of (5.3.18), we have

$$\begin{aligned}
&\mathrm{E}\left|\sqrt{p}\sum_{j=1}^n \mathrm{E}_j(\beta_j(\sigma) - \bar{b}(\sigma))\gamma_j(\sigma)\right|^2 \\
&\le p\sum_{j=1}^n\left(\mathrm{E}|\beta_j(\sigma) - \bar{b}(\sigma)|^4\mathrm{E}|\gamma_j(\sigma)|^4\right)^{1/2} \\
&= o(n^{-3/2})\cdot(\max_j \mathrm{E}|\mathrm{tr}\mathbf{A}_j^{-1}(\sigma) - \mathrm{E}\mathrm{tr}\mathbf{A}^{-1}(\sigma)|^4)^{1/2} = o(n^{-1/2}),
\end{aligned}$$

where the last step follows from applying the martingale decomposition and Burkholder inequality and from the facts that

$$|\mathrm{tr}A_j^{-1}(\sigma) - \mathrm{tr}A^{-1}(\sigma)| \le 1$$
$$\mathrm{E}|\mathrm{tr}A^{-1}(\sigma) - \mathrm{Etr}A^{-1}(\sigma)|^4 = O(n^2).$$

Thus, we have

$$\sqrt{p}(\mathbf{x}_n^*\mathbf{A}^{-1}(\sigma)\mathbf{y}_n - \mathrm{E}\mathbf{x}_n^*\mathbf{A}^{-1}(\sigma)\mathbf{y}_n) = \sqrt{p}\sum_{j=1}^{n} \bar{b}\mathrm{E}_j\gamma_j + o_p(1). \tag{5.3.19}$$

Step 2. Asymptotic Covariances

To compute Σ, by the limiting property in (5.3.19), we only need to compute the limit

$$\nu_{i,j} = \lim p\sum_{k=1}^{n} \bar{b}(\sigma_i)\bar{b}(\sigma_j)\mathrm{E}_{k-1}\mathrm{E}_k\gamma_k(\mathbf{t}_{i,1}, \mathbf{t}_{i2}, \sigma_i)\mathrm{E}_k\gamma_k(\mathbf{t}_{j1}, \mathbf{t}_{j2}, \sigma_j)$$

in which, for $i, k = 1, \cdots, K$, we have

$$\begin{aligned}&\gamma_k(\mathbf{t}_{i1}, \mathbf{t}_{i2}, \sigma_i)\\ &= \mathbf{r}_k^*\mathbf{A}_k^{-1}(\sigma_i)\mathbf{x}(\mathbf{t}_{i2})\mathbf{x}^*(\mathbf{t}_{i1})\mathbf{A}_k^{-1}(\sigma_i)\mathbf{r}_k - \frac{1}{n}\mathbf{x}^*(\mathbf{t}_{i1})\mathbf{A}_k^{-1}(\sigma_i)\mathbf{A}_k^{-1}(\sigma_i)\mathbf{x}(\mathbf{t}_{i2}).\end{aligned}$$

Note that by Lemma 5.25 $\bar{b}(\sigma) \to b(\sigma) = 1/(1 + ym(\sigma))$, a.s., we have to calculate

$$\nu_{i,j} = \lim_n p\sum_{k=1}^{n} b(\sigma_i)b(\sigma_j)\mathrm{E}_{k-1}\mathrm{E}_k\gamma_k(\mathbf{t}_{i1}, \mathbf{t}_{i2}, \sigma_i)\mathrm{E}_k\gamma_k(\mathbf{t}_{j1}.\mathbf{t}_{j2}, \sigma_j). \tag{5.3.20}$$

For simplicity, we will use $\mathbf{x}, \mathbf{y}, \mathbf{u}, \mathbf{v}, \sigma_1, \sigma_2$ to denote $\mathbf{x}(\mathbf{t}_{i1})$, $\mathbf{x}(\mathbf{t}_{i2})$, $\mathbf{x}(\mathbf{t}_{j1})$, $\mathbf{x}(\mathbf{t}_{j2})$, σ_i and σ_j. For $\mathbf{X} = (X_1, \cdots, X_p)'$ of iid entries with mean 0 and variance 1, and $\mathbf{A} = (A_{ij})$ and $\mathbf{B} = (B_{ij})$ are Hermitian matrices, the following equality holds:

$$\begin{aligned}&\mathrm{E}(\mathbf{X}^*\mathbf{A}\mathbf{X} - \mathrm{tr}\mathbf{A})(\mathbf{X}^*\mathbf{B}\mathbf{X} - \mathrm{tr}\mathbf{B})\\ &= \mathrm{tr}\mathbf{A}\mathbf{B} + |\mathrm{E}X_1^2|^2\mathrm{tr}\mathbf{A}\mathbf{B}^T + \sum A_{ii}B_{ii}(\mathrm{E}|X_1|^4 - 2 - |\mathrm{E}X_1^2|^2).\end{aligned}$$

Utilizing this equality, we have

$$\nu = \lim \frac{pb(\sigma_1)b(\sigma_2)}{n^2}\sum_{k=1}^{n} \mathrm{tr}\Big(\mathrm{E}_k\mathbf{A}_k^{-1}(\sigma_1)\mathbf{x}\mathbf{y}^*\mathbf{A}_k^{-1}(\sigma_1)\mathrm{E}_k\mathbf{A}_k^{-1}(\sigma_2)\mathbf{u}\mathbf{v}^*\mathbf{A}_k^{-1}(\sigma_2)\Big) \tag{5.3.21}$$

for the complex case. For the real case, we have

$$\nu = \lim \frac{pb(\sigma_1)b(\sigma_2)}{n^2} \times \sum_{k=1}^{n} \mathrm{tr}\left(\mathrm{E}_k \mathbf{A}_k^{-1}(\sigma_1)\mathbf{x}\mathbf{y}^* \mathbf{A}_k^{-1}(\sigma_1)\mathrm{E}_k \mathbf{A}_k^{-1}(\sigma_2)(\mathbf{u}\mathbf{v}^* + \bar{\mathbf{v}}\mathbf{u}')\mathbf{A}_k^{-1}(\sigma_2) \right). \quad (5.3.22)$$

To calculate the limit in (5.3.21), we apply the same method used in Bai, Miao and Pan (2006) to prove their (4.7). Thus, we only need to calculate the limit of

$$\frac{yb(\sigma_1)b(\sigma_2)}{n} \sum_{k=1}^{n} \mathrm{tr}\mathrm{E}_k(\mathbf{A}_k^{-1}(\sigma_1)\mathbf{x}\mathbf{y}^* \mathbf{A}_k^{-1}(\sigma_1))\mathrm{E}_k(\mathbf{A}_k^{-1}(\sigma_2)\mathbf{u}\mathbf{v}^* \mathbf{A}_k^{-1}(\sigma_2))$$
$$= \frac{yb(\sigma_1)b(\sigma_2)}{n} \sum_{k=1}^{n} \mathrm{E}_k(\mathbf{y}^* \mathbf{A}_k^{-1}(\sigma_1)\breve{\mathbf{A}}_k^{-1}(\sigma_2)\mathbf{u})(\mathbf{v}^* \breve{\mathbf{A}}_k^{-1}(\sigma_2)\mathbf{A}_k^{-1}(\sigma_1)\mathbf{x}) \quad (5.3.23)$$

where $\breve{\mathbf{A}}_k^{-1}(z_2)$ is similarly defined as $\mathbf{A}_k^{-1}(\sigma_2)$ by $(\mathbf{r}_1, \cdots, \mathbf{r}_{k-1}, \breve{\mathbf{s}}_{k+1}, \cdots, \breve{\mathbf{s}}_n)$ and $\breve{\mathbf{s}}_{k+1}, \cdots, \breve{\mathbf{s}}_n$ are iid copies of $\mathbf{r}_{k+1}, \cdots, \mathbf{r}_n$.

Following the arguments in Bai, Miao and Pan (2006), we only have to change their vectors $\mathbf{x}_n$, $\mathbf{x}_n^*$ connected with $\mathbf{A}_k^{-1}(\sigma_1)$ by $\mathbf{x}$ and $\mathbf{y}^*$ and those connected with $\mathbf{A}_k^{-1}(\sigma_2)$ by $\mathbf{u}$ and $\mathbf{v}^*$ respectively. Going along the lines from their (4.7) to (4.23), we obtain

$$\begin{aligned}
&\mathrm{E}_k \mathbf{y}^* \mathbf{A}_k^{-1}(\sigma_1)\breve{\mathbf{A}}_k^{-1}(\sigma_2)\mathbf{u}\mathbf{v}^* \breve{\mathbf{A}}_k^{-1}(\sigma_2)\mathbf{A}_k^{-1}(\sigma_1)\mathbf{x} \\
&\times[1 - \frac{k-1}{n}\bar{b}(\sigma_1)\bar{b}(\sigma_2)\frac{1}{n}\mathrm{tr}\mathbf{T}^{-1}(\sigma_2)\mathbf{T}^{-1}(\sigma_1)] \\
&= \mathbf{y}^*\mathbf{T}^{-1}(\sigma_1)\mathbf{T}^{-1}(\sigma_2)\mathbf{u}\mathbf{v}^*\mathbf{T}^{-1}(\sigma_2)\mathbf{T}^{-1}(\sigma_1)\mathbf{x} \\
&(1 + \frac{j-1}{n}\bar{b}(\sigma_1)\bar{b}(\sigma_2)\frac{1}{n}\mathrm{E}_{k-1}\mathrm{tr}(\mathbf{A}_k^{-1}(\sigma_1)\breve{\mathbf{A}}_k^{-1}(\sigma_2))) + o_p(1) \quad (5.3.24)
\end{aligned}$$

and

$$\mathrm{E}_k\mathrm{tr}(\mathbf{A}_k^{-1}(\sigma_1)\breve{\mathbf{A}}_k^{-1}(\sigma_2)) = \frac{\mathrm{tr}(\mathbf{T}^{-1}(\sigma_1)\mathbf{T}^{-1}(\sigma_2)) + o_p(1)}{1 - \frac{j-1}{n^2}b(\sigma_1)b(\sigma_2)\mathrm{tr}(\mathbf{T}^{-1}(\sigma_1)\mathbf{T}^{-1}(\sigma_2))},$$

where

$$\mathbf{T}(\sigma) = (\sigma + \frac{n-1}{n}b(\sigma))\mathbf{I}.$$

We then obtain

$$\begin{aligned}
d(\sigma_1, \sigma_2) &:= \lim \bar{b}(\sigma_1)\bar{b}(\sigma_2)\frac{1}{n}\mathrm{tr}(\mathbf{T}^{-1}(\sigma_1)\mathbf{T}^{-1}(\sigma_2)) \\
&= \frac{yb(\sigma_1)b(\sigma_2)}{(\sigma_1 + b(\sigma_1))(\sigma_2 + b(\sigma_2))} \quad (5.3.25)
\end{aligned}$$

and

$$h(\sigma_1,\sigma_2) := b(\sigma_1)b(\sigma_2)\mathbf{y}^*\mathbf{T}^{-1}(\sigma_1)\mathbf{T}^{-1}(\sigma_2)\mathbf{u}\mathbf{v}^*\mathbf{T}^{-1}(\sigma_2)\mathbf{T}^{-1}(\sigma_1)\mathbf{x}$$
$$= \frac{\mathbf{y}^*\mathbf{u}\mathbf{v}^*\mathbf{x}b(\sigma_1)b(\sigma_2)}{(\sigma_1+b(\sigma_1))^2(\sigma_2+b(\sigma_2))^2}. \tag{5.3.26}$$

From (5.3.25) and (5.3.26), we get

$$(5.3.23) \overset{a.s.}{\to} yh(\sigma_1,\sigma_2)\left(\int_0^1 \frac{1}{(1-td(\sigma_1,z_2))}dt + \int_0^1 \frac{td(\sigma_1,\sigma_2)}{(1-td(\sigma_1,\sigma_2))^2}dt\right)$$
$$= \frac{yh(\sigma_1,\sigma_2)}{1-d(\sigma_1,\sigma_2)}$$
$$= \frac{y\mathbf{y}^*\mathbf{u}\mathbf{v}^*\mathbf{x}b(\sigma_1)b(\sigma_2)}{(\sigma_1+b(\sigma_1))(\sigma_2+b(\sigma_2))[(\sigma_1+b(\sigma_1))(\sigma_2+b(\sigma_2))-yb(\sigma_1)b(\sigma_2)]}.$$

From (5.3.7), one can easily derive the following identities

$$\frac{1}{\sigma+b(\sigma)} = m(\sigma) \quad \text{and} \quad \frac{b(\sigma)}{\sigma+b(\sigma)} = \frac{1}{1+\sigma+y\sigma m(\sigma)}. \tag{5.3.27}$$

Applying these identities, we could further simplify the limit of (5.3.23) to be

$$\mathbf{y}^*\mathbf{u}\mathbf{v}^*\mathbf{x}W(\sigma_1,\sigma_2)$$

where

$$W(\sigma_1,\sigma_2) = \frac{ym(\sigma_1)m(\sigma_2)}{(1+\sigma_1+\sigma_1 ym(\sigma_1))(1+\sigma_2+\sigma_2 ym(\sigma_2))-y}.$$

From the first equation in (5.3.27), we have

$$1+\sigma+y\sigma m(\sigma) = \frac{1+ym(\sigma)}{m(\sigma)}.$$

Substituting this into the expression of $W(\sigma_1,\sigma_2)$, we could express it in the following form:

$$W(\sigma_1,\sigma_2) = \frac{ym^2(\sigma_1)m^2(\sigma_2)}{(1+ym(\sigma_1))(1+ym(\sigma_2))-ym(\sigma_1)m(\sigma_2)}. \tag{5.3.28}$$

By symmetry, we obtain the limit of (5.3.22) to be

$$(\mathbf{y}^*\mathbf{u}\mathbf{v}^*\mathbf{x}+\mathbf{y}^*\bar{\mathbf{v}}\mathbf{u}'\mathbf{x})W(\sigma_1,\sigma_2).$$

That is, for the complex case, the covariance function of the process $Y(\mathbf{t},\mathbf{s},\sigma)$ is

$$\mathrm{E}Y(\mathbf{t}_1,\mathbf{r}_1,\sigma_1)Y(\mathbf{t}_2,\mathbf{r}_2,\sigma_2) = \vartheta(\mathbf{t}_1,\mathbf{r}_2)\vartheta(\mathbf{t}_2,\mathbf{r}_1)W(\sigma_1,\sigma_2)$$

while for the real case it is

$$\mathrm{E}Y(\mathbf{t}_1,\mathbf{r}_1,\sigma_1)Y(\mathbf{t}_2,\mathbf{r}_2,\sigma_2) = (\vartheta(\mathbf{t}_1,\mathbf{r}_2)\vartheta(\mathbf{t}_2,\mathbf{r}_1)+\vartheta'(\mathbf{t}_1,\mathbf{t}_2)\bar{\vartheta}'(\mathbf{r}_2,\mathbf{r}_1))W(\sigma_1,\sigma_2).$$

Remark 5.26. *When the random variables involved with S_n are real and satisfy the finite moment conditions for the real case, the time index set $\mathbb{S}_T$ can still be chosen as a subset of $\mathbb{C}_1^p$ and the covariance function of the limiting process $Y(\mathbf{u}, \mathbf{v}, \sigma)$ has the same expression as (5.3.13) in the real case.*

Step 3. Tightness

Theorem 5.27. *Under the conditions in Theorem 5.19, the sequence of $Y_n(\mathbf{u}, \sigma) - E(Y_n(\mathbf{u}, \sigma))$ is tight.*

For ease reference on the tightness, we quote a theorem from page 267 of Loeve (1978) as follows:

Proposition 5.28. (Tightness Criterion) *The sequence (P_n) is tight if and only if*

(i) $\quad \sup_n \mathrm{P}_n(x : |x(0)| > c) \to 0$ *as* $c \to \infty$

and, for every $\varepsilon > 0$, as $\delta \to 0$, we have

(ii) $\quad \mathrm{P}_n(\omega_x(\delta) > \varepsilon) \to 0,$

where δ-oscillation is defined by

$$\omega_x(\delta) = \sup_{|\mathbf{t}-\mathbf{s}|<\delta} |x(\mathbf{t}) - x(\mathbf{s})|.$$

To complete the proof of the tightness for Theorem 5.27, we note that the condition (i) in Proposition 5.28 is a consequence of finite dimensional convergence which has been proved in the previous section. To show condition (ii) in Proposition 5.28, we will use the two lemmas given below. Thus, to complete the proof of theorem 5.27, by Proposition 5.28 and Lemma 5.29, it is sufficient to verify

$$\sup_{\mathbf{u}_1, \mathbf{u}_2 \in T \times T} \mathrm{E} \left| \frac{Y_n(\mathbf{u}_1) - Y_n(\mathbf{u}_2)}{\|(\mathbf{u}_1, \sigma_1) - (\mathbf{u}_2, \sigma_2)\|} \right|^{4m+2} < \infty. \tag{5.3.29}$$

This will be proved in Lemma 5.30.

Lemma 5.29. *Suppose that $X_n(\mathbf{t})$ is a sequence of stochastic processes whose paths are continuous and Lipschitz; that is, there is a random variable $M = M_n$ such that*

$$|X_n(\mathbf{t}) - X_n(\mathbf{s})| \leq M\|\mathbf{t} - \mathbf{s}\|.$$

We further assume that

$$\sup_{\mathbf{t} \neq \mathbf{s} \in T} \mathrm{E} \left| \frac{X_n(\mathbf{t}) - X_n(\mathbf{s})}{\|\mathbf{t} - \mathbf{s}\|} \right|^{\alpha} < \infty, \tag{5.3.30}$$

Then, for any fixed $\varepsilon > 0$, we have

$$\lim_{\delta \downarrow 0} \mathrm{P}_n(\omega_x(\delta) > \varepsilon) = 0. \tag{5.3.31}$$

Proof. Without loss of generality, we assume that $T = [0, M]^m$. Firstly, for any given $\varepsilon > 0$ and $\eta > 0$, we choose an integer K such that $2\sqrt{m}MK^{-1} < \delta$ and $K^{m-\alpha}m^{\alpha/2}\varepsilon^{-\alpha}2^{\alpha} < \eta/2$. Thereafter, we choose an integer a such that $a^{m-\alpha}2^{\alpha} < 1/2$. For each $\ell = 1, 2, \cdots$, we define

$$t_{K,i}(j,\ell) = \frac{jM}{Ka^{\ell}+1}, j = 1, \cdots, Ka^{\ell},$$

where M is the longest edge of T. Taking conventional $t_i(j) = t_{K,i}(j, 0)$ and, similarly, defining $\mathbf{t}_K(\mathbf{j}, \ell)$ as the vector with its ith component to be $t_{K,i}(j_i, \ell)$ for $\mathbf{j} = (j_1, \cdots, j_m)'$, we have

$$\begin{aligned}
&\mathrm{P}_n(\omega_x(\delta) \geq \varepsilon) \\
&\leq \mathrm{P}\Big(\sup_{\mathbf{j}} \sup_{|\mathbf{t}-\mathbf{t}_K(\mathbf{j})|\leq\delta} |X_n(\mathbf{t}) - X_n(\mathbf{t}_K(\mathbf{j}))| \geq \varepsilon/2\Big) \\
&\leq \sum_{\ell=0}^{L} \sum_{\mathbf{t}_K(\mathbf{j},\ell)} \mathrm{P}\Big(|X_n(\mathbf{t}_K(\mathbf{j}^*, \ell-1)) - X_n(\mathbf{t}_K(\mathbf{j},\ell))| \geq 2^{-\ell-1}\varepsilon\Big) \\
&\quad + \mathrm{P}\Big(\sup_{\mathbf{t}_K(\mathbf{j},L+1)} \sup_{\|\mathbf{t}-\mathbf{t}_K(\mathbf{j},L+1)\|\leq a^{-L-2}\delta} |X_n(\mathbf{t}) - X_n(\mathbf{t}_K(\mathbf{j}, L+1))| \geq 2^{-L-2}\varepsilon\Big) \\
&\leq \sum_{\ell=1}^{\infty} (Ka^{\ell})^m \left(\frac{\sqrt{m}}{\varepsilon 2^{-\ell-1}(Ka^{\ell})}\right)^{\alpha} < \frac{\eta}{2} \sum_{\ell=0}^{\infty} \left(\frac{1}{2}\right)^{\ell} \\
&= \eta,
\end{aligned} \tag{5.3.32}$$

where the summation $\sum_{\mathbf{t}_K(\mathbf{j},\ell)}$ runs over all vectors $\mathbf{t}_K(\mathbf{j}, \ell)$, and $\mathbf{t}_K(\mathbf{j}^*, \ell 1)$ is the $\mathbf{t}_K(\mathbf{j}, \ell-1)$ vector closest to $\mathbf{t}_K(\mathbf{j}, \ell)$. Here, we have used the fact that the term in second inequality of (5.3.32) tends to 0 for all fixed n when $L \to \infty$. It is because

$$|X_n(\mathbf{t}) - X_n(\mathbf{t}_K(\mathbf{j}, L+1))| \leq Ma^{-L-2}\delta$$

which implies that

$$\begin{aligned}
&\mathrm{P}\Big(\sup_{\mathbf{t}_K(\mathbf{j},L+1)} \sup_{\|\mathbf{t}-\mathbf{t}_K(\mathbf{j},L+1)\|\leq a^{-L-2}\delta} |X_n(\mathbf{t}) - X_n(\mathbf{t}_K(\mathbf{j}, L+1))| \geq 2^{-L-2}\varepsilon\Big) \\
&\leq \mathrm{P}(M_n \geq (a/2)^{L+2}\varepsilon/\delta) \to 0.
\end{aligned}$$

The proof of the lemma is then complete.

Lemma 5.30. *Under the conditions of Theorem* 5.19, *the property in* (5.3.29) *holds for any* m.

Proof. For simplicity, we shall prove the lemma for a general m instead of $4m + 2$. Since,

$$
\begin{aligned}
&E\left|\frac{Y_n(\mathbf{u}_1,\sigma_1)-Y_n(\mathbf{u}_2,\sigma_2)-\mathrm{E}(Y_n(\mathbf{u}_1,\sigma_1)-Y_n(\mathbf{u}_2,\sigma_2))}{\|\mathbf{u}_1-\mathbf{u}_2\|+|\sigma_1-\sigma_2|}\right|^m\\
&\asymp p^{m/2}\mathrm{E}\left|\frac{\mathbf{x}_n(\mathbf{t}_1)^*\mathbf{A}^{-1}(\sigma_1)\mathbf{x}_n(\mathbf{t}_2)-E\mathbf{x}_n(\mathbf{t}_1)^*\mathbf{A}^{-1}(\sigma_1)\mathbf{x}_n(\mathbf{t}_2)}{\|\mathbf{t}_1-\mathbf{t}_3\|+\|\mathbf{t}_2-\mathbf{t}_4\|+|\sigma_1-\sigma_2|}\right.\\
&\quad\left.-\frac{(\mathbf{x}_n(\mathbf{t}_3)^*\mathbf{A}^{-1}(\sigma_2)\mathbf{x}_n(\mathbf{t}_4)-E\mathbf{x}_n(\mathbf{t}_3)^*\mathbf{A}^{-1}(\sigma_2)\mathbf{x}_n(\mathbf{t}_4))}{\|\mathbf{t}_1-\mathbf{t}_3\|+\|\mathbf{t}_2-\mathbf{t}_4\|+|\sigma_1-\sigma_2|}\right|^m\\
&\le Ln^{m/2}\left\{\mathrm{E}\left|\frac{(\mathbf{x}_n(\mathbf{t}_1)-\mathbf{x}_n(\mathbf{t}_3))^*\mathbf{A}^{-1}(\sigma_1)\mathbf{x}_n(\mathbf{t}_2)-E(\mathbf{x}_n(\mathbf{t}_1)-\mathbf{x}_n(\mathbf{t}_3))^*\mathbf{A}^{-1}(\sigma_1)\mathbf{x}_n(\mathbf{t}_2)}{\|\mathbf{t}_1-\mathbf{t}_3\|}\right|^m\right.\\
&\quad+\mathrm{E}\left|\frac{\mathbf{x}_n^*(\mathbf{t}_3)\mathbf{A}^{-1}(\sigma_1)(\mathbf{x}_n(\mathbf{t}_2)-\mathbf{x}_n(\mathbf{t}_4))-E\mathbf{x}_n^*(\mathbf{t}_3)\mathbf{A}^{-1}(\sigma_1)(\mathbf{x}_n(\mathbf{t}_2)-\mathbf{x}_n(\mathbf{t}_4))}{\|\mathbf{t}_2-\mathbf{t}_4\|}\right|^m\\
&\quad\left.+\mathrm{E}\left|\mathbf{x}_n^*(\mathbf{t}_3)\mathbf{A}^{-1}(\sigma_1)\mathbf{A}^{-1}(\sigma_2)\mathbf{x}_n(\mathbf{t}_4)-E\mathbf{x}_n^*(\mathbf{t}_2)\mathbf{A}^{-1}(\sigma_1)\mathbf{A}^{-1}(\sigma_2)\mathbf{x}_n(\mathbf{t}_4)\right|^m\right\},
\end{aligned}
$$

for a constant L, where $\mathbf{A}\asymp\mathbf{B}$ means $\mathbf{A}$ and $\mathbf{B}$ have the same order, *i.e.* $\mathbf{A}=O(\mathbf{B})$ and $\mathbf{B}=O(\mathbf{A})$.

Note that $\|\mathbf{x}_n(\mathbf{t}_1)-\mathbf{x}_n(\mathbf{t}_3)\|/\|\mathbf{t}_1-\mathbf{t}_3\|\le 1$ or bounded for the general case. By the martingale decomposition (5.3.17), Burkholder inequality, we have

$$
\begin{aligned}
&n^{m/2}\mathrm{E}\left|\frac{(\mathbf{x}_n(\mathbf{t}_1)-\mathbf{x}_n(\mathbf{t}_3))^*\mathbf{A}^{-1}(\sigma_1)\mathbf{x}_n(\mathbf{t}_2)-E(\mathbf{x}_n(\mathbf{t}_1)-\mathbf{x}_n(\mathbf{t}_3))^*\mathbf{A}^{-1}(\sigma_1)\mathbf{x}_n(\mathbf{t}_2)}{\|\mathbf{t}_1-\mathbf{t}_3\|}\right|^m\\
&\quad=O(1).
\end{aligned}
$$

Similarly, we obtain

$$
\begin{aligned}
&n^{m/2}\mathrm{E}\left|\frac{\mathbf{x}_n^*(\mathbf{t}_3)\mathbf{A}^{-1}(\sigma_1)(\mathbf{x}_n(\mathbf{t}_2)-\mathbf{x}_n(\mathbf{t}_4))-E\mathbf{x}_n^*(\mathbf{t}_3)\mathbf{A}^{-1}(\sigma_1)(\mathbf{x}_n(\mathbf{t}_2)-\mathbf{x}_n(\mathbf{t}_4))}{\|\mathbf{t}_2-\mathbf{t}_4\|}\right|^m\\
&\quad=O(1).
\end{aligned}
$$

Using the martingale decomposition and Burkholder inequality, we obtain

$$
\begin{aligned}
&n^{m/2}\mathrm{E}\left|\mathbf{x}_n^*(\mathbf{t}_3)\mathbf{A}^{-1}(\sigma_1)\mathbf{A}^{-1}(\sigma_2)\mathbf{x}_n(\mathbf{t}_4)-E\mathbf{x}_n^*(\mathbf{t}_2)\mathbf{A}^{-1}(\sigma_1)\mathbf{A}^{-1}(\sigma_2)\mathbf{x}_n(\mathbf{t}_4)\right|^m\\
&\le Ln^{m/2}\left[\sum_{k=1}^n\mathrm{E}|\mathbf{x}_n^*(\mathbf{t}_3)[\mathbf{A}^{-1}(\sigma_1)\mathbf{A}^{-1}(\sigma_2)-\mathbf{A}_k^{-1}(\sigma_1)\mathbf{A}_k^{-1}(\sigma_2)]\mathbf{x}_n(\mathbf{t}_4)|^m\right.\\
&\quad\left.+\mathrm{E}\left(\sum_{k=1}^n\mathrm{E}_{k-1}|\mathbf{x}_n^*(\mathbf{t}_3)[\mathbf{A}^{-1}(\sigma_1)\mathbf{A}^{-1}(\sigma_2)-\mathbf{A}_k^{-1}(\sigma_1)\mathbf{A}_k^{-1}(\sigma_2)]\mathbf{x}_n(\mathbf{t}_4)|^2\right)^{m/2}\right]\\
&=O(1),
\end{aligned}
$$

which follows from applying the following decomposition:

$$
\begin{aligned}
&A^{-1}(\sigma_1)A^{-1}(\sigma_2)-A_k^{-1}(\sigma_1)A_k^{-1}(\sigma_2)\\
&=\beta(\sigma_1)A_k^{-1}(\sigma_1)\mathbf{r}_k\mathbf{r}_k^*A_k^{-1}(\sigma_1)A_k^{-1}(\sigma_2)+\beta(\sigma_2)A_k^{-1}(\sigma_1)A_k^{-1}(\sigma_2)\mathbf{r}_k\mathbf{r}_k^*A_k^{-1}(\sigma_2)\\
&\quad+\beta(\sigma_k)\beta(\sigma_2)A_k^{-1}(\sigma_1)\mathbf{r}_k\mathbf{r}_k^*A_k^{-1}(\sigma_1)A_k^{-1}(\sigma_2)\mathbf{r}_k\mathbf{r}_k^*A_k^{-1}(\sigma_2).
\end{aligned}
$$

Thereafter, condition (5.3.29) is verified.

5.3.3 Proof of Corollary 5.23

Applying the quadratic equation (5.3.7), we have

$$\sigma = \frac{1}{m} - \frac{1}{1+ym}. \tag{5.3.33}$$

Making a difference of σ_1 and σ_2, we obtain

$$\sigma_1 - \sigma_2 = \frac{m(\sigma_2) - m(\sigma_1)}{m(\sigma_1)m(\sigma_2)} - \frac{y(m(\sigma_2) - m(\sigma_1))}{(1+ym(\sigma_1))(1+ym(\sigma_2))}.$$

Also, we get

$$\frac{m(\sigma_2) - m(\sigma_1)}{\sigma_1 - \sigma_2} = \frac{m(\sigma_1)m(\sigma_2)(1+ym(\sigma_1))(1+ym(\sigma_2))}{(1+ym(\sigma_1))(1+ym(\sigma_2)) - ym(\sigma_1)m(\sigma_2)}. \tag{5.3.34}$$

Finally, we obtain

$$\begin{aligned}&\frac{m(\sigma_2) - m(\sigma_1)}{\sigma_1 - \sigma_2} - m(\sigma_1)m(\sigma_2)\\ &= \frac{ym^2(\sigma_1)m^2(\sigma_2)}{(1+ym(\sigma_1))(1+ym(\sigma_2)) - ym(\sigma_1)m(\sigma_2)} = W(\sigma_1, \sigma_2).\end{aligned}$$

Note that the left side of the above equation is

$$\int_a^b \frac{dF_y(x)}{(x+\sigma_1)(x+\sigma_2)} - \int_a^b \frac{dF_y(x)}{x+\sigma_1} \int_a^b \frac{dF_y(x)}{x+\sigma_2}.$$

By the unique extension of analytic functions, we have

$$W(z_1, z_2) = \int_a^b \frac{dF_y(x)}{(x-z_1)(x-z_2)} - \int_a^b \frac{dF_y(x)}{x-z_1} \int_a^b \frac{dF_y(x)}{x-z_2}.$$

Substituting this into Corollary 5.8, we obtain Corollary 5.23.

6

Wireless Communications

6.1 Introduction

In the past decade, RMT has found wide applications in wireless communications. In fact, random matrices are related to the propagation channels of two important wireless communications systems: the multiple-input multiple-output (MIMO) antenna system and the direct-sequence code division multiple access (DS-CDMA) system. In a MIMO antenna system, multiple antennas are used at the transmitter side for simultaneous data transmission, and at the receiver side for simultaneous reception. For a rich multipath environment, the channel responses between the transmit antennas and the receive antennas can be simply modeled as independent and identically distributed (i.i.d.) random variables. Thus the wireless channel for such communication scenario can be described by a random matrix. DS-CDMA is a multiple access scheme supporting multiple users to communicate with a single base station using the same time and frequency resources, but different spreading codes. CDMA is the key physical layer air-interface in third generation (3G) cellular mobile communications. In a frequency-flat, synchronous DS-CDMA uplink system with random spreading codes, the channel can also be described by a random matrix.

Using RMT, Foschini [96] and Telatar [249] have proven that, for a given power budget and a given bandwidth, the ergodic capacity of MIMO Rayleigh fading channel increases with the minimum of the numbers of transmit antennas and receive antennas. This can be considered as the first important application of RMT in wireless communications. Furthermore, it is this promising result that makes MIMO an attractive solution to achieve high-speed wireless connections over limited bandwidth [204].

Another set of early results applying RMT to wireless communications comes from [112], [192], [260], [264]. Using limiting spectral properties of large random matrices, it is proven that for a frequency-flat synchronous DS-CDMA uplink with random spreading codes, the output signal-to-interference-plus-noise ratios (SINRs) using well-known linear receivers such as matched filter, decorrelator and minimum mean-square-error (MMSE) receiver, converge to deterministic values for large systems, i.e., when both spreading gain and number of users go to infinity, with their ratio being a constant. The limiting results provide us a fundamental guideline in designing the system parameters and predicting the system performance, without requiring the exact knowledge of the specific spreading codes for each user.

Inspired by these fundamental discoveries, in the past decade, researchers have exploited various applications of RMT in wireless communications, including, but not limited to:

(i) limiting capacity and asymptotic capacity distribution for random MIMO channels [264], [262];
(ii) asymptotic SINR distribution analysis for random channels [261], [164];
(iii) limiting SINR analysis for linearly precoded systems, such as the multicarrier CDMA using linear receivers [73];
(iv) limiting SINR analysis for random channels with interference cancellation receivers [124], [259], [165];
(v) asymptotic performance analysis for reduced rank receivers [125], [200];
(vi) limiting SINR analysis for coded multiuser systems [60];
(vii) design of receivers, such as the reduced-rank minimum mean-square-error (MMSE) receiver [157]; and
(viii) the asymptotic normality study for multiple access interference (MAI) [288] and linear receiver output [117].

An excellent overview of the above and other related applications of RMT in wireless communications has been documented by Tulino and Verdú in [262].

More recently, RMT has been applied to an emerging area in wireless communications, which is called "cognitive radio" networks [285], [286], [160]. One of the usage models in cognitive radio networks is "opportunistic spectrum access" for which the secondary user opportunistically utilizes the frequency bands of the primary user when these bands are not used. By doing so, the spectrum utilization efficiency can be increased drastically [191], [123].

To support opportunistic spectrum access, the secondary user is required to periodically sense the radio environment [168], [169], i.e., to detect whether the primary user is active or not. If the primary user is detected to be inactive, the secondary user can use the frequency band. On the other hand, while using the band, if the primary user is found to be active, the secondary user is required to vacate the channel within a certain amount of time. Therefore, spectrum sensing is of significant importance in cognitive radio networks.

There are two parameters associated with the performance of spectrum sensing: probability of detection and probability of false alarm. The physical meanings of these probabilities are quite interesting in a cognitive radio setup. In fact, the probability of detection defines the protection to the primary user. The probability of false alarm, on the other hand, reflects the missed opportunity for the secondary user to use the band when the primary user is inactive.

In [284], [285], [286], [287], sensing methods have been proposed based on the maximum eigenvalue of the covariance matrix of the received signals. The covariance matrix can be formed using multiple receive antennas or through oversampling the received signal. If the secondary user uses multiple antennas for spectrum sensing and when the primary user is inactive, the measurements the receivers collect form a random matrix. Since the maximum eigenvalue of the covariance matrix follows Tracy-Widom distribution [256], the probability of false alarm can thus be determined using RMT [285], [286]. Alternatively, to keep a target probability of false alarm, RMT can be used to determine the detection threshold used in the sensing algorithms.

In this and next chapters, our main objective is to show that there are indeed many wireless channels and problems which can be modeled using random matrices, and to review some of the typical applications of RMT in wireless communications. Since there are new results published out every year, we are not in a position to review all of the results. Instead, we concentrate on some of the representing examples and main results. Hopefully this part of the book can help the readers from both mathematical and engineering backgrounds to see the link between the two different societies, to identify new problems to work on, and to promote interdisciplinary collaborations.

The following notations are used in this and next chapters: $(\cdot)^T$ for transpose; $(\cdot)^*$ for conjugate transpose; $\mathsf{E}[\cdot]$ for expectation; $\mathrm{tr}(\cdot)$ for trace; $\mathsf{Var}(\cdot)$ for variance; $\|\cdot\|$ for the spectral norm of a matrix or the Euclidean norm of a vector; $\mathrm{diag}(\mathbf{x})$ for a diagonal matrix with diagonal elements being vector $\mathbf{x}$; $\mathbf{I}_k$ for $k \times k$ identity matrix.

6.2 Channel Models

6.2.1 Basics of Wireless Communication Systems

In this subsection, we provide a brief introduction on the basics of wireless communications. In particular, we are interested in the physical layer of a wireless communication system.

Figure 6.1 shows the block diagram of a wireless communication system, which consists of three basic parts: transmitter, channel and receiver. The objective of the transmitter design is to transform the information bits into a signal format which is suitable for transmission over the wireless channels. The key components in the transmitter side include channel coding, modulation, and linear or nonlinear precoding. When the signal passes through the channel, the signal strength will be attenuated due to propagation loss, shadowing and multipath fading, and the received signal waveform will be different from the transmitted signal waveform due to multipath delay, time/frequency

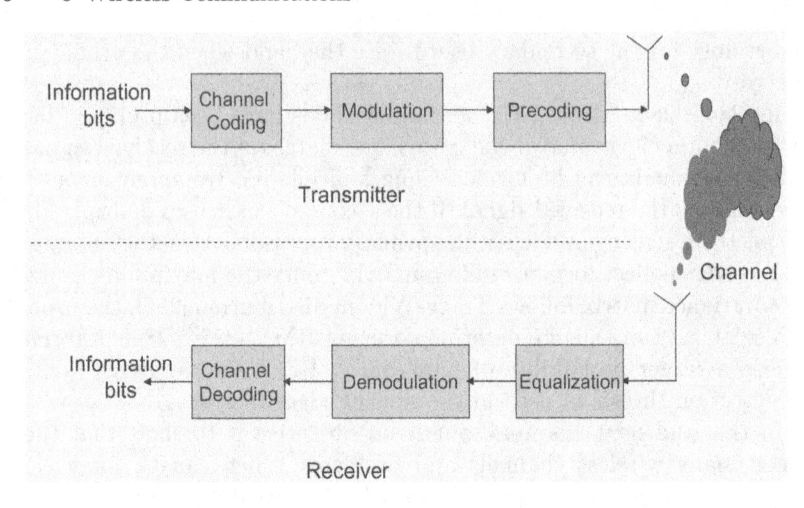

Fig. 6.1. Block diagram of wireless communication system.

selectivity of the channel, and the addition of noise and unwanted interference. Finally, at the receiver side, the transmitted information bits are to be recovered through the operations of equalization, demodulation and channel decoding.

With channel coding, the information bits are converted into coded bits with redundancy so that the effect of the channel noise and multipath fading is minimized. The modulation operation transforms the coded bits into modulated symbols with the purpose of achieving efficient transmission of the signal over the channel given a limited amount of bandwidth. The objective of the precoding operation is to provide robustness over the fading channel with multipath delay, or to compensate for the unwanted interference.

The equalization operation estimates the modulated symbols by removing the effect of the channel. With the help of the precoding operation, equalization sometimes becomes very simple. The demodulation operation converts the estimated symbols into a bit format, which is then used to recover the information bits through the channel decoding operation.

6.2.2 Matrix Channel Models

In this and next chapters, we consider the following input-output model arising from wireless communication systems:

$$\mathbf{x}(n) = \sum_{i=1}^{K} \mathbf{h}_i s_i(n) + \mathbf{u}(n) \tag{6.2.1}$$

$$= \mathbf{Hs}(\mathbf{n}) + \mathbf{u}(\mathbf{n}), \tag{6.2.2}$$

where

$$\mathbf{s}(n) = [s_1(n), s_2(n), \cdots, s_K(n)]^T \tag{6.2.3}$$

represents the transmitted signal vector of dimension $K \times 1$; $\mathbf{h}_i$ represents the channel vector of dimension $N \times 1$, corresponding to symbol $s_i(n)$;

$$\mathbf{x}(n) = [x_1(n), x_2(n), \cdots, x_N(n)]^T, \tag{6.2.4}$$

$$\mathbf{u}(n) = [u_1(n), u_2(n), \cdots, u_N(n)]^T, \tag{6.2.5}$$

denote the received signal vector, and received noise vector, respectively, both with dimension $N \times 1$; and

$$\mathbf{H} = [\mathbf{h}_1, \mathbf{h}_2, \cdots, \mathbf{h}_K] \tag{6.2.6}$$

is the $N \times K$ channel matrix. In (6.2.2), K and N are referred to as the signal dimension and observation dimension, respectively. The matrix model of 6.2.1 and (6.2.2) can be derived either in the time, frequency, space, or code domain, or any combinations of them. In the following subsections, we describe two popular matrix models in wireless communications: random matrix channels and linearly precoded channels.

6.2.3 Random Matrix Channels

The following three communications systems can be modeled using random matrix channels directly: DS-CDMA uplink, MIMO antenna systems and spatial division multiple access (SDMA) uplink.

DS-CDMA Uplink

In a DS-CDMA system, all users within the same cell communicate with a single base station using the same time and frequency resources. The transmission from the users to the base station is called uplink, while the transmission from the base station to the users is referred to as downlink. The block diagram of DS-CDMA uplink is illustrated in Figure 6.2. In order to achieve user differentiation, each user is assigned a unique spreading sequence. The matrix model in 6.2.1 and 6.2.2 directly represents the frequency-flat synchronous DS-CDMA uplink, where $s_i(n)$ and $\mathbf{h}_i$ represent the transmitted symbol and spreading sequence of user i, respectively. In this case, K and N denote the number of active users and processing gain, respectively. In the third generation (3G) wideband CDMA system, which is one of the 3G physical layer standards, the uplink spreading codes are designed as random codes, thus the equivalent propagation channel from the users to the base station can be modeled as a random matrix channel.

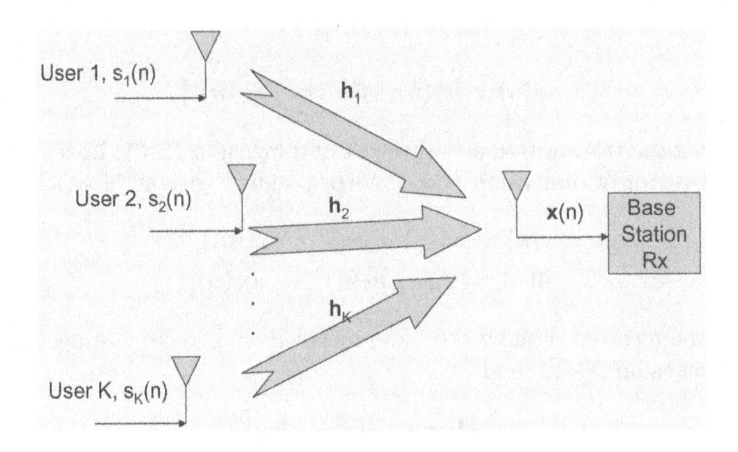

Fig. 6.2. Block diagram of DS-CDMA uplink.

MIMO Antenna Systems

Figure 6.3 shows the block diagram of a MIMO antenna system. The transmitter has K transmit antennas and the receiver has N receive antennas. The matrix model in 6.2.1 and 6.2.2 can also be used to represent such a system, where $s_i(n)$ and $\mathbf{h}_i$ denote respectively the transmitted symbol from the ith transmit antenna, and the channel responses from that transmit antenna to all receive antennas. While the MIMO channel modeling is related to the antenna configurations at both the transmitter and receiver sides, as well as to the multipath environment [204], when there are rich local scatters surrounding both sides, $\mathbf{h}_i$ can be simply modeled as an i.i.d. vector, thus the channel in (6.2.2) becomes a random matrix channel.

SDMA Uplink

In a SDMA system, the base station supports multiple users for simultaneous transmission using the same time and frequency resources [102], [213] [212]. This is achieved by equipping the base station with multiple antennas, and by doing so, the spatial channels for different users are different, which allows the signals from different users to be distinguishable. Figure 6.4 shows the block diagram of SDMA uplink, where K users communicate with the same base station equipped with N antennas. The matrix channel in 6.2.1 and 6.2.2 can be used to represent the uplink scenario where $s_i(n)$ and $\mathbf{h}_i$ denote respectively the transmitted symbol from user i and the channel responses from this user to all receive antennas at the base station. When the base station antennas and the users are surrounded with rich scatters, the channel matrix can be modeled as an i.i.d. matrix. Note SDMA and CDMA can be further combined to generate SDMA-CDMA systems [163], [162].

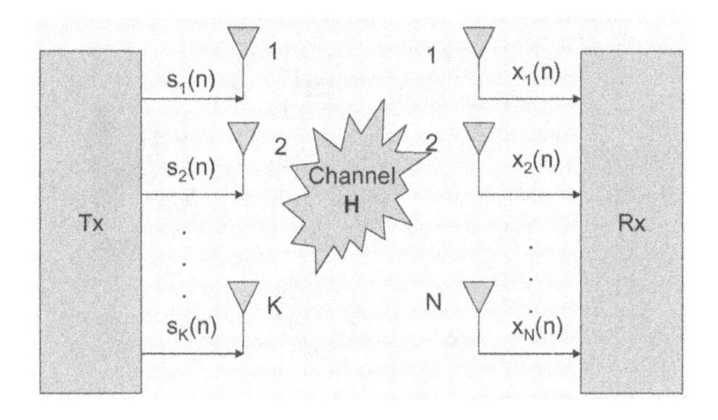

Fig. 6.3. Block diagram of MIMO antenna system.

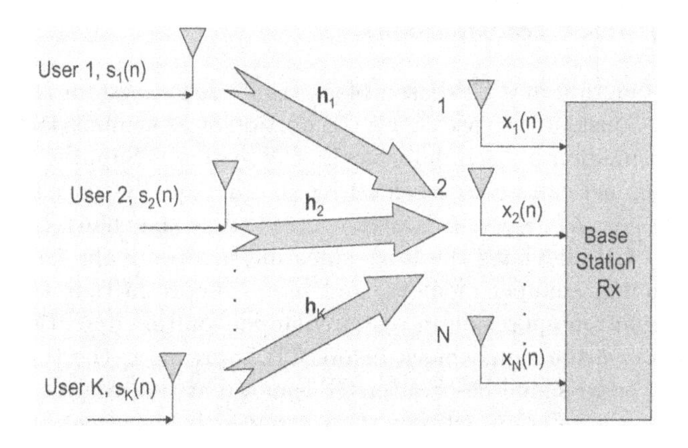

Fig. 6.4. Block diagram of SDMA uplink.

6.2.4 Linearly Precoded Systems

In broadband communications, the wireless channels usually have memories, thus are frequency selective. The frequency selectiveness of the channel introduces inter-symbol interferences (ISI) at the receiver side. Linear or nonlinear equalizers are designed to suppress the ISI. To simplify the complexity of the equalizers, linear precoding at the transmitter side can be applied. The matrix channel in 6.2.1 and 6.2.2 can be used to represent wireless channels of cyclic-prefix (CP) based block transmissions, which include, e.g., orthogonal frequency division multiplexing (OFDM), single carrier CP (SCCP) systems, multi-carrier CDMA (MC-CDMA) and CP-CDMA.

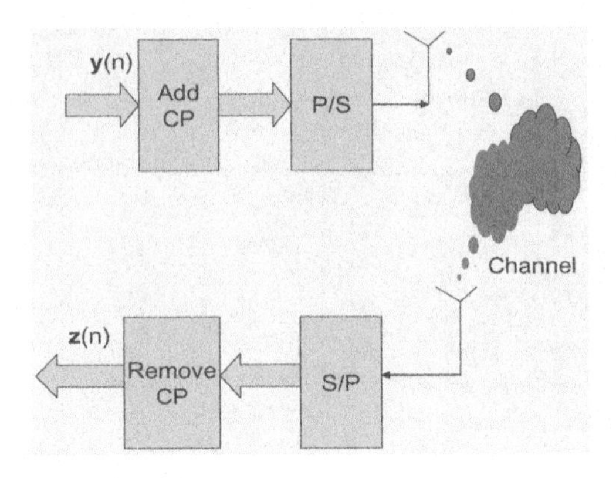

Fig. 6.5. Block diagram of CP-based block transmission system.

CP-based Block Transmissions

The block diagram of a CP-based block transmission system is illustrated in Figure 6.5. Consider the case that a CP portion of M symbols is inserted prior to the transmission of each data block of N symbols. Suppose the frequency selective channel can be represented by $(L+1)$ equally-spaced time domain taps, $h_0, h_1, \cdots, h_L$. Here, L is also referred to as the channel memory. The insertion of CP alleviates the inter-block interference if the CP length M is larger than the channel memory, and more importantly, it transforms the linear convolution operation into a circular convolution operation.

Let $\mathbf{y}(n)$ be the signal block before CP insertion at the transmitter and $\mathbf{z}(n)$ the received signal block after CP removal at the receiver, with circular convolution, the relation between $\mathbf{y}(n)$ and $\mathbf{z}(n)$ is given by

$$\mathbf{z}(n) = \mathbf{W}_N^* \mathbf{\Lambda}_N \mathbf{W}_N \mathbf{y}(n) + \tilde{\mathbf{u}}(n), \tag{6.2.7}$$

where $\mathbf{W}_N \in \mathcal{C}^{N \times N}$ is the $N \times N$ discrete Fourier transform matrix,

$$\mathbf{W} = \frac{1}{\sqrt{N}} \begin{bmatrix} 1 & 1 & \cdots & 1 \\ 1 & e^{\frac{-j2\pi}{N}} & \cdots & e^{\frac{-j2\pi \times (N-1)}{N}} \\ \cdots & & & \\ 1 & e^{\frac{-j2\pi \times (N-1)}{N}} & \cdots & e^{\frac{-j2\pi \times (N-1)(N-1)}{N}} \end{bmatrix} \tag{6.2.8}$$

and $\mathbf{\Lambda}_N = \text{diag}\{[f_0, \cdots, f_{N-1}]\}$ is the $N \times N$ diagonal matrix with $f_k = \sum_{l=0}^{L} h_l e^{-j\frac{2\pi kl}{N}}$.

The data block $\mathbf{y}(n)$ is the linear transform of the modulated block $\mathbf{s}(n)$ with size $K \times 1$ and is described as follows: $\mathbf{y}(n) = \mathbf{W}_N^* \mathbf{Q}_N \mathbf{s}(n)$, where $\mathbf{Q}_N \in \mathcal{C}^{N \times K}$ is the first precoding matrix. Performing discrete Fourier transform (DFT) on the CP-removed block $\mathbf{z}(n)$, we then have the following input-output relation

$$\mathbf{x}(n) = \mathbf{\Lambda}_N \mathbf{Q}_N \mathbf{s}(n) + \mathbf{u}(n), \tag{6.2.9}$$

where $\mathbf{x}(n) = \mathbf{W}_N \mathbf{z}(n)$ and $\mathbf{u}(n) = \mathbf{W}_N \tilde{\mathbf{u}}(n)$.

In (6.2.9), if we choose $\mathbf{Q}_N^* \mathbf{Q}_N = \mathbf{I}_K$, the system is referred to as an *isometrically precoded system*. If $\mathbf{Q}_N$ is chosen as an i.i.d. matrix, then the system is called *randomly precoded system*. Finally, if $\mathbf{\Lambda}_N = \mathbf{I}_N$ and $\mathbf{Q}_N$ is an i.i.d. matrix, the system is equivalent to a *random MIMO system* [260], [264].

Orthogonal Frequency Division Multiplexing (OFDM) Systems

In OFDM systems, we choose $K = N$ and $\mathbf{Q}_N = \mathbf{I}_N$. Thus, we have

$$\mathbf{y}(n) = \mathbf{W}_N^* \mathbf{s}(n), \tag{6.2.10}$$

$$\mathbf{x}(n) = \mathbf{\Lambda}_N \mathbf{s}(n) + \mathbf{u}(n). \tag{6.2.11}$$

In 6.2.11, since $\mathbf{\Lambda}_N$ is a diagonal matrix, the received signals have been completely decoupled, i.e.,

$$x_i(n) = f_i s_i(n) + u_i(n) \tag{6.2.12}$$

for $i = 0, \cdots, N-1$, where $x_i(n)$ and $u_i(n)$ are the ith elements of $\mathbf{x}(n)$ and $\mathbf{u}(n)$, respectively. Thus the signal detection problem for recovering the transmitted signals becomes very simple for implementation. Therefore, OFDM has become the most popular scheme to handle the ISI issue, and has been adopted in various wireless standards, e.g., IEEE802.11 wireless local area networks, IEEE802.16 wireless metropolitan area networks, and third generation long term evolution (3G-LTE).

Single Carrier CP (SCCP) Systems

In an OFDM system, from 6.2.10, $\mathbf{y}(n)$ is the IDFT output of the modulated symbols $\mathbf{s}(n)$, thus the signals transmitted out may suffer from high peak-to-average power ratio (PAPR), which causes difficulty in practical implementation. In a SCCP system, we directly choose $K = N$ and $\mathbf{y}(n) = \mathbf{s}(n)$, thus we have

$$\mathbf{x}(n) = \mathbf{\Lambda}_N \mathbf{W}_N \mathbf{s}(n) + \mathbf{u}(n). \tag{6.2.13}$$

Obviously, SCCP is an isometrically precoded system. From 6.2.13, it can be seen that the transmitted symbols are mixed up together, thus the equalization for SCCP is more complicated as compared to that for OFDM system. However, due to its simplicity in transmitter side and lower PAPR, SCCP has been adopted in IEEE802.16, and its multiuser version, interleaved frequency division multiple access (IFDMA) has been adopted in 3G-LTE uplink.

MC-CDMA

Single user scenario is considered in OFDM and SCCP systems. In order to support multiple users simultaneously, in the next two subsections, we will introduce downlink models for MC-CDMA and CP-CDMA. To do so, we use the following common notations: G for processing gain common to all users; T for the number of active users; $\mathbf{D}(n)$ for the long scrambling codes used at the nth block, where

$$\mathbf{D}(n) = \operatorname{diag}\{[d(n;0), \cdots, d(n;N-1)]\}, \tag{6.2.14}$$

with $|d(n;k)| = 1$; $\mathbf{c}_i$ for the short codes of user i, where

$$\mathbf{c}_i = [c_i(0), \cdots, c_i(G-1)]^T \tag{6.2.15}$$

with $\mathbf{c}_i^* \mathbf{c}_j = 1$ for $i = j$, and $\mathbf{c}_i^* \mathbf{c}_j = 0$ for $i \neq j$. From now on, we look at the channel model from the base station to one particular mobile user.

MC-CDMA performs frequency domain spreading by transmitting the chip signals associated with each modulated symbol over different sub-carriers within the same time block [122]. Denote Q as the number of symbols transmitted in one block for each user, G as the processing gain, T as the number of users, and $N = QG$ as the total number of sub-carriers. There are then $K = TQ$ multiuser symbols in each block.

At the receiver side, the nth received block after CP removal and FFT operation can be represented as

$$\mathbf{x}(n) = \mathbf{\Lambda}_N \mathbf{D}(n) \mathbf{C} \mathbf{s}(n) + \mathbf{u}(n), \tag{6.2.16}$$

where

$$\mathbf{x}(n) = [x(n;0), \cdots, x(n;N-1)]^T, \tag{6.2.17}$$

$$\mathbf{s}(n) = \left[\bar{\mathbf{s}}_1^T(n), \cdots, \bar{\mathbf{s}}_Q^T(n)\right]^T, \tag{6.2.18}$$

$$\mathbf{u}(n) = [u(n;0), \cdots, u(n;N-1)]^T, \tag{6.2.19}$$

$$\mathbf{C} = \operatorname{diag}\{\bar{\mathbf{C}}, \cdots, \bar{\mathbf{C}}\}, \tag{6.2.20}$$

with $\bar{\mathbf{s}}_i(n) = [s_0(n;i), \cdots, s_{T-1}(n;i)]^T$, and $\bar{\mathbf{C}} = [\mathbf{c}_0, \cdots, \mathbf{c}_{T-1}]$.

CP-CDMA

CP-CDMA is the single carrier dual of MC-CDMA. The $Q = N/G$ symbols of each user are first spread out with user-specific spreading codes, then the chip sequence for all users are summed up; the total chip signal of size N is then passed to CP inserter, which adds a CP. Using the duality between CP-CDMA and MC-CDMA, from (6.2.16), the nth received block of CP-CDMA after FFT can be written as

$$\mathbf{x}(n) = \mathbf{\Lambda}_N \mathbf{W}_N \mathbf{D}(n) \mathbf{C} \mathbf{s}(n) + \mathbf{u}(n), \tag{6.2.21}$$

where $\mathbf{x}(n)$, $\mathbf{C}$, $\mathbf{s}(n)$ and $\mathbf{u}(n)$ are defined same as in MC-CDMA case. Again, there are $P = TQ$ multiuser symbols in each block.

For MC-CDMA downlink, $\mathbf{Q}_N = \mathbf{D}(n)\mathbf{C}$, and for CP-CDMA downlink, $\mathbf{Q}_N = \mathbf{W}_N \mathbf{D}(n)\mathbf{C}$. Since $\mathbf{Q}_N^* \mathbf{Q}_N = \mathbf{I}_K$, both systems belong to the isometrically precoded category.

6.3 Channel Capacity for MIMO Antenna Systems

Channel capacity is a fundamental performance indicator used in communication theory study; it describes the maximum rate of data transmission that the channel can support with an arbitrarily small probability of error incurred due to the channel impairment. The channel capacity for additive white Gaussian noise channels was derived by Claude Shannon in 1948 [71]. For single-input single-output systems, the capacity limits for fading channels have been well-documented in, e.g., [111], [45], [59], [44]. In this section, we consider the channel capacity of MIMO antenna systems in the fading channel environment.

6.3.1 Single-Input Single-Output Channels

Let us first consider the following AWGN channel:

$$x(n) = s(n) + u(n), \tag{6.3.22}$$

and assume that: (i) the transmitted signal $s(n)$ is zero-mean i.i.d. Gaussian with $\mathsf{E}[|s(n)|^2] = E_s$; (ii) the noise $u(n)$ is zero-mean i.i.d. Gaussian with $\mathsf{E}[|u(n)|^2] = \sigma^2$, and denote $\Gamma = E_s/\sigma^2$ as the signal-to-noise ratio (SNR) of the channel.

The capacity of the channel is determined by the mutual information between the input and output, which is given by

$$C = \log_2(1 + \Gamma). \tag{6.3.23}$$

Here the unit of capacity is bits per second per Hertz (bits/sec/Hz). In the high SNR regime, the channel capacity increases by 1 bit/sec/Hz for every 3 dB increase in SNR. Note the channel capacity determines the maximum rate of codes that can be transmitted over the channel and recovered with arbitrarily small error.

Next, we consider the following SISO block fading channel:

$$x(n) = hs(n) + u(n) \tag{6.3.24}$$

and assume that: (i) the transmitted signal $s(n)$ is zero-mean i.i.d. Gaussian with $\mathsf{E}[|s(n)|^2] = E_s$; (ii) the noise $u(n)$ is zero-mean i.i.d. Gaussian with

$\mathsf{E}[|u(n)|^2] = \sigma^2$, and denote $\Gamma = E_s/\sigma^2$; (iii) the fading state h is a random variable with $\mathsf{E}[|h|^2] = 1$.

Let us first introduce the concept of a "block fading channel." A block fading channel refers to a slow fading channel whose coefficient is constant over an interval of large time T, and changes to another independent value which is again constant over an interval of time T, and so on. The *instantaneous* mutual information between $s(n)$ and $x(n)$ of channel (6.3.24) conditional on channel state h is given by

$$I(s; x|h) = \log_2(1 + |h|^2 \Gamma). \tag{6.3.25}$$

Since h is a random variable, the instantaneous mutual information is also a random variable. Thus if the distribution of $|h|^2$ is known, the distribution of $I(s; x|h)$ can be calculated accordingly.

The channel capacity of a fading channel can be quantified either in an ergodic sense or in an outage sense, yielding *ergodic capacity* and *outage capacity.*

The ergodic capacity of the SISO fading channel (6.3.24) is defined as

$$C = \mathsf{E}[\log_2(1 + |h|^2 \Gamma)], \tag{6.3.26}$$

where the expectation is taken over the channel state variable h. Physically speaking, the ergodic capacity defines the maximum (constant) rate of codes which can be transmitted over the channel and recovered with arbitrarily small probability of error when the codes are long enough to cover the all possible channel states.

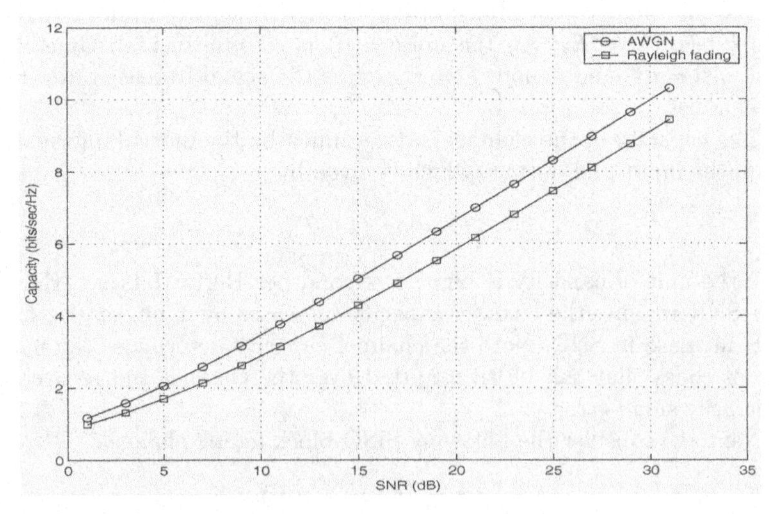

Fig. 6.6. The capacity comparison for AWGN channels and SISO Rayleigh fading channels

In Figure 6.6, we compare the capacities of the AWGN channel and the SISO Rayleigh fading channel with respect to the received SNR. Here, for the fading channel case, we have used the average received SNR. It can be seen that at high SNR, the capacity of the fading channel increases by 1 bit/sec/Hz for every 3 dB increase in SNR, which is the same as the AWGN channel.

Since the instantaneous mutual infirmation is a random variable, if a code with constant rate C_0 is transmitted over the fading channel, this code cannot be correctly recovered at the receiver at a fading block whose instantaneous mutual information is lower than the code rate C_0, thus causing an outage event. We define the *outage probability* as the probability that the instantaneous mutual information is less than the rate of C_0, i.e.,

$$P_{\text{out}}(C_0) = \Pr(I(s;x|h) < C_0) \tag{6.3.27}$$

Based on this, the $\mathsf{q}\%$ *outage capacity* $C_{\text{out},\mathsf{q}\%}$ is defined as the maximum information rate of codes transmitted over the fading channel for which the outage probability does not exceed $\mathsf{q}\%$.

6.3.2 MIMO Fading Channels

In order to analyze the capacity of MIMO fading channels, we make the following assumptions:

(A1) The channel vectors $\mathbf{h}_i$'s can be represented as

$$\mathbf{h}_i = [X_{1i}, X_{2i}, \cdots, X_{Ni}]^T, \tag{6.3.28}$$

for $i = 1, \cdots, K$, where X_{ki}'s are i.i.d. random variables with zero mean and unit variance, i.e., $\mathsf{E}[|X_{ki}|^2] = 1$ for all k's and i's.

(A2) The K symbols constituting the transmitted signal vector are drawn from a *Gaussian codebook* with $\mathbf{R}_x = \mathsf{E}[\mathbf{x}(n)\mathbf{x}^*(n)]$ and $\mathsf{Tr}(\mathbf{R}_x) = E_s$. This is the total power constraint.

(A3) The elements of $\mathbf{u}(n)$ are zero-mean, circularly symmetric complex Gaussian with $\mathbf{R}_u = \mathsf{E}[\mathbf{u}(n)\mathbf{u}^*(n)] = \sigma^2\mathbf{I}$.

The following two cases need to be considered separately when we study the MIMO channel capacity. In the first case, the channel state information (CSI) $\mathbf{H}$ is available at the transmitter side. This case is called *CSI-known case*. In the second case, the CSI is unavailable at the transmitter side. This case is referred to as *CSI-unknown case*. In both cases, we assume that the CSI is perfectly known at the receiver side.

Let us first look at the instantaneous mutual information under one CSI realization $\mathbf{H}$. Based on the distribution of $\mathbf{H}$, the distribution of the instantaneous mutual information can be derived. Similar to the SISO fading channel case, we study the ergodic capacity and outage capacity of MIMO fading channels.

Using singular value decomposition (SVD), the $N \times K$ channel matrix $\mathbf{H}$ can be represented as follows:

$$\mathbf{H} = \mathbf{U}\mathbf{\Sigma}\mathbf{V}^*,$$

where matrix $\mathbf{U}$ is a $N \times N$ unitary matrix ($\mathbf{U}^*\mathbf{U} = \mathbf{I}_N$), and is called the left singular vector matrix of $\mathbf{H}$; matrix $\mathbf{V}$ is a $K \times K$ unitary matrix ($\mathbf{V}^*\mathbf{V} = \mathbf{I}_K$), which is referred to as the right singular vector matrix of $\mathbf{H}$; and $\mathbf{\Sigma}$ is the $N \times K$ singular value matrix, the elements of which are zeros except that $(\mathbf{\Sigma})_{i,i} = \sigma_i \geq 0$, where $\sigma_1 \geq \ldots \geq \sigma_M \geq 0$ ($M = \min(K, N)$) are the singular values of $\mathbf{H}$. Note $\mathbf{H}\mathbf{H}^* = \mathbf{U}\mathbf{\Sigma}\mathbf{\Sigma}^*\mathbf{U}^*$, thus $\lambda_i = \sigma_i^2$, $\forall\ i$ are the non-zero eigenvalues of $\mathbf{H}\mathbf{H}^*$.

MIMO Fading Channels for CSI-Known Case

When CSI is available at the transmitter, joint transmit and receive beamforming can be used to decouple the MIMO fading channel into M SISO fading channels. Figure 6.7 shows the block diagram of the decoupling process.

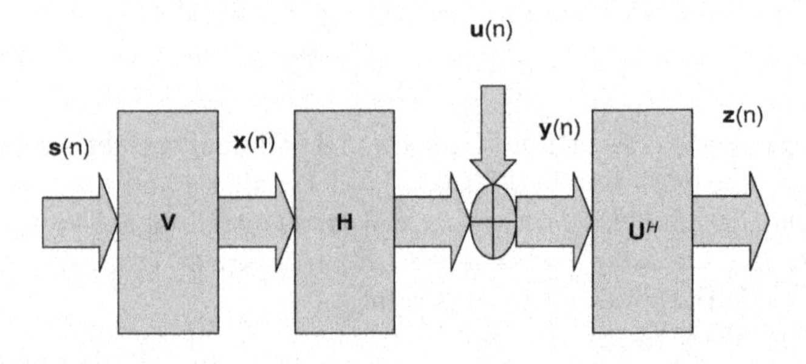

Fig. 6.7. Block diagram of joint transmit and receive eigen-beamforming.

At the transmitter side, we precode the transmitted signals using transmit beamforming:

$$\mathbf{x}(n) = \mathbf{V}\mathbf{s}(n). \tag{6.3.29}$$

Then the received signal vector is given by

$$\begin{aligned}\mathbf{y}(n) &= \mathbf{U\Sigma V}^*\mathbf{V}\mathbf{s}(n) + \mathbf{u}(n) \\ &= \mathbf{U\Sigma s}(n) + \mathbf{u}(n).\end{aligned} \tag{6.3.30}$$

At the receiver side, if we pre-multiply $\mathbf{y}(n)$ with $\mathbf{U}^*$ (this processing is called *receive eigen-beamforming*), we then have

$$\begin{aligned}\mathbf{z}(n) &= \mathbf{\Sigma s}(n) + \mathbf{U}^*\mathbf{u}(n) \\ &= \mathbf{\Sigma s}(n) + \tilde{\mathbf{u}}(n).\end{aligned} \tag{6.3.31}$$

Note $\tilde{\mathbf{u}}(n) = \mathbf{U}^*\mathbf{u}(n)$, thus $\mathbf{R}_{\tilde{u}} = \mathsf{E}[\tilde{\mathbf{u}}(n)\tilde{\mathbf{u}}^*(n)] = \mathsf{E}[\mathbf{U}^*\mathbf{u}(n)\mathbf{u}^*(n)\mathbf{U}] = \sigma^2\mathbf{I}$. Therefore, we have

$$z_i(n) = \sigma_i s_i(n) + \tilde{u}_i(n), \quad i = 1, \ldots, M. \tag{6.3.32}$$

From the above, it can be seen that with CSI-known, the MIMO channel has been decoupled into M SISO channels through joint transmit and receive beamforming. That is to say, M data streams can be transmitted in a parallel manner.

Suppose the transmission power to the ith data stream is $\gamma_i = \mathsf{E}[|s_i(n)|^2]$, then the SNR for this data stream is given by:

$$SNR_i = \frac{\sigma_i^2 \mathsf{E}[|s_i(n)|^2]}{\mathsf{E}[|\tilde{u}_i(n)|^2]} = \frac{\sigma_i^2\gamma_i}{\sigma^2}. \tag{6.3.33}$$

Note the total power over the M data streams has to be less than or equal to E_s, i.e., $\sum_{i=1}^M \gamma_i \le E_s$.

The capacity of the MIMO channel under channel state $\mathbf{H}$ is equal to the sum of the individual SISO channel's capacity:

$$I_{\mathbf{H}} = \sum_{i=1}^M \log_2\left(1 + \frac{\sigma_i^2\gamma_i}{\sigma^2}\right) = \sum_{i=1}^M \log_2\left(1 + \frac{\lambda_i\gamma_i}{\sigma^2}\right) \tag{6.3.34}$$

under the power constraint: $\sum_{i=1}^M \gamma_i \le E_s$.

If equal power is allocated to each data stream, i.e., $\gamma_i = \frac{E_s}{M}$, then

$$I_{\mathbf{H}}^{(\mathsf{EP})} = \sum_{i=1}^M \log_2\left(1 + \frac{\lambda_i E_s}{M\sigma^2}\right). \tag{6.3.35}$$

In order to achieve maximum capacity, we can allocate different power to the data streams using the Lagrangian method. Define the objective function:

$$J(\gamma_1, \cdots, \gamma_M) = \sum_{i=1}^M \log_2\left(1 + \frac{\lambda_i\gamma_i}{\sigma^2}\right) - \mu_1\left(\sum_{i=1}^M \gamma_i - E_s\right). \tag{6.3.36}$$

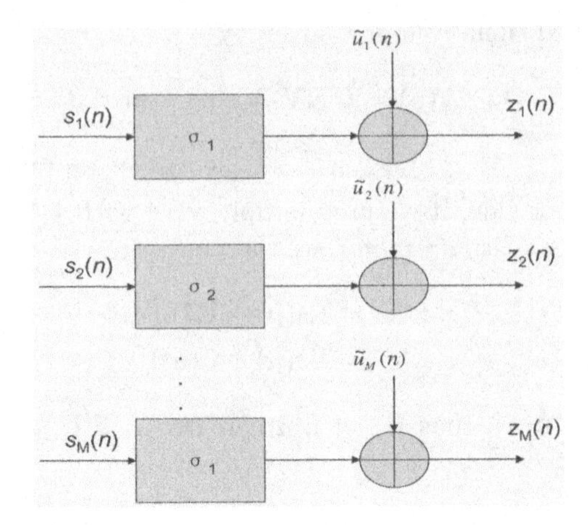

Fig. 6.8. A MIMO channel is equivalent to a set of parallel SISO channels.

Calculating $\frac{\partial J(\gamma_1,\cdots,\gamma_M)}{\partial \gamma_i}$ and setting $\frac{\partial J(\gamma_1,\cdots,\gamma_M)}{\partial \gamma_i} = 0$ for all i's, we obtain the following water-filling solution:

$$\gamma_i^* = \left(\mu - \frac{\sigma^2}{\lambda_i}\right)^+, \quad i = 1, \cdots, M, \tag{6.3.37}$$

where μ is the water level for which the equality power constraint is satisfied, and $(x)^+ = x$ for $x \geq 0$, and $(x)^+ = 0$ for $x < 0$. With the water-filling power allocation, the channel capacity can then be calculated as follows:

$$I_{\mathbf{H}}^{(\mathtt{WF})} = \sum_{i=1}^{M} \log_2 \left(1 + \frac{\lambda_i \gamma_i^*}{\sigma^2}\right). \tag{6.3.38}$$

For block fading channels, both $I_{\mathbf{H}}^{(\mathtt{EP})}$ and $I_{\mathbf{H}}^{(\mathtt{WF})}$ are random variables. If the distribution of the eigenvalues is known, we can then calculate the distributions of $I_{\mathbf{H}}^{(\mathtt{EP})}$ and $I_{\mathbf{H}}^{(\mathtt{WF})}$.

Taking expectations on $I_{\mathbf{H}}^{(\mathtt{EP})}$ and $I_{\mathbf{H}}^{(\mathtt{WF})}$ over the random matrix $\mathbf{H}$, we obtain the ergodic capacities of the MIMO fading channels as follows:

$$C^{(\mathtt{EP})} = \mathsf{E}_{\mathbf{H}}[I_{\mathbf{H}}^{(\mathrm{EP})}] = \mathsf{E}_{\mathbf{\Lambda}}\left[\sum_{i=1}^{M} \log_2 \left(1 + \frac{\lambda_i E_s}{M\sigma^2}\right)\right], \tag{6.3.39}$$

$$C^{(\mathtt{WF})} = \mathsf{E}_{\mathbf{H}}[I_{\mathbf{H}}^{(\mathrm{WF})}] = \mathsf{E}_{\mathbf{\Lambda}}\left[\sum_{i=1}^{M} \log_2 \left(1 + \frac{\lambda_i \gamma_i^*}{\sigma^2}\right)\right]. \tag{6.3.40}$$

Since $I_{\mathbf{H}}^{(\mathtt{WF})} \geq I_{\mathbf{H}}^{(\mathtt{EP})}$, thus $C^{(\mathtt{WF})} \geq C^{(\mathtt{EP})}$. In the high SNR regime, however, these two capacities tend to be equal.

The expressions of the ergodic capacities can be derived based on the eigenvalue distributions [166]. Alternatively, we may quantify the performance gain of using multiple antennas by looking at lower-bound of the ergodic capacity. In fact, for CSI-known case with equal power allocation, a lower-bound of the ergodic capacity of MIMO Rayleigh fading channels is given by ([199]):

$$C = C(\Gamma) \geq M \log_2 \left[1 + \frac{\Gamma}{M} \exp\left(\frac{1}{M} \sum_{j=1}^{M} \sum_{p=1}^{Q-j} \frac{1}{p} - \gamma \right) \right], \quad (6.3.41)$$

where $\gamma \approx 0.57721566$ is the Euler's constant, $\Gamma = \frac{E_s}{\sigma^2}$ and $Q = \max(K, N)$.

Let us define spatial multiplexing gain as: $r = \lim_{\Gamma \to \infty} \frac{C(\Gamma)}{\log_2(\Gamma)}$. From (6.3.41), we can see that $r = \lim_{\Gamma \to \infty} \frac{C(\Gamma)}{\log_2(\Gamma)} = M$. Therefore, in the high SNR regime, the ergodic capacity increases by M bits/sec/Hz for every 3 dB increase in the average SNR, Γ. Recall that for SISO AWGN channels and SISO Rayleigh fading channels, the capacity increases by 1 bit/sec/Hz for every 3 dB increase of SNR, in the high SNR regime. This shows the tremendous capacity gain by using MIMO antenna systems.

MIMO Fading Channels for CSI-Unknown Case

In the previous subsection, we have proved that, when the CSI is known at the transmitter side, the spatial multiplexing gain for a $N \times K$ MIMO channel is equal to M, which is the minimum of the numbers of transmit and receive antennas. In practice, the CSI may not be available at the transmitter, can we still achieve a spatial multiplexing gain of M? We deal with this question in this subsection.

For MIMO flat fading channels, when the input signals are i.i.d. Gaussian, the instantaneous mutual information between $\mathbf{x}(n)$ and $\mathbf{y}(n)$ under channel state $\mathbf{H}$ is given by ([96], [249]):

$$I_{\mathbf{H}} = \log_2 \{\det(\pi e \mathbf{R}_y)\} - \log_2 \{\det(\pi e \mathbf{R}_u)\}. \quad (6.3.42)$$

Since $\mathbf{R}_y = \mathsf{E}[\mathbf{y}(n)\mathbf{y}^*(n)] = \mathbf{R}_u + \mathbf{H}\mathbf{R}_x\mathbf{H}^*$ and $\mathbf{R}_u = \sigma^2 \mathbf{I}_N$, we have

$$I_{\mathbf{H}} = \log_2 \left\{ \det \left(\mathbf{I}_N + \frac{\mathbf{H}\mathbf{R}_x\mathbf{H}^*}{\sigma^2} \right) \right\}. \quad (6.3.43)$$

For CSI-unknown case, we will choose the transmitted signal vector $\mathbf{x}$ such that $\mathbf{R}_x = \frac{E_s}{K}\mathbf{I}_K$. Thus if $K \geq N$[1],

---

[1] Suppose matrices $\mathbf{A}$ and $\mathbf{B}$ are of dimensions $m \times n$ and $n \times m$, respectively, we have the equality $\det(\mathbf{I}_m + \mathbf{AB}) = \det(\mathbf{I}_n + \mathbf{BA})$.

$$
\begin{aligned}
I_{\mathbf{H}} &= \log_2\left\{\det\left(\mathbf{I}_N + \frac{E_s}{K\sigma^2}\mathbf{H}\mathbf{H}^*\right)\right\} \\
&= \log_2\left\{\det\left(\mathbf{I}_N + \frac{E_s}{K\sigma^2}\mathbf{U}\boldsymbol{\Sigma}\boldsymbol{\Sigma}^*\mathbf{U}^*\right)\right\} \\
&= \log_2\left\{\det\left(\mathbf{I}_N + \frac{E_s}{K\sigma^2}\boldsymbol{\Sigma}\boldsymbol{\Sigma}^*\mathbf{U}^*\mathbf{U}\right)\right\} \\
&= \sum_{i=1}^{M}\log_2\left(1+\frac{E_s\sigma_i^2}{K\sigma^2}\right) = \sum_{i=1}^{M}\log_2\left(1+\frac{E_s\lambda_i}{K\sigma^2}\right) = I_{\boldsymbol{\Lambda}}, \qquad (6.3.44)
\end{aligned}
$$

where again, σ_i is the ith singular value of $\mathbf{H}$ and λ_i is the ith eigenvalue of $\mathbf{H}\mathbf{H}^*$, respectively. On the other hand, if $K < N$,

$$
\begin{aligned}
I_{\mathbf{H}} &= \log_2\left\{\det\left(\mathbf{I}_N + \frac{E_s}{K\sigma^2}\mathbf{H}\mathbf{H}^*\right)\right\} \\
&= \log_2\left\{\det\left(\mathbf{I}_K + \frac{E_s}{K\sigma^2}\mathbf{H}^*\mathbf{H}\right)\right\} \\
&= \log_2\left\{\det\left(\mathbf{I}_K + \frac{E_s}{K\sigma^2}\mathbf{V}\boldsymbol{\Sigma}^*\boldsymbol{\Sigma}\mathbf{V}^*\right)\right\} \\
&= \log_2\left\{\det\left(\mathbf{I}_K + \frac{E_s}{K\sigma^2}\boldsymbol{\Sigma}^*\boldsymbol{\Sigma}\mathbf{V}^*\mathbf{V}\right)\right\} \\
&= \sum_{i=1}^{M}\log_2\left(1+\frac{E_s\lambda_i}{K\sigma^2}\right) = I_{\boldsymbol{\Lambda}}. \qquad (6.3.45)
\end{aligned}
$$

Combining (6.3.44) with (6.3.45) yields

$$
I_{\mathbf{H}} = \sum_{i=1}^{M}\log_2\left(1+\frac{E_s\lambda_i}{K\sigma^2}\right). \qquad (6.3.46)
$$

When $K \leq N$, the above formula is the same as (6.3.35). That is to say, if the number of transmit antennas is not greater than the number of receive antennas, even when CSI is unknown at the transmitter, the capacity of the MIMO channel is the same as that for CSI-known case when equal power allocation is applied.

Taking expectation on $I_{\mathbf{H}}$ over the random matrix $\mathbf{H}$, we obtain the ergodic capacity of the MIMO fading channel as follows

$$
C = \mathsf{E}_{\mathbf{H}}[I_{\mathbf{H}}] = \mathsf{E}_{\boldsymbol{\Lambda}}\left[\sum_{i=1}^{M}\log_2\left(1+\frac{E_s\lambda_i}{K\sigma^2}\right)\right]. \qquad (6.3.47)
$$

For MIMO Rayleigh fading channels, similar to the CSI-known case, we have the following inequality

$$C \geq M \log_2 \left[1 + \frac{E_s}{K\sigma^2} \exp\left(\frac{1}{M} \sum_{j=1}^{M} \sum_{p=1}^{Q-j} \frac{1}{p} - \gamma \right) \right]. \tag{6.3.48}$$

Thus a spatial multiplexing gain of M can be achieved for MIMO Rayleigh fading channels even when the CSI is unavailable at the transmitter side.

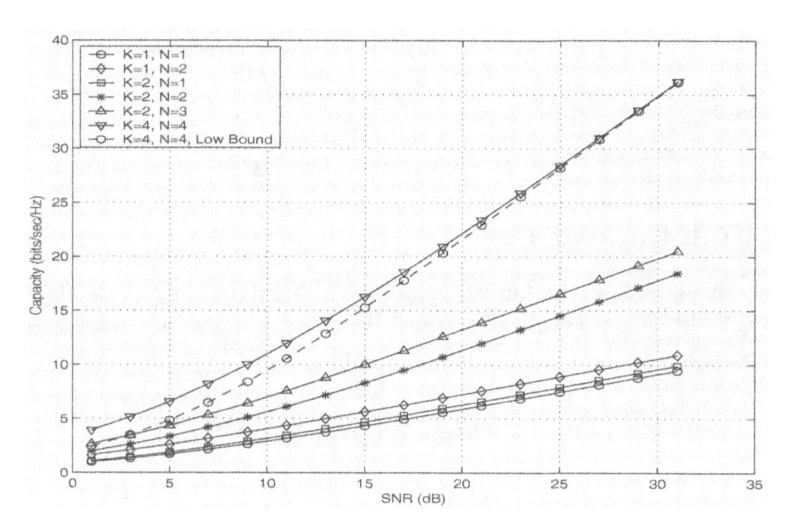

Fig. 6.9. The ergodic capacity comparison for MIMO Rayleigh fading channels with different numbers of antennas.

Figure 6.9 illustrates the ergodic capacities of MIMO fading channels with different antenna configurations for CSI-unknown case. Here we also plot the lower bound of the ergodic capacity for 4×4 MIMO channels. It is seen that this lower bound is tight when the SNR is greater than 25dB. Further, for asymmetric antenna configurations, for a given M and at high SNR regime, there exists a fixed SNR loss when the transmit antenna number is larger than the receive antenna number.

6.4 Limiting Capacity of Random MIMO Channels

In the previous section, we have studied the ergodic capacity of MIMO fading channels with limited dimensions of K and N, and have shown that for random matrix channel with Rayleigh fading coefficients, the MIMO channel achieves a spatial multiplexing gain of M, which is the minimum of the transmit antenna number and receive antenna number.

In this section, we are interested in the limiting performance of the instantaneous mutual information for any given random channel realization, when

$K \to \infty$, $N \to \infty$ with $\frac{K}{N} \to \alpha$ (constant). For the sake of brevity, we assume that K and N scale up with the same speed, i.e., $\frac{K}{N} = \alpha$ for every N.

According to the spectral theory of large random matrices, under assumption (A1) in Section 6.3.2 and for the limiting case, the empirical distribution of the eigenvalues of $\frac{1}{N}\mathbf{HH}^*$ converges almost surely to Marčenko-Pastur law, whose density function is given by

$$f_\alpha(x) = (1-\alpha)^+\delta(x) + \frac{\sqrt{(x-a)^+(b-x)^+}}{2\pi x}, \tag{6.4.49}$$

where

$$a = (1-\sqrt{\alpha})^2, \quad b = (1+\sqrt{\alpha})^2. \tag{6.4.50}$$

6.4.1 CSI-Unknown Case

Now, let us look at the limiting capacity of MIMO fading channels with CSI unavailable at the transmitter side. When $\alpha \geq 1$, we are interested in the following normalized mutual information

$$\begin{aligned}\tilde{I}_{\mathbf{H}} &= \frac{1}{N}\log_2\left\{\det\left(\mathbf{I}_N + \frac{E_s}{K\sigma^2}\mathbf{HH}^*\right)\right\} \\ &= \frac{1}{N}\sum_{i=1}^{N}\log_2\left(1+\frac{E_s\lambda_i}{K\sigma^2}\right).\end{aligned} \tag{6.4.51}$$

In the previous section we have defined λ_i's as the non-zero eigenvalues of $\mathbf{HH}^*$. In the limiting case, the empirical distributions of $\tilde{\lambda}_i = \frac{\lambda_i}{N}$, $i = 1, \cdots, N$, converges to $f_\alpha(x)$ almost surely for $x \in (a,b)$. Thus we have the following ([264]):

$$\begin{aligned}\tilde{I}_{\mathbf{H}} &= \frac{1}{N}\sum_{i=1}^{N}\log_2\left(1+\frac{NE_s\tilde{\lambda}_i}{K\sigma^2}\right) \\ &\to \int_a^b \log_2(1+\bar{\Gamma}x)f_\alpha(x)dx, \qquad (6.4.52) \\ &\triangleq \tilde{I}_1(\Gamma,\alpha), \qquad (6.4.53)\end{aligned}$$

where $\bar{\Gamma} = \frac{1}{\alpha}\frac{E_s}{\sigma^2}$. The above limit is termed the Shannon transform of the Marcenko-Pastur law, and the closed-form expression for the limit can be found in [262]. Note $\tilde{I}_1(\Gamma,\alpha)$ is defined for the case $\alpha \geq 1$.

Analogously, when $\alpha < 1$, we have the following

$$\begin{aligned}\tilde{I}_{\mathbf{H}} &= \frac{1}{K}\log_2\left\{\det\left(\mathbf{I}_N + \frac{E_s}{K\sigma^2}\mathbf{HH}^*\right)\right\} \\ &= \frac{1}{K}\log_2\left\{\det\left(\mathbf{I}_K + \frac{E_s}{K\sigma^2}\mathbf{H}^*\mathbf{H}\right)\right\}\end{aligned}$$

$$= \frac{1}{K}\sum_{i=1}^{K}\log_2\left(1+\frac{E_s\lambda_i}{K\sigma^2}\right). \tag{6.4.54}$$

When $K < N$, the nonzero eigenvalues of $\mathbf{HH}^*$ are the same as the eigenvalues of $\mathbf{H}^*\mathbf{H}$. Thus in the limiting case, the empirical distributions of the eigenvalues of $\frac{1}{K}\mathbf{H}^*\mathbf{H}$, i.e., $\tilde{\lambda}_i = \frac{\lambda_i}{K}$ for $i = 1, \cdots, K$, converges to $f_{\tilde{\alpha}}(x)$ almost surely for $x \in (\tilde{a}, \tilde{b})$, where $\tilde{\alpha} = \frac{1}{\alpha}$, $\tilde{a} = (1-\sqrt{\tilde{\alpha}})^2$ and $\tilde{b} = (1+\sqrt{\tilde{\alpha}})^2$. Thus we have the following ([264]):

$$\tilde{I}_{\mathbf{H}} \to \int_{\tilde{a}}^{\tilde{b}} \log_2(1+\Gamma x) f_{\tilde{\alpha}}(x)dx, \tag{6.4.55}$$

$$\triangleq \tilde{I}_2(\Gamma, \alpha), \tag{6.4.56}$$

where $\Gamma = \frac{E_s}{\sigma^2}$. Note $\tilde{I}_2(\Gamma, \alpha)$ is defined for the case $\alpha < 1$.

Let us compare the limiting capacities of MIMO random channels with two asymmetric antenna configurations: one is the case with $K/N = \alpha_1 \geq 1$, the other is the case with $K/N = \alpha_2 = 1/\alpha_1 \leq 1$. From 6.4.52 and 6.4.55, for a given pair of (α_1, α_2), we have the following equality:

$$\tilde{I}_1(\Gamma, \alpha_1) = \tilde{I}_2(\Gamma/\alpha_1, \alpha_2). \tag{6.4.57}$$

That is to say, if there are more antennas at transmitter side, a SNR loss α_1 exists due to the unavailability of CSI at the transmitter side.

6.4.2 CSI-Known Case

When $\alpha \geq 1$ ($K \geq N$ and $M = N$), we are interested in the following normalized mutual information

$$\begin{aligned}
\tilde{I}_{\mathbf{H}} &= \frac{1}{N}\log_2\left\{\det\left(\mathbf{I}_N + \frac{E_s}{N\sigma^2}\mathbf{HH}^*\right)\right\} \\
&= \frac{1}{N}\sum_{i=1}^{N}\log_2\left(1+\frac{E_s\lambda_i}{N\sigma^2}\right) \\
&\to \int_a^b \log_2(1+\Gamma x) f_\alpha(x)dx && (6.4.58) \\
&\triangleq \tilde{I}_3(\Gamma, \alpha). && (6.4.59)
\end{aligned}$$

Here, $\tilde{I}_3(\Gamma, \alpha)$ is defined for the case $\alpha \geq 1$.

Analogously, when $\alpha < 1$, we have the following

$$\begin{aligned}
\tilde{I}_{\mathbf{H}} &= \frac{1}{K}\log_2\left\{\det\left(\mathbf{I}_N + \frac{E_s}{K\sigma^2}\mathbf{HH}^*\right)\right\} \\
&= \frac{1}{K}\log_2\left\{\det\left(\mathbf{I}_K + \frac{E_s}{K\sigma^2}\mathbf{H}^*\mathbf{H}\right)\right\}
\end{aligned}$$

$$= \frac{1}{K}\sum_{i=1}^{K}\log_2\left(1+\frac{E_s\lambda_i}{K\sigma^2}\right)$$

$$\to \int_{\tilde{a}}^{\tilde{b}} \log_2(1+\Gamma x) f_{\tilde{\alpha}}(x)dx \tag{6.4.60}$$

$$\triangleq \tilde{I}_4(\Gamma,\alpha). \tag{6.4.61}$$

Note $\tilde{I}_4(\Gamma,\alpha)$ is defined for the case $\alpha < 1$.

From 6.4.58 and 6.4.60, for a given pair of (α_1,α_2) with $\alpha_1 \geq 1$ and $\alpha_2 = \frac{1}{\alpha_1}$, we have the following equality:

$$\tilde{I}_3(\Gamma,\alpha_1) = \tilde{I}_4(\Gamma,\alpha_2). \tag{6.4.62}$$

This means, if CSI is available at the transmitter side, the limiting capacities of MIMO random channels are the same for the two asymmetric antenna configurations considered earlier. This is different from the case when CSI is unavailable at the transmitter side.

6.5 Concluding Remarks

In this chapter, we first provided a review on the applications of RMT in wireless communications. Then, we briefly introduced the basics of wireless communication systems, and presented the input-output model for matrix channels which are very popular from the physical layer perspective of wireless communications. Examples for random matrix channels and linearly precoded channels are given, including MIMO antenna systems, DS-CDMA, MC-CDMA and CP-CDMA systems. We then turned our attention to the derivation of capacities of MIMO fading channels under two different scenarios: CSI-known and CSI-unknown at the transmitter side. For both cases, considering the randomness of the channel realization, ergodic channel capacities are derived for MIMO systems with limited antenna dimensions. Finally, we studied the limiting capacities of MIMO fading channels. It was shown that while the instantaneous mutual information of the MIMO channel is a function of the channel realization, for the limiting case and under some mild assumptions, the normalized mutual information of the MIMO channel converges almost surely to a deterministic value, which depends only on the average received SNR and the ratio between the number of transmit antennas and the number of receive antennas.

7

Limiting Performances of Linear and Iterative Receivers

7.1 Introduction

In this chapter, we are interested in the receiver design for communication systems with the following matrix channel model:

$$\mathbf{x}(n) = \sum_{i=1}^{K} \mathbf{h}_i s_i(n) + \mathbf{u}(n) \tag{7.1.1}$$

$$= \mathbf{H}\mathbf{s}(n) + \mathbf{u}(n), \tag{7.1.2}$$

where $\mathbf{s}(n) = [s_1(n), s_2(n), \cdots, s_K(n)]^T$ represents the transmitted signal vector of dimension $K \times 1$; $\mathbf{h}_i$ represents the channel vector of dimension $N \times 1$, corresponding to symbol $s_i(n)$; $\mathbf{x}(n) = [x_1(n), x_2(n), \cdots, x_N(n)]^T$ denotes the received signal vector of dimension $N \times 1$; $\mathbf{u}(n) = [u_1(n), u_2(n), \cdots, u_N(n)]^T$ represents the received noise vector of dimension $N \times 1$; and $\mathbf{H} = [\mathbf{h}_1, \mathbf{h}_2, \cdots, \mathbf{h}_K]$ is the $N \times K$ channel matrix. In (7.1.2), K and N are referred to as the signal dimension and observation dimension, respectively.

The fundamental objective of the receiver design in a communication system is to recover the information bits dedicated to this particular receiver. Referring back to the matrix channel model given in 7.1.1 and 7.1.2, when the transmitted signals, $s_1(n), s_2(n), \cdots, s_K(n)$, arrive at the receiver side, they are generally interfering with each other. Thus we need to design signal processing techniques to separate and recover these signals. These techniques are called equalizers, and they can be implemented either in a linear processing manner or in a nonlinear processing manner.

In this chapter, both linear and iterative receivers are considered. We are interested in the limiting performance of these receivers when $K \to \infty$, $N \to \infty$ with $K/N \to \alpha$ (constant).

7.2 Linear Equalizers

In this section, linear equalizers are reviewed for the purpose of recovering the transmitted symbols. For the sake of brevity, we refer each of the transmitted signals to one user's signal, and consider a single signal block only. Thus in the following, the block index (n) is dropped.

To describe the linear receivers, we make the following assumptions.

- The transmitted symbols s_i's are i.i.d. random variables with zero mean and average power $\mathsf{E}[|s_i|^2] = P_i$, for $i = 1, \cdots, K$.
- The noise vector $\mathbf{u}$ is i.i.d., zero mean, circularly symmetric complex Gaussian and with covariance matrix $\mathsf{E}[\mathbf{u}\mathbf{u}^*] = \sigma^2 \mathbf{I}_N$.
- The transmitted symbols are statistically independent of the noises.

Denote $\mathbf{q}_k$ as the linear equalizer for recovering the signal s_k, and let

$$\mathbf{Q} = [\mathbf{q}_1, \mathbf{q}_2, \cdots, \mathbf{q}_K] \tag{7.2.3}$$

be the equalization matrix for all K users. The linear equalization output for user k is given by

$$\hat{s}_k = \mathbf{q}_k^* \mathbf{x}. \tag{7.2.4}$$

Using 7.1.1, $\hat{s}_k$ can be further represented as

$$\hat{s}_k = \mathbf{q}_k^* \mathbf{h}_k s_k + \sum_{j \neq k} \mathbf{q}_k^* \mathbf{h}_j s_j + \mathbf{q}_k^* \mathbf{u}. \tag{7.2.5}$$

On the right hand side (RHS) of 7.2.5, $\mathbf{q}_k^* \mathbf{h}_k s_k$ is the desired signal component for user k, and the rest are the residual interference and noise after equalization. The signal-to-interference-plus-noise ratio (SINR) of the linear equalization output at $\hat{s}_k$ can be written as

$$SINR_k = \frac{|\mathbf{q}_k^* \mathbf{h}_k|^2 P_k}{\sum_{j \neq k} P_j |\mathbf{q}_k^* \mathbf{h}_j|^2 + \sigma^2 \|\mathbf{q}_k\|^2} \tag{7.2.6}$$

$$= \frac{|\mathbf{q}_k^* \mathbf{h}_k|^2 P_k}{\mathbf{q}_k^* \left(\mathbf{H}_k \mathbf{D}_k \mathbf{H}_k^* + \sigma^2 \mathbf{I}_N \right) \mathbf{q}_k} \tag{7.2.7}$$

where

$$\mathbf{H}_k = [\mathbf{h}_1, \cdots, \mathbf{h}_{k-1}, \mathbf{h}_{k+1}, \cdots, \mathbf{h}_K], \tag{7.2.8}$$

and

$$\mathbf{D}_k = \mathrm{diag}\{[P_1, \cdots, P_{k-1}, P_{k+1}, \cdots, P_K]\}. \tag{7.2.9}$$

We first study three types of well-known linear equalizers: zero-forcing (ZF) equalizer, matched-filter (MF) equalizer and minimum mean-square-error (MMSE) equalizer. A suboptimal MMSE receiver will also be considered.

7.2.1 ZF Equalizer

The ZF equalizer tries to null out the interferences from the other users, and is designed as follows:

$$\mathbf{Q}^{(\mathtt{ZF})} = (\mathbf{H}\mathbf{H}^*)^{-1}\mathbf{H}. \tag{7.2.10}$$

Note that the ZF equalizer usually suffers from noise enhancement because of the nulling purpose. In (7.2.10), the inverse should be replaced by pseudo-inverse when $N > K$.

7.2.2 Matched Filter (MF) Equalizer

The MF receiver is given by

$$\mathbf{Q}^{(\mathtt{MF})} = \mathbf{H}. \tag{7.2.11}$$

This receiver maximizes the power gain at the desired user's directions, but ignores the interferences from the other users. As such, MF receiver suffers from strong interferences.

7.2.3 MMSE Equalizer

The MMSE receiver is the optimal linear receiver which, for the purpose of recovering s_k, minimizes the following mean-square-error (MSE)

$$J(\mathbf{q}_k) = \mathsf{E}[|\mathbf{q}_k^*\mathbf{x} - s_k|^2]. \tag{7.2.12}$$

The optimal solution to minimize $J(\mathbf{q}_k)$ is given by

$$\begin{aligned}\mathbf{q}_k^{(\mathtt{MMSE})} &= (\mathsf{E}[\mathbf{x}\mathbf{x}^*])^{-1}\,\mathsf{E}[\mathbf{x}s_k^*] \\ &= \left(\mathbf{H}\mathbf{D}\mathbf{H}^* + \sigma^2\mathbf{I}_N\right)^{-1}\mathbf{h}_kP_k,\end{aligned} \tag{7.2.13}$$

where $\mathbf{D} = \mathrm{diag}\{[P_1, P_2, \cdots, P_K]\}$. Using the matrix inversion lemma, we can easily verify that

$$\mathbf{q}_k^{(\mathtt{MMSE})} = c_k\left(\mathbf{H}_k\mathbf{D}_k\mathbf{H}_k^* + \sigma^2\mathbf{I}_N\right)^{-1}\mathbf{h}_kP_k, \tag{7.2.14}$$

where c_k is a non-zero scalar. Thus, plugging the above MMSE equalizer solution to (7.2.7), we obtain the output SINR of the MMSE equalizer for user k as follows

$$SINR_k^{(\mathtt{MMSE})} = \mathbf{h}_k^*(\mathbf{H}_k\mathbf{D}_k\mathbf{H}_k^* + \sigma^2\mathbf{I}_N)^{-1}\mathbf{h}_kP_k. \tag{7.2.15}$$

The MMSE receiver in fact maximizes the output SINR by compromising between the interference nulling and noise enhancement.

7.2.4 Suboptimal MMSE Equalizer

To implement the optimal MMSE receiver shown in (7.2.14), it is required for the receiver to have the perfect knowledge of the instantaneous received power profile. When this power profile changes, the MMSE equalizer vector has to be recalculated, which results in high computational complexity. While the ZF and MF receivers do not have such requirement, their performances are generally much worse than that of the MMSE receiver. In the next subsection, we introduce a suboptimal MMSE (SMMSE) receiver, which does not require the exact knowledge of the received powers and is much less complex compared to the MMSE receiver.

The suboptimal MMSE (SMMSE) receiver is designed by treating all users as having equal received power. With this design, the equalization vector for user k can be represented as

$$\mathbf{q}_k^{(\text{SMMSE})} = (\mathbf{H}_k\mathbf{H}_k^* + \mu^2\mathbf{I}_N)^{-1}\mathbf{h}_k. \tag{7.2.16}$$

In 7.2.16, if we choose $\mu^2 = 0$, the receiver becomes the ZF receiver. On the other hand, if a large positive μ^2 is used, the SMMSE receiver is equivalent to the MF receiver. It is thus expected that there exists an optimal μ^2 which maximizes the output SINR of the suboptimal equalizer.

7.3 Limiting SINR Analysis for Linear Receivers

When designing the linear equalizers, we notice that the equalizer coefficients are dependent on the instantaneous channel coefficients, and similarly for the output SINRs for each of the linear receivers. If the channel coefficients are random variables, the output SINRs are also random. In [260] and [264], it is proven that for the limiting case, i.e., when $K \to \infty$, $N \to \infty$, and $K/N \to \alpha$ where α is a constant, the output SINR of the MMSE equalizer converges to a fixed value, regardless of the instantaneous channel realizations. In [164], the asymptotic SINR performance of the SMMSE receiver is analyzed for random matrix channels. The limiting results for linearly precoded systems are derived in [73].

7.3.1 Random Matrix Channels

To assist the limiting performance analysis, the following assumptions are made for the *random matrix channels*.

(A1) The channel vector $\mathbf{h}_i$ can be represented as

$$\mathbf{h}_i = \frac{1}{\sqrt{N}}[X_{1i}, X_{2i}, \cdots, X_{Ni}]^T, \tag{7.3.17}$$

for $i = 1, \cdots, K$, where X_{ki}'s are i.i.d. random variables with zero mean and unit variance, i.e., $\mathsf{E}[|X_{ki}|^2] = 1$ for all k's and i's.

(A2) The transmitted symbols s_i's are i.i.d. random variables with zero mean and average power $\mathsf{E}[|s_i|^2] = P_i$, for $i = 1, \cdots, K$.
(A3) The noise vector $\mathbf{u}$ is i.i.d., zero mean, circularly symmetric complex Gaussian and with covariance matrix $\mathsf{E}[\mathbf{u}\mathbf{u}^*] = \sigma^2 \mathbf{I}_N$.

Remark 7.1. *The channel coefficients can be real-valued or complex-valued. For complex-valued case, we assume that the real and imaginary parts of each channel coefficient are independent, zero mean and with equal variance.*

Remark 7.2. *In (7.3.17), a normalization factor $1/\sqrt{N}$ has been included, which implies that the signal-to-noise ratio (SNR) for user i, $\gamma_i = \frac{P_i}{\sigma^2}$, is defined as the average received SNR over all observation dimensions.*

In this section, we introduce the limiting performance of the MMSE and SMMSE receivers for *random matrix channels*, assuming that $K \to \infty$, $N \to \infty$, and $K/N \to \alpha$, where α is a fixed constant. Here, the fixed constant α can be greater than one.

MMSE Receiver

Without loss of generality, we consider the detection of user 1, whose output SINR after MMSE equalization is given by

$$SINR_1^{(\mathtt{MMSE})} = \mathbf{h}_1^*(\mathbf{H}_1\mathbf{D}_1\mathbf{H}_1^* + \sigma^2\mathbf{I}_N)^{-1}\mathbf{h}_1 P_1. \tag{7.3.18}$$

When the dimensions become large, we assume that the empirical distribution of the received powers of the users converges to a fixed distribution. The limiting SINR of MMSE receiver for random matrix channels is derived in [260].

Theorem 7.3. *Consider the random MIMO channel 7.1.2 satisfying assumptions (A1), (A2) and (A3), and suppose the empirical distribution function of $\mathbf{D}$ converges to a probability distribution function $F_p(x)$. Then $SINR_1^{(\mathtt{MMSE})}$ converges to β_1 in probability as $K \to \infty$, $N \to \infty$, and $K/N \to \alpha$, where β_1 is the positive solution (which is unique) to the following equation*

$$\beta_1 = \frac{P_1}{\sigma^2 + \alpha \int \frac{pP_1}{P_1 + p\beta_1} dF_p(x)}. \tag{7.3.19}$$

There is in general no explicit solution to the limiting SINR β_1 in (7.3.19). However, when all users have same received power P and thus same received SNR, $\gamma = \frac{P}{\sigma^2}$, the limiting SINR has a simple expression, which is described by the following theorem [260], [264].

Theorem 7.4. *Consider the random MIMO channel model 7.1.2 satisfying assumptions (A1), (A2) and (A3). If all users have the same received SNR γ,* *then the SINR output $SINR_k^{(\mathrm{MMSE})}$ for user k converges to $\beta(\gamma)$ almost surely as $K \to \infty$, $N \to \infty$, and $K/N \to \alpha$, where $\beta(\gamma)$ is the positive solution of the following equation*

$$\beta(\gamma) = \frac{\gamma}{1 + \frac{\alpha\gamma}{1+\beta(\gamma)}}. \tag{7.3.20}$$

The explicit expression for $\beta(\gamma)$ is further given by

$$\beta(\gamma) = \frac{(1-\alpha)\gamma}{2} - \frac{1}{2} + \sqrt{\frac{(1-\alpha)^2\gamma^2}{4} + \frac{(1+\alpha)\gamma}{2} + \frac{1}{4}}. \tag{7.3.21}$$

SMMSE Receiver

Next, we consider the limiting performance of the SMMSE receiver given in (7.2.16). With this receiver, from (7.2.7), the output SINR for user 1 can be written as

$$SINR_1^{(\mathrm{SMMSE})} = \frac{p_1|\mathbf{h}_1^*(\mathbf{H}_1\mathbf{H}_1^* + \mu^2\mathbf{I}_N)^{-1}\mathbf{h}_1|^2}{\mathbf{h}_1^*(\mathbf{H}_1\mathbf{H}_1^* + \mu^2\mathbf{I}_N)^{-1}(\mathbf{H}_1\tilde{\mathbf{D}}_1\mathbf{H}_1^* + \tilde{\sigma}^2\mathbf{I}_N)(\mathbf{H}_1\mathbf{H}_1^* + \mu^2\mathbf{I}_N)^{-1}\mathbf{h}_1}, \tag{7.3.22}$$

where $p_1 = P_1/\bar{P}$, $\tilde{\mathbf{D}}_1 = \mathrm{diag}\{[P_2/\bar{P}, ..., \cdots, P_K/\bar{P}]\}$ with $\bar{P} = \frac{1}{K}\sum_{i=1}^K P_i$ and $\tilde{\sigma}^2 = \sigma^2/\bar{P}$.

Without loss of generality, we assume that $p_1 = 1$. The limiting theorem for the output SINR of the SMMSE receiver is stated as follows [164].

Theorem 7.5. *Consider the random MIMO channel 7.1.2 satisfying assumptions (A1), (A2) and (A3), and assume that the normalized powers of all users are i.i.d., with a fixed distribution and are uniformly bounded, that is, there exists a constant R such that $p_i \le R$. Then $SINR_1^{(\mathrm{SMMSE})}$ in (7.3.22) converges almost surely to a deterministic constant $\beta^{(\mathrm{S})}$ given by*

$$\beta^{(\mathrm{S})} = \beta_\mu \mathcal{K}, \tag{7.3.23}$$

where

$$\beta_\mu = \frac{-\mu^2 - \alpha + 1 + \sqrt{(\mu^2+\alpha+1)^2 - 4\alpha}}{2\mu^2}, \tag{7.3.24}$$

and

$$\mathcal{K} = \frac{\alpha + \mu^2(1+\beta_\mu)^2}{\alpha + \tilde{\sigma}^2(1+\beta_\mu)^2}. \tag{7.3.25}$$

Theorem 7.6. *Under the conditions of Theorem 7.5, the limiting SINR $\beta^{(\mathrm{S})}$ is maximized when $\mu^2 = \tilde{\sigma}^2$.*

From Theorem 7.5, it is seen that the limiting SINR is independent of the actual power distribution of the users. This is due to the fact that the SMMSE receiver is designed by assuming equal powers for all users. The parameter $\mathcal{K}$ in (7.3.25) can be thought of as a mismatch factor defining the difference between the true noise variance $\tilde{\sigma}^2$ and selected noise variance μ^2. If the users are indeed with the same received powers, the SMMSE receiver in (7.2.16) becomes optimal if we choose $\mu^2 = \tilde{\sigma}^2$. In this case, $\mathcal{K} = 1$ and $\beta^{(\mathtt{S})} = \beta_\mu$. This result is consistent with the limiting result of Theorem 7.4 with $\gamma = \frac{1}{\mu^2}$.

7.3.2 Linearly Precoded Systems

In the previous chapter, the input-output model for linearly precoded systems is defined as follows:

$$\mathbf{x}(n) = \mathbf{\Lambda}_N \mathbf{Q}_N \mathbf{s}(n) + \mathbf{u}(n), \tag{7.3.26}$$

where $\mathbf{\Lambda}_N = \text{diag}([f_0, \cdots, f_{N-1}])$. If we choose $\mathbf{Q}_N^* \mathbf{Q}_N = \mathbf{I}_K$, the system is referred to as an *isometrically precoded system*. If $\mathbf{Q}_N$ is chosen as an i.i.d. matrix, then the system is called *randomly precoded system*.

Asymptotic performance of the MMSE receiver for isometrically precoded systems have been analyzed in [73]. As mentioned in the previous chapter, the CP-based CDMA systems (MC-CDMA downlink and CP-CDMA downlink) belong to isometric precoded systems, thus the asymptotic SINR performances provide some guidelines in designing the system parameters for various CDMA systems, without requiring the exact knowledge of the spreading codes.

Isometrically precoded systems

To assist the asymptotic performance analysis, we make the following assumptions [73]:

(B1) $\mathbf{\Lambda}_N = \text{diag}([f_0, \cdots, f_{N-1}])$ has identically distributed centered random diagonal entries. $|f_i|^2$ has a probability density $p(t)$ with finite moments of all orders, and $p(t)$ has a compact support in a positive interval.
(B2) For each $l \geq 1$, $\lim_{N\to+\infty} \frac{1}{N} \sum_{k=0}^{N-1} |f_k|^{2l} = \mathsf{E}[|f_i|^{2l}]$ almost surely.
(B3) The matrix $\mathbf{Q}_N$ is generated by extracting K columns from an $N \times N$ Haar unitary random matrix independent of $\mathbf{\Lambda}_N$.

The assumption (B2) implies that the channel contains a large number of independent delay paths, so that the channel responses in each sub-carrier are sufficiently uncorrelated.

Theorem 7.7. *Consider the MIMO channel model 7.1.2 and suppose* $\mathbf{H} = \mathbf{\Lambda}_N \mathbf{Q}_N$, *and the matrices* $\mathbf{\Lambda}_N$ *and* $\mathbf{Q}_N$ *are chosen according to assumptions (B1), (B2) and (B3). If all users have the same received SNR* γ, *when* $N \to \infty$, $K \to \infty$, *and* $K/N \to \alpha \leq 1$, *then* $SINR_k^{(\mathtt{MMSE})}$ *of the MMSE receiver for user*

k converges almost surely to a deterministic value $\beta^*(\gamma)$, which is the unique positive solution of the equation:

$$\int_0^{\infty} \frac{t}{\alpha t + (1/\gamma)(1-\alpha)\beta^*(\gamma) + 1/\gamma} p(t)dt = \frac{\beta^*(\gamma)}{\beta^*(\gamma)+1}. \tag{7.3.27}$$

Remark 7.8. *Let us consider Rayleigh fading case where $p(t) = e^{-t}$. Let $E_i(x) = \int_x^{\infty} \frac{e^{-s}}{s} ds$, and denote $\mu = \frac{(1-\alpha)\beta^*(\gamma)+1}{\gamma\alpha}$. Then the limiting SINR $\beta^*(\gamma)$ is the positive solution of*

$$1 - \mu e^{\mu} E_i(\mu) = \frac{\alpha\beta^*(\gamma)}{\beta^*(\gamma)+1}. \tag{7.3.28}$$

Further, if the system is fully loaded, i.e, $\alpha = 1$, then $\mu = 1/\gamma$, thus

$$\beta^*(\gamma) = \frac{1}{(1/\gamma)e^{1/\gamma}E_i(1/\gamma)} - 1. \tag{7.3.29}$$

Random Precoded Systems

For random precoded systems, we make the following assumptions.

(C1) $\mathbf{\Lambda}_N = \text{diag}([f_0, \cdots, f_{N-1}])$ has identically distributed centered random diagonal entries. $|f_i|^2$ has a probability density $p(t)$ with finite moments of all orders, and $p(t)$ has a compact support in a positive interval.
(C2) For each bounded continuous function π: $\lim_{N\to+\infty} \frac{1}{N}\sum_{i=1}^{N} \pi(1/|f_i|^2) = \mathsf{E}(\pi(1/|f_i|^2))$ almost surely.
(C3) $\mathbf{Q}_N$ is a i.i.d. matrix, with each element being zero-mean i.i.d. with variance $1/N$.

Theorem 7.9. *Consider the MIMO channel model 7.1.2 and suppose $\mathbf{H} = \mathbf{\Lambda}_N\mathbf{Q}_N$, and the matrices $\mathbf{\Lambda}_N$ and $\mathbf{Q}_N$ are chosen according to assumptions (C1), (C2) and (C3). If all users have the same received SNR γ, when $N \to \infty$, $K \to \infty$, and $K/N \to \alpha \leq 1$, then $SINR_k^{(\mathtt{MMSE})}$ of the MMSE receiver for user k converges almost surely to a deterministic value $\beta^*(\gamma)$, which is the positive solution of the equation:*

$$\int_0^{\infty} \frac{t}{\alpha t + (1/\gamma)\beta^*(\gamma) + 1/\gamma} p(t)dt = \frac{\beta^*(\gamma)}{\beta^*(\gamma)+1}. \tag{7.3.30}$$

Remark 7.10. *For Rayleigh fading, the limiting SINR $\beta^*(\gamma)$ is still determined by (7.3.28), but with μ being replaced by $\mu = \frac{\beta^*(\gamma)+1}{\gamma\alpha}$.*

7.3.3 Asymptotic SINR Distribution

In the previous section, we have derived the limiting SINR results for the optimal MMSE receiver and the SMMSE receiver, when $K \to \infty$, $N \to \infty$ and $K/N \to \alpha$. In practical systems, both the signal dimension K and observation dimension N are limited numbers, thus the output SINR of any linear receivers should be a random variable depending on the instantaneous random channel matrix. In this section, we are interested in the asymptotic distribution of the output SINR when the dimensions are large but with limited values.

Tse and Zeitouni [261] first studied the asymptotic distribution of SINR for the ZF and MMSE receivers using RMT for random matrix channels under the assumption that all users have the same received powers and the channel coefficients are real-valued. In [164], assuming that all users may have different received powers, Liang, Pan and Bai studied the asymptotic SINR distribution of SMMSE receivers for random matrix channels with either real-valued or complex valued coefficients. Since the results of [164] include the one of [261] as a special case, here, we only review the results of [164].

Theorem 7.11. *Consider the random MIMO channel 7.1.2 satisfying assumptions (A1), (A2) and (A3), and assume that the normalized powers of all users are i.i.d., with a fixed distribution and are uniformly bounded, that is, there exists a constant R such that $p_i \leq R$. Furthermore, we assume that the empirical second order moment of the users' powers tends to a limit ϑ_2, i.e.,*

$$\lim_{K\to\infty} \frac{1}{K} \sum_{k=2}^{K} p_k^2 = \vartheta_2. \tag{7.3.31}$$

then as $N \to \infty$,

$$\sqrt{N}(SINR_1^{(\mathsf{SMMSE})} - \beta^{(\mathsf{S})}) \xrightarrow{\mathcal{D}} \mathcal{N}(0, l^2), \tag{7.3.32}$$

where $\xrightarrow{\mathcal{D}}$ denotes convergence in distribution, and $\mathcal{N}(0, l^2)$ denotes the Gaussian distribution with zero mean and variance l^2. The variance l^2 can be calculated as follows.

(i) If X_{11} is real, then

$$l^2 = l_1 \mathsf{E}[X_{11}^4 - 3] + 2l_2 \triangleq l_r^2, \tag{7.3.33}$$

(ii) If X_{11} is complex and $\mathsf{E}(X_{11}^2) = 0$, then the variance becomes

$$l^2 = l_1 \mathsf{E}[|X_{11}|^4 - 2] + l_2 \triangleq l_c^2, \tag{7.3.34}$$

where

$$l_1 = \mathcal{K}^2 \beta_\mu^2, \tag{7.3.35}$$

and

$$l_2 = \mathcal{K}^2 \left[(1+\beta_\mu)^2 \Gamma_2 + 2\alpha^2(\mathcal{K}-1)^2\Gamma_2^3 + 2\alpha(\mathcal{K}-1)\beta\Gamma_2^2 + \alpha\Gamma_2^2(\mathcal{K}^2\vartheta_2 - 1)\right]. \tag{7.3.36}$$

In the above equation, $\Gamma_j = (1+\beta_\mu)^{-j} \int \frac{dF_y(x)}{(x+\mu^2)^j}$ *for* $j \geq 2$, *where* $F_y(x)$ *is the limiting spectral distribution function of the random matrix* HH^*.

From Theorem 7.11, if the received powers for all users are equal, then $\vartheta_2 = 1$, thus if we choose $\mathcal{K} = 1$, we obtain

$$l_1 = \beta_\mu^2, \tag{7.3.37}$$

$$l_2 = (1+\beta_\mu)^2 \Gamma_2 = \frac{\beta_\mu(1+\beta_\mu)^2}{y + \tilde{\sigma}^2(1+\beta_\mu)^2}. \tag{7.3.38}$$

For the real-valued channel case, the above results are consistent with the results derived in [261]. For the complex-valued channel case, the variance expression of SINR distribution is different from the conjecture made in [261].

We now consider the relationship between the asymptotic SINR distributions for real-valued and complex-valued channel cases. For real-valued channels, we denote X_{11} as $X_{11}^{(\mathtt{RL})}$. For the corresponding complex-valued channels, X_{11} is denoted as $X_{11}^{(\mathtt{CX})}$, which can be further represented as $\frac{1}{\sqrt{2}}(X_{11}^{(r)} + jX_{11}^{(i)})$ with $X_{11}^{(\mathtt{r})}$ and $X_{11}^{(\mathtt{i})}$ being independent of each other but both following the same distribution as the random variable $X_{11}^{(\mathtt{RL})}$. If we assume that $X_{11}^{(\mathtt{RL})}$ is with zero mean and unit variance, according to the assumptions of Theorem 7.11, it can be easily verified that

$$\mathsf{E}\left[\left|X_{11}^{(\mathtt{CX})}\right|^4 - 2\right] = \frac{1}{2}\mathsf{E}\left[\left(X_{11}^{(\mathtt{RL})}\right)^4 - 3\right].$$

Thus from (7.3.31) and (7.3.32), we have $l_c^2 = \frac{l_r^2}{2}$.

Remark 7.12. *If* $X_{11}^{(\mathtt{RL})}$ *is assumed to follow the Gaussian distribution, then we obtain* $l_{c,Gaussian}^2 = \frac{l_{r,Gaussian}^2}{2} = l_2$. *For CDMA systems, comparing the real-valued channel with BPSK random codes and the complex-valued channel with QPSK random codes, we have* $l_{QPSK}^2 = \frac{l_{BPSK}^2}{2} = l_2 - l_1$. *This implies that using QPSK spreading can help to reduce the SINR estimation variance when the MMSE receiver is used.*

One possible problem in using the Gaussian distribution to model the SINR distribution is that the SINR may appear to be negative, which is not feasible in practice. In order to predict the SINR distribution more accurately for a system with a small size, and to prevent the SINR from being negative, we turn to look at the distribution of SINR in dB. The following theorem directly comes from Theorem 7.11 using the Taylor series expansion.

Theorem 7.13. *Under the conditions of Theorem 7.11, as $N \to \infty$,*

$$\sqrt{N}\left(10\log_{10}(SINR_1^{(\text{SMMSE})}) - 10\log_{10}(\beta^{(\text{S})})\right) \xrightarrow{\mathcal{D}} \mathcal{N}\left(0, \left(\frac{10\log_{10}(e)}{\beta_N^{(s)}}\right)^2 l^2\right), \quad (7.3.39)$$

where the variance l^2 is given in Theorem 7.11 for both real-valued and complex-valued channels, and e is the natural logarithm base.

7.4 Iterative Receivers

In a matrix channel, the data streams from different users interfere with each other. While the linear receivers are simple to implement, they usually produce a detection performance far away from the maximum likelihood (ML) bound, which defines the optimal performance we could possibly achieve. The complexity of the ML detector however is infeasible for large system implementation as its complexity grows exponentially with the signal dimension. In this section, non-linear receivers are designed to recover the transmitted symbols with improved detection performance.

In the past decade, much research effort has been devoted to designing low-complexity receivers to achieve near-ML performance. For example, when the signal dimension is small and moderate, near ML detection can be achieved through closest lattice point search (CPS) techniques [1], [72], with much lower complexity than brute force ML detector. The computational complexity and memory requirement of CPS techniques grow dramatically with the increase of the signal dimension. Thus, it becomes impractical again to implement CPS when the signal dimension becomes large.

For matrix channels with large signal dimensions, two receivers have been proposed to achieve near-ML performance with much reduced complexity: iterative MMSE receiver with soft interference cancellation (MMSE-SIC) [268], [60] and block-iterative generalized decision feedback equalizer (BI-GDFE) [65], [165]. In this section, we first review the iterative processes of these two receivers. It is noticed that in every iterative step, both of them are designed based on the MMSE criterion, thus we can easily derive and compare the limiting performances of these receivers.

7.4.1 MMSE-SIC

In this subsection, we review the MMSE-SIC receiver in details. Let us assume that all users have the same received power P, and thus the same received SNR $\gamma = \frac{P}{\sigma^2}$. Denote $\mathbf{q}_{k,\ell}$ as the MMSE equalization vector of user k at the ℓth iteration. In the first iteration, the conventional MMSE receiver for user k is applied, which is given by

$$\mathbf{q}_{k,1} = \left(\mathbf{h}_k \mathbf{h}_k^* + \mathbf{H}_k \mathbf{H}_k^* + \frac{1}{\gamma} \mathbf{I}_N \right)^{-1} \mathbf{h}_k. \tag{7.4.40}$$

Applying the above equalizer to the received signal vector $\mathbf{x}$, and removing the bias [68], we obtain

$$\check{s}_{k,1} = s_k + e_{k,1}, \tag{7.4.41}$$

where

$$e_{k,1} = \frac{1}{\alpha_{k,1}} \left(\sum_{j \neq k} \mathbf{q}_{k,1}^* \mathbf{h}_j s_j + \mathbf{q}_{k,1}^* \mathbf{u} \right), \tag{7.4.42}$$

with $\alpha_{k,1} = \mathbf{q}_{k,1}^* \mathbf{h}_k$. The variance of $e_{k,1}$ can be written as

$$\check{\sigma}_{k,1}^2 = \frac{P}{\mathbf{h}_k^* \left[\mathbf{H}_k \mathbf{H}_k^* + (1/\gamma) \mathbf{I}_N \right]^{-1} \mathbf{h}_k}. \tag{7.4.43}$$

Thus the output SINR after the MMSE equalization is given by

$$\beta_{k,1} = \mathbf{h}_k^* \left[\mathbf{H}_k \mathbf{H}_k^* + (1/\gamma) \mathbf{I}_N \right]^{-1} \mathbf{h}_k. \tag{7.4.44}$$

Once the output of the MMSE receiver is obtained, the log-likelihood ratios (LLRs) of the modulation candidates can be derived. Let $a_0, \cdots, a_{M-1}$ be the constellation points, we calculate the soft estimate of user k's symbol at the first iteration as follows

$$\begin{aligned} \tilde{s}_{k,1} &= \mathsf{E}\{s_k | \check{s}_{k,1}\} \\ &= \sum_{i=0}^{M-1} a_i p(s_k = a_i | \check{s}_{k,1}) \\ &= \frac{\sum_{i=0}^{M-1} a_i p(\check{s}_{k,1} | s_k = a_i)}{\sum_{i=0}^{M-1} p(\check{s}_{k,1} | s_k = a_i)}. \end{aligned} \tag{7.4.45}$$

Here $p(x|y)$ denotes the probability of event x occurring under condition y. When the signal dimension is large, the multiple access interference (MAI) after the MMSE equalization is asymptotically Gaussian [288], [117] with mean zero and variance $\check{\sigma}_{k,1}^2$, thus we have

$$p(\check{s}_{k,1} | s_k = a_i) = \frac{1}{\sqrt{2\pi} \check{\sigma}_{k,1}} \exp \left(-\frac{|\check{s}_{k,1} - a_i|^2}{2 \check{\sigma}_{k,1}^2} \right). \tag{7.4.46}$$

At this point, we assume that we have obtained the soft estimates $\tilde{s}_{k,\ell}$'s for all users at the ℓth iteration. Our next step is to carry out soft interference cancellation (SIC):

$$\mathbf{y}_{k,\ell+1} = \mathbf{x} - \sum_{j \neq k} \mathbf{h}_j \tilde{s}_{j,\ell}. \tag{7.4.47}$$

The above output after SIC can be written as follows:

$$\mathbf{y}_{k,\ell+1} = \mathbf{h}_k s_k + \sum_{j \neq k} \mathbf{h}_j (s_j - \tilde{s}_{j,\ell}) + \mathbf{n}. \tag{7.4.48}$$

Given the above output, subsequent iterations shall be carried out with a new optimal MMSE receiver for every user k. By minimizing the MSE between $\mathbf{q}_{k,\ell+1}^* \mathbf{y}_{k,\ell+1}$ and s_k, the optimal MMSE vector $\mathbf{q}_{k,\ell+1}$ is given by

$$\mathbf{q}_{k,\ell+1} = \left(\mathbf{h}_k \mathbf{h}_k^* + \mathbf{H}_k \mathbf{D}_{k,\ell} \mathbf{H}_k^* + \frac{1}{\gamma} \mathbf{I}_N \right)^{-1} \mathbf{h}_k, \tag{7.4.49}$$

where

$$\mathbf{D}_{k,\ell} = \text{diag}\{[d_{1,\ell}, \cdots, d_{k-1,\ell}, d_{k+1,\ell}, \cdots, d_{K,\ell}]\} \tag{7.4.50}$$

with

$$d_{j,\ell} = \frac{1}{P} \mathsf{E} \Big\{ |s_j - \tilde{s}_{j,\ell}|^2 | \check{s}_{j,\ell} \tag{7.4.51}$$

$$= \frac{1}{P} \left[\mathsf{E}\{|s_j|^2 | \check{s}_{j,\ell}\} - |\tilde{s}_{j,\ell}|^2 \right], \tag{7.4.52}$$

and $\mathsf{E}\{|s_j|^2 | \check{s}_{j,\ell}\} = \sum_{i=0}^{M-1} |a_i|^2 p(s_j = a_i | \check{s}_{j,\ell})$. In (7.4.52), $d_{j,\ell}$ defines the normalized residual interference power for user j, which varies for every symbol interval and every iteration, thus matrix $\mathbf{D}_{k,\ell}$ is termed instantaneous interference power matrix.

Let us replace the instantaneous interference power matrix $\mathbf{D}_{k,\ell}$ in (7.4.50) with a scaled version of identity matrix:

$$\bar{\mathbf{D}}_{k,\ell} = (1 - \delta_{k,\ell}^2) \mathbf{I}_{K-1}, \tag{7.4.53}$$

where

$$\delta_{k,\ell}^2 = 1 - \frac{1}{(K-1)P} \sum_{j \neq k} \left[\mathsf{E}\{|s_j|^2 | \check{s}_{j,\ell}\} - |\tilde{s}_{j,\ell}|^2 \right]. \tag{7.4.54}$$

Here, $\bar{\mathbf{D}}_{k,\ell}$ is an approximation of $\mathbf{D}_{k,\ell}$ and $\delta_{k,\ell}^2$ can be considered as the average reliability for all users except for user k, at iteration ℓ. Thus we can then design the weight vector as

$$\mathbf{q}_{k,\ell+1} = \left(\mathbf{h}_k \mathbf{h}_k^* + \mathbf{H}_k \bar{\mathbf{D}}_{k,\ell} \mathbf{H}_k^* + \frac{1}{\gamma} \mathbf{I}_N \right)^{-1} \mathbf{h}_k. \tag{7.4.55}$$

Note from (7.4.52) and (7.4.54), we have $\mathrm{Tr}(\mathbf{D}_{k,\ell}) = \mathrm{Tr}(\bar{\mathbf{D}}_{k,\ell})$. Thus in (7.4.53), we in fact use the average residual interference power to replace the instantaneous residual interference powers for all interferers.

Using the above weight vector to obtain the output $\check{s}_{k,\ell+1} = \mathbf{q}_{k,\ell+1}^* \mathbf{y}_{k,\ell+1}$, the corresponding variance $\check{\sigma}_{k,\ell+1}^2$ of noise-plus-interference after bias removing can be generated as

$$\check{\sigma}_{k,\ell+1}^2 = \frac{\mathbf{q}_{k,\ell+1}^* \left(\mathbf{H}_k \mathbf{D}_{k,\ell} \mathbf{H}_k^* \sigma_s^2 + \sigma_n^2 \mathbf{I}_N\right) \mathbf{q}_{k,\ell+1}}{\alpha_{k,\ell+1}^2}. \tag{7.4.56}$$

Using matrix inversion lemma, we derive the output SINR as follows:

$$\beta_{k,\ell+1} = \frac{\left| \mathbf{h}_k^* \left[\mathbf{H}_k \bar{\mathbf{D}}_{k,\ell} \mathbf{H}_k^* + \frac{1}{\gamma}\mathbf{I}_N\right]^{-1} \mathbf{h}_k \right|^2}{\mathbf{h}_k^* \left[\mathbf{H}_k \bar{\mathbf{D}}_{k,\ell} \mathbf{H}_k^* + \frac{1}{\gamma}\mathbf{I}_N\right]^{-1} \left[\mathbf{H}_k \mathbf{D}_{k,\ell} \mathbf{H}_k^* + \frac{1}{\gamma}\mathbf{I}_N\right] \left[\mathbf{H}_k \bar{\mathbf{D}}_{k,\ell} \mathbf{H}_k^* + \frac{1}{\gamma}\mathbf{I}_N\right]^{-1} \mathbf{h}_k}. \tag{7.4.57}$$

The LLRs of the modulation candidates are then computed and used in determining the soft estimates used in the next iteration.

For every iteration, the soft symbols $\tilde{s}_{k,\ell}$ and the residual interference matrix $\mathbf{D}_{k,\ell}$ will vary. Thus, the MMSE-SIC receiver requires the calculation of the MMSE weights for each user, each symbol interval, and each detection iteration.

7.4.2 BI-GDFE

In the MMSE-SIC receiver, the soft estimates of the transmitted symbols need to be calculated for each user, each symbol interval, and each detection iteration. BI-GDFE [165], on the other hand, estimates the transmitted symbols in a block-iterative manner. In each iteration, the hard decisions for all users are estimated, and these decisions are used for interference cancellation in the next iteration, with the assistance of the input-decision correlation (IDC), which is a measure for the statistical reliability of the earlier-made decisions relative to the transmitted symbols.

At $(\ell+1)$th iteration, the received signal $\mathbf{x}$ is passed to the feed-forward equalizer (FFE) and at the same time, the decision-directed symbols $\hat{\mathbf{s}}_\ell$ from the ℓth iteration go through the feedback equalizer (FBE). The output from the FFE and that of the FBE in BI-GDFE are then combined and utilized to generate the filter output $\mathbf{z}_{\ell+1}$ for the $(\ell+1)$th iteration as

$$\mathbf{z}_{\ell+1} = \mathbf{F}_{\ell+1}^* \mathbf{x} - \mathbf{B}_{\ell+1} \hat{\mathbf{s}}_\ell, \tag{7.4.58}$$

where $\mathbf{F}_{\ell+1}$ represents the FFE weight and $\mathbf{B}_{\ell+1}$ represents the FBE weight at the $(\ell+1)$th iteration. The underlying design of the BI-GDFE is to preserve and recover the desirable signal component $\mathbf{s}$ in the observation vector $\mathbf{x}$ by the FFE, and to cancel out the interference by the FBE.

Let $\mathbf{A}_{\ell+1}$ be the diagonal matrix whose diagonal elements are the same those of matrix $\mathbf{F}_{\ell+1}^*\mathbf{H}$. Denote ρ_ℓ as the IDC for the ℓth iteration, i.e., $\mathsf{E}\{\mathbf{s}\hat{\mathbf{s}}^*\} = \rho_\ell P\mathbf{I}_N$. It has been shown in [165] that the optimal FFE and FBE for $(\ell+1)$th iteration which maximizes the output SINR are given by

$$\mathbf{F}_{\ell+1} = \left[(1-\rho_\ell^2)\mathbf{H}\mathbf{H}^* + (1/\gamma)\mathbf{I}_N\right]^{-1}\mathbf{H}, \tag{7.4.59}$$

$$\mathbf{B}_{\ell+1} = \rho_\ell(\mathbf{F}_{\ell+1}^*\mathbf{H} - \mathbf{A}_{\ell+1}). \tag{7.4.60}$$

The output SINR for symbol s_k at $(\ell+1)$th iteration is given by [165]

$$\theta_{k,\ell+1} = \frac{|\mathbf{f}_{k,\ell+1}^*\mathbf{h}_k|^2}{(1-\rho_\ell^2)\sum_{j\neq k}|\mathbf{f}_{k,\ell+1}^*\mathbf{h}_j|^2 + (1/\gamma)\|\mathbf{f}_{k,\ell+1}\|^2} \tag{7.4.61}$$

$$= \frac{1}{(1-\rho_\ell^2)}\frac{|\mathbf{f}_{k,\ell+1}^*\mathbf{h}_k|^2}{\sum_{j\neq k}|\mathbf{f}_{k,\ell+1}^*\mathbf{h}_j|^2 + \frac{1}{\gamma(1-\rho_\ell^2)}\|\mathbf{f}_{k,\ell+1}\|^2}$$

$$= \frac{1}{(1-\rho_\ell^2)}\frac{|\mathbf{f}_{k,\ell+1}^*\mathbf{h}_k|^2}{\mathbf{f}_{k,\ell+1}^*\left[\mathbf{H}_k\mathbf{H}_k^* + \frac{1}{\gamma(1-\rho_\ell^2)}\mathbf{I}_N\right]\mathbf{f}_{k,\ell+1}}. \tag{7.4.62}$$

where $\mathbf{f}_{k,\ell+1}$ is the kth column of $\mathbf{F}_{\ell+1}$. Using the matrix inversion lemma, we have the following

$$\mathbf{f}_{k,\ell+1} = c_{k,\ell+1}\left[\mathbf{H}_k\mathbf{H}_k^* + \frac{1}{\gamma(1-\rho_\ell^2)}\mathbf{I}_N\right]^{-1}\mathbf{h}_k, \tag{7.4.63}$$

where $c_{k,\ell+1}$ is a nonzero scalar. Thus

$$\theta_{k,\ell+1} = \frac{1}{(1-\rho_\ell^2)}\mathbf{h}_k^*\left[\mathbf{H}_k\mathbf{H}_k^* + \frac{1}{\gamma(1-\rho_\ell^2)}\mathbf{I}_N\right]^{-1}\mathbf{h}_k. \tag{7.4.64}$$

Since the diagonal elements of the FBE $\mathbf{B}_{\ell+1}$ are all zeros, the previous decisions $\hat{\mathbf{s}}_\ell$ are only used in the cancellation of the interferences caused by the transmitted symbols. This will not affect the signal components contained at $\mathbf{z}_{\ell+1}$.

The IDC ρ_0 is set to 0 and thus the BI-GDFE functions as the conventional MMSE equalizer for the first iteration. When all the symbols have been recovered correctly, the IDC coefficient becomes 1, and the FBE cancels out all the interferences and FFE becomes single user MF receiver.

7.5 Limiting Performance of Iterative Receivers

Both the MMSE-SIC and BI-GDFE are designed using the MMSE criterion, thus the limiting performances of these receivers can be derived using RMT. In this section, we study the limiting performances of MMSE-SIC and BI-GDFE receivers, assuming that $K \to \infty$, $N \to \infty$, and $K/N \to \alpha$, where α is a fixed constant. We also assume the transmitted signals are QPSK modulated, and are with zero-mean and unit received power.

7.5.1 MMSE-SIC Receiver

For the MMSE-SIC receiver, there are two important parameters: δ_ℓ and the output SINRs for each iteration. Here, we will show that, for the limiting case, both of them converge to their respective deterministic values.

Since the transmitted signals are QPSK modulated and with unit received power, for the limiting case, according to Theorem 7.4, Theorem 7.7 and Theorem 7.9, after the first iteration (using the MMSE receiver), the output SINRs for all users converge to a deterministic value, $\tilde{\beta}_1^{(\mathtt{MMSE-SIC})}(\gamma)$, depending on the received SNR γ and parameter α only. Further, the equalization outputs for all users can be modelled as the outputs from an AWGN channel with SNR $\tilde{\beta}_1^{(\mathtt{MMSE-SIC})}(\gamma)$. Therefore, for the limiting case, $K \to \infty$ and $\delta_{k,1} \to \bar{\delta}_1 = \mathsf{E}\{|\tilde{s}_{k,1}|^2\}$, which is dependent on the limiting SINR $\tilde{\beta}_1^{(\mathtt{MMSE-SIC})}(\gamma)$. In the next iteration, Theorem 7.5 needs to be used to derive the limiting SINR.

Random MIMO Channels

The limiting result of the MMSE-SIC receiver for random MIMO channels is stated as follows [161].

Theorem 7.14. *Consider the MIMO channel model 7.1.2. Under the assumptions (A1), (A2) and (A3), the output SINR $\beta_{1,\ell+1}$ in (7.4.57) of user 1 using the MMSE-SIC receiver at the $(\ell+1)$th iteration converges to $\tilde{\beta}_{\ell+1}^{(\mathtt{MMSE-SIC})}(\gamma)$ almost surely as $K \to \infty$, $N \to \infty$, and $K/N \to \alpha$, where*

$$\tilde{\beta}_{\ell+1}^{(\mathtt{MMSE-SIC})}(\gamma) = \frac{1}{(1-\bar{\delta}_\ell^2)}\beta_{\bar{\sigma}}, \tag{7.5.65}$$

for $\bar{\delta}_\ell < 1$, $\bar{\sigma}^2 = \frac{1}{\gamma(1-\bar{\delta}_\ell^2)}$, and $\tilde{\beta}_{\ell+1}^{(\mathtt{MMSE-SIC})}(\gamma) = \gamma$ for $\bar{\delta}_\ell = 1$. In 7.5.65, $\beta_{\bar{\sigma}}$ is calculated from β_μ in 7.3.24 by replacing μ with $\bar{\sigma}$.

Finally, for the $(\ell+1)$th iteration, $\delta_{k,\ell} \to \bar{\delta}_\ell = \mathsf{E}\{|\tilde{s}_{k,\ell}|^2\}$, which is dependent on the limiting SINR $\tilde{\beta}_{\ell+1}^{(\mathtt{MMSE-SIC})}(\gamma)$.

Isotropically Precoded Systems

The limiting result of the MMSE-SIC receiver for isotropically precoded systems is stated as follows.

Theorem 7.15. *Consider the MIMO channel model 7.1.2 and suppose* $\mathbf{H} = \mathbf{\Lambda}_N \mathbf{Q}_N$*, and the matrices* $\mathbf{\Lambda}_N$ *and* $\mathbf{Q}_N$ *are chosen according to assumptions (B1), (B2) and (B3). The output SINR* $\beta_{1,\ell+1}$ *in (7.4.57) of user* 1 *using the MMSE-SIC receiver at the* $(\ell+1)$*th iteration converges to* $\tilde{\beta}_{\ell+1}^{(\mathtt{MMSE-SIC})}(\gamma)$ *almost surely as* $K \to \infty$, $N \to \infty$, *and* $K/N \to \alpha$, *where*

$$\tilde{\beta}_{\ell+1}^{(\mathtt{MMSE-SIC})}(\gamma) = \frac{1}{(1-\bar{\delta}_\ell^2)}\beta^*((1-\bar{\delta}_\ell^2)\gamma) \tag{7.5.66}$$

for $\bar{\delta}_\ell < 1$, *and* $\tilde{\beta}_{\ell+1}^{(\mathtt{MMSE-SIC})}(\gamma) = \gamma$ *for* $\bar{\delta}_\ell = 1$. *In 7.5.66,* $\beta^*((1-\bar{\delta}_\ell^2)\gamma)$ *is calculated from 7.3.27.*

Randomly Precoded Systems

The limiting result of the MMSE-SIC receiver for randomly precoded systems is stated as follows.

Theorem 7.16. *Consider the MIMO channel model 7.1.2 and suppose* $\mathbf{H} = \mathbf{\Lambda}_N \mathbf{Q}_N$*, and the matrices* $\mathbf{\Lambda}_N$ *and* $\mathbf{Q}_N$ *are chosen according to assumptions (C1), (C2) and (C3). The output SINR* $\beta_{1,\ell+1}$ *in (7.4.57) of user* 1 *using the MMSE-SIC receiver at the* $(\ell+1)$*th iteration converges to* $\tilde{\beta}_{\ell+1}^{(\mathtt{MMSE-SIC})}(\gamma)$ *almost surely as* $K \to \infty$, $N \to \infty$, *and* $K/N \to \alpha$, *where*

$$\tilde{\beta}_{\ell+1}^{(\mathtt{MMSE-SIC})}(\gamma) = \frac{1}{(1-\bar{\delta}_\ell^2)}\beta^*((1-\bar{\delta}_\ell^2)\gamma) \tag{7.5.67}$$

for $\bar{\delta}_\ell < 1$, *and* $\tilde{\beta}_{\ell+1}^{(\mathtt{MMSE-SIC})}(\gamma) = \gamma$ *for* $\bar{\delta}_\ell = 1$. *In 7.5.67,* $\beta^*((1-\bar{\delta}_\ell^2)\gamma)$ *is calculated from 7.3.30.*

7.5.2 BI-GDFE Receiver

Same as for the MMSE-SIC receiver, for the BI-GDFE receiver, after the first iteration using the MMSE receiver, the output SINRs for all users converge to a deterministic value. Furthermore, the equalization outputs for all users can then be modelled as the outputs from an AWGN channel with SNR $\tilde{\beta}_1^{(\mathtt{BI-GDFE})}(\gamma)$. Therefore, the IDC, ρ_1, used for next iteration can be calculated as follows [65], [165]:

$$\rho_1 = 1 - 2P_e(\tilde{\beta}_1^{(\mathtt{BI-GDFE})}(\gamma)), \tag{7.5.68}$$

where $P_e(\gamma_0) = \mathcal{Q}(\sqrt{\gamma_0})$ with $\mathcal{Q}(u) = \frac{1}{\sqrt{2\pi}}\int_u^\infty \exp(-\frac{t^2}{2})dt$.

Random MIMO Channels

The limiting SINR of the BI-GDFE receiver for random MIMO channels is stated as follows [165].

Theorem 7.17. *Consider the MIMO channel model 7.1.2. Under the assumptions (A1), (A2) and (A3), the output SINR $\theta_{1,\ell+1}$ of user 1 using the BI-GDFE receiver at the $(\ell+1)$th iteration converges to $\tilde{\beta}_{\ell+1}^{(\mathtt{BI-GDFE})}(\gamma)$ almost surely as $K \to \infty$, $N \to \infty$, and $K/N \to \alpha$, where*

$$\tilde{\beta}_{\ell+1}^{(\mathtt{BI-GDFE})}(\gamma) = \frac{1}{(1-\rho_\ell^2)}\beta_{\bar{\sigma}}, \tag{7.5.69}$$

for $\rho_\ell < 1$, $\bar{\sigma}^2 = \frac{1}{\gamma(1-\rho_\ell^2)}$, and $\tilde{\beta}_{\ell+1}^{(\mathtt{BI-GDFE})}(\gamma) = \gamma$ for $\rho_\ell = 1$. In 7.5.69, $\beta_{\bar{\sigma}}$ is calculated from β_μ in 7.3.24 by replacing μ with $\bar{\sigma}$.

Isotropically Precoded Systems

The limiting result of the BI-DGFE receiver for isotropically precoded systems is stated as follows.

Theorem 7.18. *Consider the MIMO channel model 7.1.2 and suppose $\mathbf{H} = \mathbf{\Lambda}_N \mathbf{Q}_N$, and the matrices $\mathbf{\Lambda}_N$ and $\mathbf{Q}_N$ are chosen according to assumptions (B1), (B2) and (B3). The output SINR $\theta_{1,\ell+1}$ of user 1 using the BI-GDFE receiver at the $(\ell+1)$th iteration converges to $\tilde{\beta}_{\ell+1}^{(\mathtt{BI-GDFE})}(\gamma)$ almost surely as $K \to \infty$, $N \to \infty$, and $K/N \to \alpha$, where*

$$\tilde{\beta}_{\ell+1}^{(\mathtt{BI-GDFE})}(\gamma) = \frac{1}{(1-\rho_\ell^2)}\beta^*((1-\rho_\ell^2)\gamma) \tag{7.5.70}$$

for $\rho_\ell < 1$, and $\tilde{\beta}_{\ell+1}^{(\mathtt{MMSE-SIC})}(\gamma) = \gamma$ for $\rho_\ell = 1$. In 7.5.71, $\beta^*((1-\rho_\ell^2)\gamma)$ is calculated from 7.3.27.

Randomly Precoded Systems

The limiting result of the BI-GDFE receiver for randomly precoded systems is stated as follows.

Theorem 7.19. *Consider the MIMO channel model 7.1.2 and suppose $\mathbf{H} = \mathbf{\Lambda}_N \mathbf{Q}_N$, and the matrices $\mathbf{\Lambda}_N$ and $\mathbf{Q}_N$ are chosen according to assumptions (C1), (C2) and (C3). The output SINR $\theta_{1,\ell+1}$ of user 1 using the BI-GDFE receiver at the $(\ell+1)$th iteration converges to $\tilde{\beta}_{\ell+1}^{(\mathtt{BI-GDFE})}(\gamma)$ almost surely as $K \to \infty$, $N \to \infty$, and $K/N \to \alpha$, where*

$$\tilde{\beta}_{\ell+1}^{(\mathtt{BI-GDFE})}(\gamma) = \frac{1}{(1-\rho_\ell^2)}\beta^*((1-\rho_\ell^2)\gamma) \tag{7.5.71}$$

for $\rho_\ell < 1$, and $\tilde{\beta}_{\ell+1}^{(\mathtt{MMSE-SIC})}(\gamma) = \gamma$ for $\rho_\ell = 1$. In 7.5.71, $\beta^*((1-\rho_\ell^2)\gamma)$ is calculated from 7.3.30.

7.6 Numerical Results

In this section, numerical results are presented to illustrate the asymptotic convergence of the BI-GDFE receiver, for both isometrically and randomly precoded systems. The conventional MMSE receiver and the BI-GDFE receiver are simulated. Single user AWGN bound (the case when all interferences have been perfectly cancelled out, thus the AWGN bound is determined by the average received SNR) is used as the lower-bound for performance comparison. The distribution $p(t)$ is chosen to be exponential. QPSK modulation is employed, thus the IDC values are calculated by the theoretical formula shown in (7.5.68).

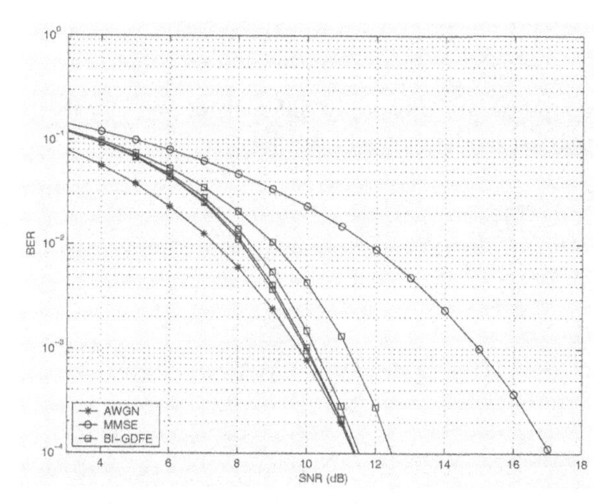

Fig. 7.1. Asymptotic performance of different receivers for isometrically precoded systems with QPSK modulation and $\alpha = 1$.

Figure 7.1 and Figure 7.2 illustrate the dynamics of BER vs SNR of different receivers for isometrically precoded systems with $\alpha = 1$ and 0.5, respectively. In each figure, the performance curves for the BI-GDFE receiver lie between the conventional MMSE receiver and the single user bound, and for a given received SNR, the BER moves down with an increase in the number of iterations. Here 4 iterations are used. From the results, the convergence of the BI-GDFE receiver to the single user AWGN bound is observed when SNR$\geq$ 11dB for all cases. Also, the larger the α, more iterations are required to achieve the single user lower-bound.

Figures 7.3 and Figure 7.4 illustrate the dynamics of BER vs SNR of different receivers for random precoded systems with $\alpha = 1.5$ and 1, respectively. Here 20 iterations are carried out for $\alpha = 1.5$ case and 10 iterations for the others. Note for random precoded system, α can be greater than 1, just like in random MIMO channels [260]. From the results, the convergence

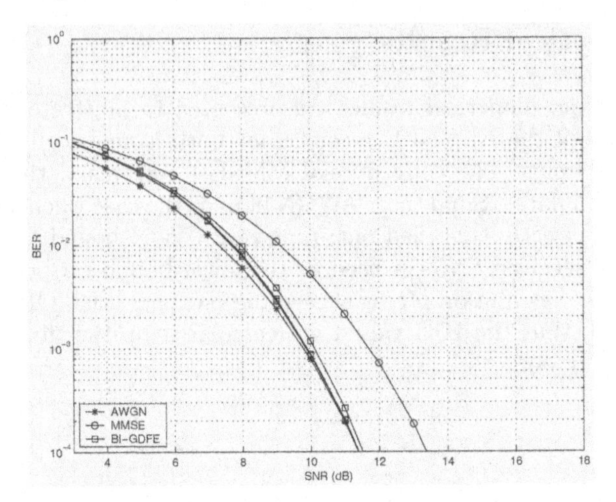

Fig. 7.2. Asymptotic performance of different receivers for isometrically precoded systems with QPSK modulation and $\alpha = 0.5$.

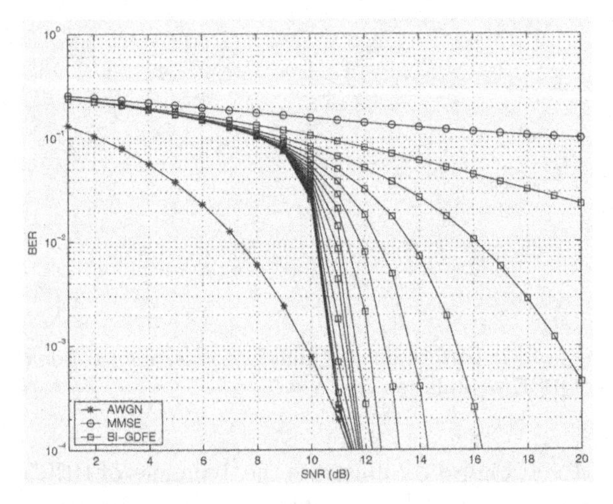

Fig. 7.3. Asymptotic performance of different receivers for randomly precoded systems with QPSK modulation and $\alpha = 1.5$.

of the BI-GDFE receiver to the single user AWGN bound is still observed when SNR$\geq$ 11dB for all cases, even thought it takes about 13 iterations for $\alpha = 1.5$ case. Finally, for the same α, the performance of the random precoded systems is worse than that of the isometric systems, for the same number of iterations. This is expected since isometric systems use orthogonal codes, while random systems employ random codes, there exists stronger interference for the random precoded systems. This also illustrates the importance of using short orthogonal codes for the downlink transmissions.

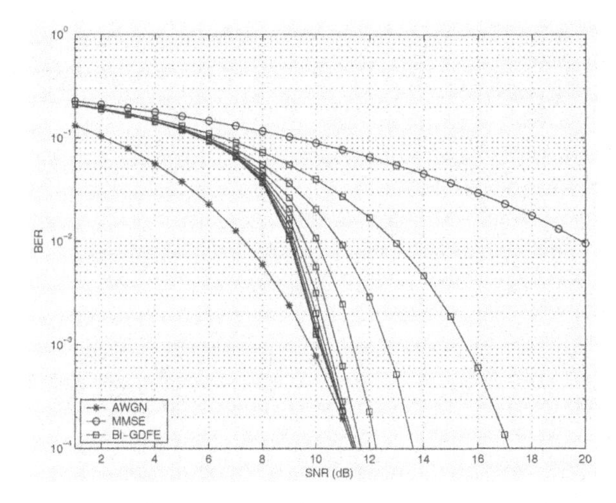

Fig. 7.4. Asymptotic performance of different receivers for randomly precoded systems with QPSK modulation and $\alpha = 1$.

7.7 Concluding Remarks

In this chapter, we started with a review on the linear equalizers for matrix channels. Then the limiting performances of the MMSE and SMMSE receivers were analyzed for random MIMO channels. We also analyzed the limiting performance of the the MMSE receiver for linearly precoded systems, including isometrically precoded systems and random precoded systems. Next, we reviewed two iterative receivers: MMSE-SIC and BI-GDFE, and studied the limiting performance of these receivers under various channel assumptions. Numerical results were presented to illustrate the asymptotic convergence behavior of the BI-GDFE receiver, and it was observed that BI-GDFE can achieve single user AWGN bound when the received SNR is high enough.

8

Application to Finance

It is a popular understanding that the financial environment today is riskier than it was used to. From the second half of the 20th century the financial environment has changed. The price indices have gone up and the volatility of foreign exchange rates, interest rates and commodity prices all has been increasing. All firms and financial institutes are facing uncertainty. The markets for risk management products have grew dramatically since 1980's. Risk management becomes a key and basic technique for all market participants. Risk should be carefully measured before well managed. Var(Value at risk) matrix and credit matrix become popular tools in Banks and Fund management company. People using correlation matrix and factor models to work on internal or external measure for financial risk.

8.1 Portfolio and Risk Management

8.1.1 Markowitz's Portfolio Selection

Portfolio theory is a very useful tool to the investment, whereas is very helpful in estimating risk of an investor's portfolio. Investor selects portfolios in accordance with the mean and variance or the mean and semi-variance of the gains. The use of these criteria are defined in terms of the theory of rational behavior under risk and uncertainty as developed by Von Neumanm, J. and Morgenstern and L. J. Savage, and the relationship between many-period and single-period utility analysis as explained by R. Bellman. Algorithms were provided to compute portfolios which minimize variance or semi-variance for various levels of expected returns one requisite estimates concerning securities are provided.

Portfolio theory refers to an investment strategy that seeks to construct an optimal portfolio by considering the relationship between risk and return. The fundamental issue of capital investment should no longer be to pick out good stocks but to diversify the wealth among different assets. The success of investment not only depends on return, but also on the risk. Risk is influenced by correlations between different assets, that means, the portfolio selection represents as a optimization problem.

A couple of popularly used selection models are listed as follows:

Mean-Variance Model

Suppose there are p assets with returns $R_1, \cdots, R_p$, all of which are random variables, and $\mathrm{E}R_i$ and $Cov(R_i, R_j)$ are assume to be known. Let $\mathbf{R} = (R_1, \cdots R_N)'$, $\mathbf{r} = \mathrm{E}\mathbf{R} = (r_1, \cdots r_p)'$,
$\Sigma = Var\mathbf{R} = \mathrm{E}(\mathbf{R} - \mathbf{r})(\mathbf{R} - \mathbf{r})' = (\sigma_{ij})$. Consider $\mathcal{P}$, a portfolio, with a vector of weights (ratio of different stocks in a portfolio, or loadings) $\mathbf{w} = (w_1, \cdots, w_p)'$. We impose a budget constraint

$$\sum_{i=1}^{p} w_i = \mathbf{w}'\mathbf{1} = 1,$$

where $\mathbf{1}$ is a vector of ones. If additionally $\forall i, \quad w_i \geq 0$, the short sell is excluded.

The return of a whole portfolio $\mathcal{P}$ is denoted by $R_{\mathcal{P}}$, then

$$R_{\mathcal{P}} = \sum_{i=1}^{p} w_i R_i = \mathbf{w}'\mathbf{R}$$

and

$$r_{\mathcal{P}} = \mathrm{E}R_{\mathcal{P}} = \sum_i w_i \mathrm{E}R_i = \sum_i w_i r_i = \mathbf{w}'\mathbf{r}.$$

The variance (or risk) of return is $\sigma_{\mathcal{P}}^2 = \mathbf{w}'\Sigma\mathbf{w}$.

According to Markowitz, a rational investor always searches for $\mathbf{w}$ which minimizes the risk with a given expected return level not less than R_0, i.e.

$$\min\left\{\mathbf{w}'\Sigma\mathbf{w} \;\middle|\; \mathbf{w}'\mathbf{r} \geq R_0, \mathbf{w}'\mathbf{1} = 1, \ \mathbf{w} \geq 0\right\}$$

or it's dual version of maximizing expected return under a given risk level σ_0^2, i.e.

$$\max\left\{\mathbf{w}'\mathbf{r} \;\middle|\; \mathbf{w}'\Sigma\mathbf{w} \leq \sigma_0^2, \ \mathbf{w}'\mathbf{1} = 1, \mathbf{w} \geq 0\right\}.$$

When we use absolute deviation to measure the risk, we get the mean-absolute deviation model to minimize $\mathrm{E}\left|\mathbf{w}'\mathbf{R} - \mathbf{w}'\mathbf{r}\right|$. In this case, we minimize $\mathbf{w}'V_-\mathbf{w}$, where

$$V_- = \Big(Cov((R_i - r_i)_-, (R_j - r_j)_-)\Big)$$

$$(R_i - r_i)_- = [-(R_i - r_i)] \vee 0.$$

Sometimes utility functions are used to evaluate the investment performance, say, $\ln x$. The utility of portfolio $\mathcal{P}$ is $\sum_{i=1}^{p} \ln r_i$. Denote $\widetilde{\Sigma} = (\widetilde{\sigma}_{ij})$, where

$$\widetilde{\sigma}_{ij} = Cov((\ln R_i - \ln r_i)_- , (\ln R_j - \ln r_j)_-).$$

We then come to the **log-utility model**:

$$\min \left\{ \mathbf{w}'\widetilde{\Sigma}\mathbf{w} \,\middle|\, \sum_{i=1}^{p} w_i \ln r_i \geq R_0 \ \mathbf{w}'\mathbf{1} = 1, \ \mathbf{w} \geq 0 \right\}.$$

A portfolio is legitimate if it satisfies constraints $\mathbf{Aw} = \mathbf{b}, \mathbf{w} \geq 0$. Denote the expected return by E of a portfolio and the variance of the return by V ($V = \mathbf{w}'\Sigma\mathbf{w}$, or $\mathbf{w}'\widetilde{\Sigma}\mathbf{w}$ in different model). An E-V combination (E_0, V_0) is obtainable if there is a legitimate portfolio $\mathbf{w}_0$ with $E_0 = \mathbf{w}_0'\mathbf{r}$ and $V_0 = \mathbf{w}_0'\Sigma\mathbf{w}_0$. An E-V combination (E_0, V_0) is efficient if (1) (E_0, V_0) is obtainable and (2) There is no other obtainable combination (E_1, V_1) such that either $E_1 > E_0$ and $V_1 \leq V_0$ or $E_1 \geq E_0$ and $V_1 < V_0$. A portfolio $\mathbf{w}$ is efficient if it is legitimate and if its E-V combination is efficient. The problem turns out to be finding the set of all efficient E-V combinations and a legitimate portfolio for each efficient E-V combination. Kuhn-Tucker's results on non-linear programming are valuable in treating the critical line method. Segments of critical line make up the piece-wise linear set of efficient portfolio. The simplex method are also useful to solve the quadratic programming for portfolio selection as shown by Philip Wolfe.

8.1.2 Financial Correlations and Information Extracting

To evaluate the risk of a given portfolio, one needs to find the population covariance matrix. To this end, it is natural to employ the sample covariance matrix. Suppose $\{\mathbf{r}_t, \ t = 1, \cdots, n\}$ is an observation of $\mathbf{R}$ at discrete time instants $t = 1, \ldots n$, then the sample covariance matrix $\mathbf{S} = (s_{ij})$ is given by

$$s_{ij} = \frac{1}{n} \sum_{t=1}^{n} (r_{i,t} - \bar{r}_i)(r_{j,t} - \bar{r}_j)$$

where $\bar{r}_i = \frac{1}{n} \sum_{t=1}^{n} r_{i,t}$. Covariance matrix plays an important role in the risk measurement and portfolio optimization. Theoretically, the covariance matrix can be well derived from historical data and thus considered to be known. But in real practice, it is not the case. Empirical covariance matrix from historical data should be treated as random and noisy. More importantly, the sample covariance matrix performs far different from the population covariance matrix especially when p is comparable to the sample size. That means, the future risk and return of a portfolio are not well estimated and controllable. We are facing the problems of covariance matrix cleaning in order to construct an efficient portfolio.

To estimate correlation matrix $\mathbf{C} = (C_{ij})$, recall that $\Sigma = \mathbf{D}'\mathbf{CD}$ where $\mathbf{D} = (\sigma_1, \ldots \sigma_p)$, we need to determine $p(p+1)/2$ parameters from the p-time series of length n. Denoting $y = p/n$, only when $y \ll 1$, we can accurately determine the true correlation matrix. Write $X_{i,t} = r_{i,t}/\sigma_i$, then the empirical correlation matrix(ECM) is given by $\mathbf{E} = (E_{ij})$, where

$$E_{ij} = \frac{1}{n}\sum_{t=1}^{n} X_{i,t}X_{j,t}.$$

When $n < p$, rank$(\mathbf{E}) = n < p$ and thus the matrix $\mathbf{E}$ has $p - n$ eigenvalue zero. The risk of a portfolio could be measured by

$$\frac{1}{n}\sum_{i,j,t} w_i\sigma_i Cov(X_{i,t}, X_{j,t})w_j\sigma_j$$

which is expected to close to $\sum_{i,j} w_i\sigma_j C_{ij}\sigma_j w_j$. The above estimator is unbiased and with mean square-error of the order $\frac{1}{n}$. But the portfolio is not constructed using $\mathbf{E}$, so the risk of portfolio should be carefully evaluated.

M. Potters *et al* define the In-sample, Out-sample and True minimum risk as

$$\begin{aligned}
\Sigma_{in} &= \mathbf{w}_{\mathbf{E}}'\mathbf{E}\mathbf{w}_{\mathbf{E}} = \frac{R_0^2}{\mathbf{r}'\mathbf{E}^{-1}\mathbf{r}} \\
\Sigma_{true} &= \mathbf{w}_{\mathbf{C}}'\mathbf{C}\mathbf{w}_{\mathbf{C}} = \frac{R_0^2}{\mathbf{r}'\mathbf{C}^{-1}\mathbf{r}} \\
\Sigma_{out} &= \mathbf{w}_{\mathbf{E}}'\mathbf{C}\mathbf{w}_{\mathbf{E}} = R_0^2\frac{\mathbf{r}'\mathbf{E}^{-1}\mathbf{C}\mathbf{E}^{-1}\mathbf{r}}{(\mathbf{r}'\mathbf{E}^{-1}\mathbf{r})^2}
\end{aligned}$$

where

$$\mathbf{w}_{\mathbf{C}} = R_0\frac{\mathbf{C}^{-1}\mathbf{r}}{\mathbf{r}'\mathbf{C}^{-1}\mathbf{r}}.$$

Since E$(\mathbf{E}) = \mathbf{C}$, by convexity, we have

$$(\mathbf{r}'\mathbf{E}^{-1}\mathbf{r}) \geq \mathbf{r}'\mathbf{C}^{-1}\mathbf{r}.$$

So we obtain

$$\Sigma_{in} \leq \Sigma_{true} \leq \Sigma_{out}.$$

This shows that the in-sample risk is an under-estimator of the true risk when the true correlation matrix is identify $\mathbf{I}$. Pafka *et al* showed that

$$\Sigma_{true} = \frac{R_0^2}{\mathbf{r}'\mathbf{r}}$$

$$\Sigma_{in} \simeq \Sigma_{true}\sqrt{1-y} \simeq \Sigma_{out}(1-y).$$

When $y \to 0$, they all coincide.

Denote λ_k and $\mathbf{V}_k = (V_{i,k})$ the eigenvalue and eigenvectors of the sample correlation matrix, the resulting loading weight

$$w_i \propto \sum_{k,j} \lambda_k^{-1} V_{i,k} V_{j,k} r_j = r_i + \sum_{k,j} (\lambda_k^{-1} - 1) V_{i,k} V_{j,k} r_j.$$

When $\sigma_i = 1$, the optimal portfolio should invest proportionally to the expected return r_i which is the first term on the RHS of the above equation. The second term is in fact an disturbance term caused by the inaccuracy of sample covariance matrix accoding to $\lambda > 1$ or $\lambda < 1$. It is possible that the Markowitz solution will allocate a large weight to small eigenvalue, and cause the domination of the measurement noise. To avoid the un-stability of the risk, people might use

$$w_i \propto r_i - \sum_{k \leq k^*; j} V_{i,k} V_{j,k} r_j$$

projecting out the k^* largest eigenvectors.

The methods of cleaning the noise in correlation matrix are discussed in G. Papp *et al*, S. Sharifi *et al*, T. Conlon *et al*, among others.

Shrinkage estimator is one way of cleaning. Let $\mathbf{E}_c$ denote the cleaned correlation matrix.

$$\mathbf{E}_c = \alpha \mathbf{E} + (1 - \alpha)\mathbf{I}$$

$$\lambda_{c,k} = 1 + \alpha(\lambda_k - 1)$$

where $\lambda_{c,k}$ is the k-th eigenvalue of $\mathbf{E}_c$. The parameter α is related to the expected signal to noise ratio, $\alpha \in (0, 1)$. $\alpha \to 0$ means that the noise is large. L. Laloux *et al* suggest the eigenvalue cleaning.

$$\lambda_{c,k} = \begin{cases} 1 - \delta & k > k^* \\ \lambda_k & k \leq k^* \end{cases}$$

where k^* is the number of meaningful sectors and δ is chosen to preserve the trace of the matrix. The choice of k^* is based on the Random Matrix Theory. The key point is to fix k^* such that λ_{k_*} of $\mathbf{E}$ is close to the theoretical edge of the random part of eigenvalue distribution. The spectrum discussed in Chapter 2 set up the foundation of application here.

Consider ECM $\mathbf{E}$ of p assets using n data points, both large with $y = p/n$ finite. For a Wishart correlation matrix

$$\mathbf{S} = \frac{1}{n} \mathbf{A}\mathbf{A}'$$

where $\mathbf{A}$ is a $p \times n$ matrix with entrance iid random variable with zero mean and unit variance, the asymptotic eigenvalue distribution is

$$P(\lambda) = \frac{1}{2y\lambda\pi\sigma^2} \sqrt{(\lambda_+ - \lambda)(\lambda - \lambda_-)} \qquad \lambda \in (\lambda_-, \lambda_+) \qquad (MP\ law)$$

where $\lambda_\pm = \sigma^2(1 \pm \sqrt{y})^2$ are the bounds of the limiting eigenvalue distribution. Comparing the eigenvalue of ECM with $P(\lambda)$, one can identify the deviating eigenvalue. These deviating eigenvalues are said to contain information about the system under consideration. If the correlation matrix $\mathbf{C}$ has a large eigenvalue greater than $1 + \sqrt{y}$, it has been shown by Baik *et al* that the ECM will have an eigenvalue which is asymptotically Gaussian with a deviation width $\sim \frac{1}{\sqrt{n}}$ and a center out of the MP sea. The k^* is determined by the expected edge of the bulk eigenvalue. The cleaned correlation matrix are used to construct portfolio. Other cleaning methods include clustering analysis by R. N. Mantegna. Empirical study are reported that the risk of the optimized portfolio using a cleaned correlation matrix is more reliable, (see LaLoux *et al*), and less than 5% of the eigenvalues appear to carry most of information.

To extract information from noisy time series, we need assess the degree to which an ECM is noise-dominated. By comparing the eigenspectra properties, we identify the eigenstates of ECM which contains genuine information content. Other remaining eigenstates will be noise-dominated and unstable. To analyze the structure of eigenvectors with corresponding eigenvalues outside the 'MP sea', Paul Ormerod, V. Rojkova *et al* calculate the Inverse Participation Ratio(IPR) (c.f. Plerou *et al* 1999, 2000). Given the k-th eigenvalue λ_k and the corresponding eigenvector $\mathbf{V}_k$ with components $V_{k,i}$, the IPR is then defined by

$$I_k = \sum_{i=1}^{p} V_{k,i}^4.$$

It is commonly utilized in localization theory to quantify the contribution of the different components of an eigenvector to the magnitude of the eigenvector. Two extreme cases: the eigenvector consists of identical components $V_{k,i} = \frac{1}{\sqrt{p}}$ or only one non-zero component. We get $I_k = \frac{1}{p}$ or 1 respectively. When applying it to finance, IPR is the reciprocal of the number of economics contributing to the eigenvector. By analyzing the quarterly levels of GDP over the period 1977-2000 from the OECD database for EU economics, France, Germany, Italy, Spain and the UK, Ormerod shows that the co-movement over time between the growth rates of the EU economies do contain a large amount of information.

Recently, V. Rojkova *et al* use delayed correlation to find long memory dependence, including the delayed IPR, which is

$$I_k(\tau) = \sum_{i=1}^{p} [V_{k,i}(\tau)]^4$$

and the related time-lagged correlation matrix $\mathbf{D}(\tau) = (D_{i,j}(\tau))$ with

$$D_{i,j}(\tau) = \frac{1}{2n} \sum_{t=0}^{n-\tau} (g_i(t)g_j(t+\tau) + g_j(t)g_i(t+\tau))$$

and $g_i(t)$ is the normalized $G_i(t)$, where $G_i(t) = \ln T_i(t + \triangle t) - \ln T_i(t)$, T_i is the time series of interest. These can serve as a measure of back-in-time correlation within the economic system (or traffic network in their paper). Through FFT (fast Fourier transform) to IPRs, they found new localization trend. The analysis of delayed correlation matrices would find new applications in modern finance and should be prospected.

When large eigenvalues are not large enough to lie outside the bulk of the spectral density, Guhr *et al*, based on power mapping, analyzed the correlation matrices for noise detection and identification. In his framework of correlation structure for different companies in different branches, the spectral density is

$$\rho_n(\lambda) = (K - \kappa - B)G(\lambda - \mu_B, v_B^2/n) + \sum_{b=1}^{B} \delta(\lambda - \frac{1 + \kappa_b P_b}{1 + P_P}) + \kappa \rho_{ch}(\lambda, \frac{\kappa}{n})$$

(c. f. Guhr for notations in details.) the defined power mapping for the entries of correlation matrix is

$$C_{ij}^{(q)}(n) = \text{sign}(C_{ij}(n))|C_{ij}(n)|^q$$

where n is the time, q is a positive number. Through simulation and theoretical analysis, they found best separation for noise is reached at $q = 1.5$. The methods could be used to estimate the noise and strength of the true correlation in the financial correlation matrix. It could also be considered as a generalization of shrinking techniques. It allows people to distinguish different correlation structure.

8.2 Factor Models

When we evaluate the performance of investment, the simplest model is the constant-mean-return model:

Let μ_i be the mean return for asset i, $\boldsymbol{\mu} = (\mu_1, \ldots, \mu_p)'$, the model is

$$R_{i,t} = \mu_i + \varepsilon_{it}$$

$$E\varepsilon_{it} = 0 \qquad \text{Var}(\varepsilon_{it}) = \sigma_{\varepsilon_i}^2$$

where R_{it}, the period-t return on security i, ε_{it} is the disturbance term, $\mathbf{R}_t = (R_{1t}, \ldots R_{nt})'$, $Cov(\varepsilon_{it}, \varepsilon_{jt})$ is the (i, j) elements of Σ. $\mathbf{R}_t \sim MN(\mu, \Sigma)$, iid through time t.

One step ahead, we may consider the market model. This model relates the return of any given security to the return of the market portfolio. The linearity follows from the normality of asset returns. For any security i

$$R_{it} = \alpha_i + \beta_i R_{mt} + \varepsilon_{it}$$

$$E\varepsilon_{it} = 0 \qquad \text{Var}(\varepsilon_{it}) = \sigma_{\varepsilon_i}^2$$

where R_{it} and R_{mt} are the period-t returns on security i and the market portfolio, ε_{it} is the disturbance term. $\alpha_i, \beta_i, \sigma_{\varepsilon_i}^2$ are parameters of this model.

In financial application the stock index is used for the market portfolio, such as S&P500 CRSP index. By removing variance in the market return, the variance of the abnormal return is reduced. The higher the R^2 of market model regression, the greater is the variance reduction of the abnormal return. This leads to the larger of gain. The above market model is the simplest single factor model.

The factor model originates in the work of Spearman and Hotelling in statistical area and Goldberger, Robinson and Zellner in the economic area. As mentioned in Onatski the factor model with large dimension on both cross-section and time-series has attracted considerable attention in finance. Approximate factor models, where the idiosyncratic components may be weakly correlated and the common factors non-trivially affect a large number of the cross-sectional units, are particularly useful in applications. In finance such a model is the extension of arbitrage pricing theory (*c.f.* Chamberlain and Rothschild (1983), Ingersol (1984)). In macroeconomy, the models are used to identify economy-wide and global shocks, to construct coincidence indexes, to forecast individual macroeconomic time series to study relationship between microeconomic and aggregated macroeconomic dynamics; and to augment the information in the VAR models used for monetary policy analysis (*c.f.* Forni and Reichlin (1998); Forni, Hallin, Lippi and Reichlin (2000); Stock and Watson (1999); Forni and Lippi (1999) and Bernanke, Boivin and Eliasz (2004)).

Nowadays, huge information are collected and high-capacity computers are available. Dynamic factor models are developed to do forecasting and macroeconomic analysis. With the help of dynamic factor model, the idiosyncratic movements, which possibly include measurement error and local shocks, can be eliminated. In some applications, people estimate the common factors. Another important advantage is the modelers could remain agnostic about the structure of the economy.

We will start from Principle Component Analysis (PCA) and introduce the approximate and the dynamic factor models in the sequel. Some important application areas are discussed through examples in forecasting, index construction and financial crisis.

8.2.1 From PCA to Generalized Dynamic Factor Models

Fisher (1941) in his book "Statistical Methods for Research Workers" offered the definition for statistics.

Statistics may be regarded as

(i) the study of populations,
(ii) the study of variation,
(iii) the study of methods of the reduction of data.

The Principal Components Analysis (PCA) is a useful foundation for reducing the dimension of a multivariate random sample. Consider $\mathbf{X} = (X_1, \dots X_p)'$, a random vector in R^p, with second moment finite. Let $\boldsymbol{\mu} = \mathrm{E}\mathbf{X},\ \Sigma = \mathrm{Var}\mathbf{X}$. It's linear transform

$$\mathbf{Y}_{p\times 1} = \mathbf{L}_{p\times p}\mathbf{X}_{p\times 1}$$

where

$$\mathbf{L} = \begin{pmatrix} l_{11} & \dots & l_{p1} \\ \vdots & \vdots & \vdots \\ l_{1p} & \dots & l_{pp} \end{pmatrix} = \begin{pmatrix} \boldsymbol{\ell}_1' \\ \vdots \\ \boldsymbol{\ell}_p' \end{pmatrix}$$

$\mathrm{Var}\mathbf{Y} = \mathbf{L}'\Sigma\mathbf{L}$. The problem is find Y_i contain as more as possible information in $\mathbf{X}$'s with the constraint $\boldsymbol{\ell}_i'\boldsymbol{\ell}_i = 1$. One way is to find $\mathbf{L}$, such that Y_1 takes maximum information. This lead to the first Principal Component. In other words, we have

Theorem 8.2.1 *Denote $\lambda_1 \geq \lambda_2 \ldots \geq \lambda_p \geq 0$ be the eigenvalues of Σ, $\mathbf{V}_1, \ldots, \mathbf{V}_p$ be the corresponding eigenvectors. If Σ exists, the i-th PC is $Y_i = \mathbf{V}_i'\mathbf{X}$ and $Var(Y_i) = \lambda_i,\ i = 1, \ldots p$.*

Proof. Since $\mathbf{V}_1, \ldots, \mathbf{V}_p$ form a orthnomarl basis in R^p, for any linear combination $\boldsymbol{\ell} = \sum_{i=1}^{p} \alpha_i \mathbf{V}_i$, we have $\max\limits_{\boldsymbol{\ell}\neq 0} \frac{\boldsymbol{\ell}'\Sigma\boldsymbol{\ell}}{\boldsymbol{\ell}'\boldsymbol{\ell}} = \max\limits_{\|\boldsymbol{\ell}\|=1} \boldsymbol{\ell}'\Sigma\boldsymbol{\ell} = \max\limits_{\|\boldsymbol{\ell}\|=1} \sum\limits_{i=1}^{p} \lambda_i\alpha_i' \leq \lambda_1$, since $\boldsymbol{\ell}'\Sigma\boldsymbol{\ell} = \sum\limits_{i=1}^{P} \lambda_i\alpha_i^2$. Similarly, we get that the i-th PC is $Y_i = \mathbf{V}_i'\mathbf{X}$.

Correlation of Y_k and $\mathbf{X}$ is called factor loading $\lambda_k / \sum\limits_{i=1}^{p} \lambda_i$ which is the contribution rate of Y_k, and $\sum\limits_{i=1}^{m} \lambda_i / \sum\limits_{i=1}^{p} \lambda_i,\ m < p$ is the cumulated contribution rate of $Y_i, \ldots, Y_m$.

An r-factor model is

$$\mathbf{Y} = \mathbf{F}\boldsymbol{\Lambda}' + \mathbf{U}$$

where

$$\begin{aligned} \mathbf{Y} &= [\mathbf{y}_1, \ldots, \mathbf{y}_n]' \quad \mathbf{U} = (\mathbf{u}_1, \ldots, \mathbf{u}_n)' \\ \mathbf{F} &= [\mathbf{f}_1, \ldots, \mathbf{f}_n]' \quad \boldsymbol{\Lambda} = (\boldsymbol{\lambda}_i, \ldots \boldsymbol{\lambda}_p)' \\ \boldsymbol{\lambda}_i' &= (\lambda_{i1}, \ldots \lambda_{ir}) \quad \mathbf{f}_t = [f_{1t}, \ldots f_{rt}]' \\ \mathbf{u}_t &= [u_{1t}, \ldots u_{pt}]' \quad \mathbf{y}_t = [y_{1t}, \ldots y_{pt}]' \end{aligned}$$

and

$$\begin{aligned} y_{it} &= \boldsymbol{\lambda}_i'\mathbf{f}_t + u_{it} \\ \mathbf{y}_t &= \boldsymbol{\Lambda}\mathbf{f}_t + \mathbf{u}_t \end{aligned}$$

For the static strict factor model it is assumed that $\mathrm{E}\mathbf{u}_t = 0,\ \mathrm{E}(\mathbf{u}_t\mathbf{u}_t') = \Sigma = \mathrm{diag}(\sigma_1^2, \ldots \sigma_p^2)$, $\mathbf{u}_t$ comprises p idiosyncratic components, $\mathbf{f}$ comprises r common factors and $\mathrm{E}\mathbf{f}_t = 0,\ \mathrm{E}(\mathbf{f}_t\mathbf{f}_t') = \boldsymbol{\Lambda}$ and also $\mathrm{E}(\mathbf{f}_t\mathbf{u}_t') = 0$. The loading matrix Λ could be estimated by solving

$$\min\left\{\sum_{t=1}^{n}(\mathbf{y}_t-\mathbf{B}\mathbf{f}_t)'(\mathbf{y}_t-\mathbf{B}\mathbf{f}_t)\Big|\ \mathbf{B}'\mathbf{B}=\mathbf{I}_r\right\}$$

i.e. $\widehat{\mathbf{B}}$ is the estimator of $\boldsymbol{\Lambda}$, $\widehat{\boldsymbol{\beta}}_i$ is the i-th column of $\widehat{\mathbf{B}}$, should satisfy conditions

$$(\mu\mathbf{I}_p-\mathbf{S})\widehat{\boldsymbol{\beta}}_k=0 \quad k=1,\dots r$$

where

$$\mathbf{S}=\frac{1}{n}\sum_{t=1}^{n}\mathbf{y}_t\mathbf{y}_t', \quad \mu \text{ the eigenvalue}$$

i.e. $\widehat{\mathbf{B}}$ is the PC estimator of $\boldsymbol{\Lambda}$.

If $\mathbf{y}_t$ is normally distributed and $\Sigma=\alpha^2\mathbf{I}$, then the PC-estimator is the maximum likelihood estimator. For general Σ, iterative algorithms can be used to calculate the PC estimator. For large factor models the convergence rate is slow and difficult.

If p is allowed to tend to infinity, the restrictions on above model can be relaxed (*c.f.* Stock, Watson *et al*, Chamberlain, Rothshield, Bai). Approximate factor models are discussed under some assumptions:
(1) Allow serial correlation of the idiosyncratic errors.
(2) Idiosyncratic errors could be cross-correlated and heteroscedasticity or assume all eigenvalues of $\Sigma=\mathrm{E}(\mathbf{u}_t\mathbf{u}_t')$ are bounded.
(3) $\mathbf{f}_t$ and $\mathbf{u}_t$ could be weak correlation.
(4) $\frac{1}{p}\boldsymbol{\Lambda}'\boldsymbol{\Lambda}$ converge to a positive definite limiting matrix.
Under some more assumptions on moment of $\mathbf{f}_t$ and $\mathbf{u}_t$, Bai (2003) proved the consistency and asymptotic normality of PC-estimator for $\boldsymbol{\Lambda}$.

A Dynamic factor model is

$$\mathbf{y}_t=\sum_{i=0}^{m}\boldsymbol{\Lambda}_i\mathbf{g}_{t-i}+\mathbf{u}_t \tag{8.2.1}$$

where $\boldsymbol{\Lambda}_i$ is $p\times q$ matrices and $\mathbf{g}_t$ is a vector of q stationary factors, $\mathbf{u}_t$ is assumed independent and stationary. Forni *et al* (2004) suggest that we can first estimate $\mathbf{f}_t$ by PC and then set up a Vector Autoregressive (VAR) model for the estimated $\mathbf{f}_t$, and finally estimate the innovations of dynamic factors, which could be used to drive the common factors. Here $\mathbf{f}_t=[\mathbf{g}_t',\mathbf{g}_{t-1}',\dots,\mathbf{g}_{t-m}']$, $\boldsymbol{\Lambda}^*=[\boldsymbol{\Lambda}_0,\dots,\boldsymbol{\Lambda}_m]'$, (8.2.1) could be written as

$$\mathbf{y}_t=\boldsymbol{\Lambda}^*\mathbf{f}_t+\mathbf{u}_t$$

Lag operator presentation is used in some literature for dynamic factor models. (Dynamic means considering lags of factors in the model) *c. f.* Forni *et al.*

An important problem is to determine the number r of dynamic factors from the vectors of q static factors with $r=(m+1)q$. To this end, Forni *et al* (2004) used the information criterion and Bai and Ng (2005) and Stock and Watson (2005) used the PCA approach, Breitung and Eickmeier (2005) used the test the hypothesis on canonical correlation between $\widehat{\mathbf{f}}_t$ and $\widehat{\mathbf{f}}_{t-1}$ through a regression of $\widehat{\mathbf{V}}_i'\mathbf{f}_t$ on $\widehat{\mathbf{f}}_{t-1}$ to explain R^2 (the i-th eigenvalue). We will come back to this topic in Section 8.2.3.

8.2.2 CAPM and APT

Markowitz's Portfolio Selection Theory sets up the foundation for the Capital Asset Pricing Model (CAPM) (1959, 1991). Economists were able to quantify the risks and the reward for bearing risks. The CAPM says that the expected return of an asset is linearly related to the covariance of its return with the market portfolio return.

Sharpe (1964) and Lintner (1965) showed that if investors have homogeneous expectations and optimally hold mean-variance efficient portfolios then the portfolio of all invested wealth (market portfolio) will itself be a mean-variance efficient portfolio in the absence of mark friction.

Assuming the existence of lending and borrowing at a risk-free rate of interest. Let R_m be the market portfolio return, R_f be the return on the risk-free asset, we have for asset i the expected return

$$\mathrm{E}[R_i] = R_f + \beta_{im}(\mathrm{E}(R_m) - R_f)$$

when

$$\beta_{im} = \frac{\mathrm{Cov}(R_i, R_m)}{\mathrm{Var}(R_m)}.$$

Denote $Z_i = R_i - R_f$, we get the Sharpe-Lintener version in excess returns, i.e.

$$\mathrm{E}(Z_i) = \beta_{im}\mathrm{E}(Z_m), \quad \beta_{im} = \mathrm{Cov}(Z_i, Z_m)/\mathrm{Var}(Z_m).$$

Empirical tests of the model are developed on the positivity of $\mathrm{E}(Z_m)$ and Beta completely captures the cross-sectional variation of expected excess returns as well as the intercept is zero.

Black's (1972) model generalized the CAPM in the absence of a risk-free asset. He restricts the asset-specific intercept of the real-return market model to be equal to the expected zero-beta portfolio return ($\mathrm{E}(R_{0m})$) times one minus the asset's beta. *i.e.*

$$\mathrm{E}(R_i) = \alpha_{im} + \beta_{im}\mathrm{E}(R_m)$$

$$\beta_{im} = \mathrm{Cov}(R_i, R_m)/\mathrm{Var}(R_m)$$

$$\alpha_{im} = \mathrm{E}(R_{0m})(1 - \beta_{im}) \quad \forall i.$$

The CAPM is a single-period model, the simplest case of which is to assume the the returns are iid through time and multivariate normally distributed. The applications include cost of capital estimation, portfolio performance evaluation and event-study analysis. For real data, S&P500 index serves as a proxy for the market portfolio, the US Treasury bill rate serves as risk-free return.

The cost of capital could be calculated through the excess return on market portfolio over risk-free return.

Details of testing and estimation can be found in John Y. Campbell *et al*, among others.

The CAPM beta does not explain the cross-section of expected asset return completely more factors need be considered. Two progresses are made by Ross (1976) and Merton (1973).

One is the Arbitrage Pricing Theory (APT), and the other is Intertemporal Capital Asset Pricing Model (ICAPM). The APT allows for multiple risk factors and does not require the identifications of the market portfolio. The cost is that APT provides an approximate relation for factors. The determination of numbers of factors and the testability of model become an important research issue.

APT assumes that the markets are competitive and frictionless. The asset return is

$$R_i = a_i + \mathbf{b}_i'\mathbf{f} + \varepsilon_i$$

$$\mathrm{E}[\varepsilon_i|\mathbf{f}] = 0 \qquad \mathrm{E}(\varepsilon_i^2) = \sigma_i^2 < \sigma^2 < \infty$$

where R_i is the return of asset i, a_i intercept of the factor model, $\mathbf{b}_i$ is the factor sensitivities for asset i, $\mathbf{f}$ is a vector of common factors. ε_i is the disturbance term, if $i = 1, \ldots N$, a vector or matrix version is available.

Ross (1976) shows that the absence of arbitrage in large economics implies that

$$\boldsymbol{\mu} \equiv \mathbf{1}\lambda_0 + \mathbf{B}\boldsymbol{\lambda}_k$$

where $\boldsymbol{\mu}$ is the expected return vector, λ_0 is the model zero-beta parameter (if risk-free asset exist, λ_0 equals risk-free return), $\boldsymbol{\lambda}_k$ is a vector of factor risk premia and

$$\mathbf{B} = [\mathbf{b}_1, \ldots, \mathbf{b}_p]'$$

$$\mathbf{1} = [1, \ldots, 1]'.$$

As mentioned in Campbell *et al*, the APT model may overfit the data and not be able to predict asset returns in the future nicely. People need provide a true out-of-sample check on model's performance.

8.2.3 Determine the Number of Factors

As reviewed in Breitung *et al*, when time is long enough, *i.e.* $n \gg p \gg q$, q is the true number of common factors, some estimation and test for q are applicable. For a sample correlation matrix $\mathbf{R}$, $\mathrm{tr}(\mathbf{R}) = p = \sum_{i=1}^{p} \mu_i$, where μ_i the i-th eigenvalue of $\mathbf{R}$, ordered as $\mu_1 > \mu_1 > \ldots > \mu_p$, $\sum_{i=1}^{k} \mu_i/p$ explain the fraction of variance for the first k factors. If the true number of factors is q, the idiosyncratic components $\mathbf{u}_t$ should be uncorrelated. A score test based on $\binom{p}{2}$ squared correlations could be used.

For the case of p and n tending to infinity, Bai and Ng (2002) used information criteria (IC) to estimate q consistently.

Let $V(k) = \frac{1}{pn} \sum_{t=1}^{n} \widehat{\mathbf{u}}_t' \widehat{\mathbf{u}}_t$ denote the overall sums of squared residuals for a k-factor model, where $\widehat{\mathbf{u}}_t = \mathbf{y}_t - \widehat{\mathbf{B}}\widehat{\mathbf{f}}_t$ is the estimate idiosyncratic errors. Chose $\widehat{k}$ which minimizes the

$$IC(k) = \ln[V(k)] + k(\frac{p+n}{pn}) \ln\{n \wedge p\}.$$

We have $\widehat{k} \to q$ in pr., when n and p tends to infinity. The difference with other IC is the penalty term depending also on p.

Kapetanios's ME Algorithm

From the framework of the factor model, the covariance matrix of the dataset is given by

$$\Sigma_Y = \Sigma_f + \Sigma_u$$

where rank of Σ_f is q, Σ_u is the covariance matrix of the idiosyncratic component which is assumed to have bounded eigenvalues for all p. The q largest eigenvalues of Σ_f will tend to infinity at rate p whereas the rest are zero.

When $\lim_{p,n \to \infty} \frac{p}{n} = y$, Yin *et al* (1988) (Theorem 3.5 of this book) proved that the largest eigenvalue converges almost surely to $(1+\sqrt{y})^2$ if $E(u_{i,t}^4) < \infty$, whereas the smallest (or the $p-n$-th smallest when $p > n$) eigenvalue of $\widehat{\Sigma}_u$ converges almost surely to $(1-\sqrt{y})^2$, where $\widehat{\Sigma}_u$ is the sample covariance matrix $\frac{1}{n}\mathbf{u}'\mathbf{u}$. The condition $\mathrm{E}(u_{i,t}^4) < \infty$ is crucial. If it does not hold, then the lim sup of the largest eigenvalue is infinity with probability 1. Theorem 3.16 tells us that as $p, n \to \infty$ the empirical distribution of eigenvalues of $\frac{1}{n}\mathbf{Q}_p^{\frac{1}{2}}\mathbf{u}'\mathbf{u}\mathbf{Q}_p^{\frac{1}{2}}$ has a limiting support which is almost surely determined by the limiting spectral distribution of $\mathbf{Q}_p$. These are basis for the algorithm.

The ideas behind are (1) to check weather there is a factor structure, if the number of factor is zero, the maximum eigenvalue $\mu_{\max}$ of $\widehat{\Sigma}_y$ should not exceed $(1+\sqrt{y})^2$ a. s. If the factor structure exists then $\mu_{\max} \to \infty$. Set up a bound $b = (1+\sqrt{c})^2 + d$, where $d > 0$ is chosen *a priori* (d could be the average eigenvalue of sample covariance matrix, where normalized it is equal to 1). If $\mu_{\max} > b$ it is concluded that a factor structure exists. (2) Extracting the largest PC from the data given by

$$y_{i,t}^{(1)} = y_{i,t} - \widehat{\beta}_{1,i}\widehat{f}_{1,t},$$

where $\widehat{\beta}_{1,i}$ is estimated regression coefficient and $\widehat{f}_{1t}$ is first PC of $\mathbf{y}_t$.

Remember first PC is a consistent estimate of a linear combination of the true factors. If there are more than one factors, repeate above procedure on $\mathbf{y}_t^{(j)}, j = 1, 2, \ldots q_{\max}$, until the bound b is not exceeded by the maximum eigenvalue of the sample covariance matrix of the normalized data.

These algorithm could extend to the dynamic factor model, with one more parameters of the bandwidth used for estimation of spectral density.

A set of assumptions are needed:

Assumption 1. $\mathrm{E}\|\mathbf{f}_t\|^4 \leq M < \infty$, $\frac{1}{n}\sum_{t=1}^{n} \mathbf{f}_t\mathbf{f}_t' \xrightarrow{P} \Sigma$, for some Σ positive definite.

Assumption 2. $\|\lambda_i\| \leq \lambda < \infty$ and $\left\|\frac{1}{p}\boldsymbol{\Lambda}'\boldsymbol{\Lambda} - \mathbf{D}\right\| \to 0$, for some $\mathbf{D}$ positive definite.

Assumption 3. $\mathrm{E}(\varepsilon_{i,t}) = 0$, $\mathrm{E}(ep_{i,t}^2) = \sigma_i^2$, $E|\varepsilon_{i,t}|^8 \leq M$

Assumption 4. $\varepsilon_{i,t}$, iid across i and n

Assumption 5. $\mathbf{f}_t$ and $\varepsilon_{i,t}$ are independent.

Assumption 6. $\lim_{p,n\to\infty} \frac{p}{n} \to y$, where $0 \leq y < \infty$.

ME Algorithm for estimation of the number of factors

Step 1. Demean the data $y_{i,t}$. Set $y_{i,t}^f = y_{i,t}$. Normalized $y_{i,t}^f$ by dividing every observation of each series with the estimated standard deviation of that series. Estimate the first $q_{\max}$ principal components of $y_{i,t}^f$. Denote the estimates by $\widehat{f}_{s,t}, s = 1, \ldots, q^{max}$. Set $b = (1 + \sqrt{\frac{p}{n}})^2 + 1$ and $y_{i,t}^{(0)} = y_{i,t}$.

Step 2. Normalize $y_{i,t}^{(j)}$ by dividing every observation of each series with the estimated standard deviation of that series. Denoting the resulting value by $\tilde{y}_{i,t}^{(j)}$.

Step 3. Calculate the maximum eigenvalue of the estimated covariance matrix of $\tilde{y}_{i,t}^{(j)}$ denoted by $\mu_{\max}^{(j)}$. If $\mu_{\max}^{(j)} < b$, then set $\widehat{q} = j$ and stop, otherwise go to Step 4.

Step 4. Regress $y_{i,t}^{(j)}$ for all i on $\widehat{f}_{(j+1,t)}$. Denote the estimated regression coefficients $\beta_i^{(j)}$. Let $y_{i,t}^{j+1} = y_{i,t}^{(j)} - \beta_i^{(j)}\widehat{f}_{j+1,t}$. Set $j = j+1$, If $j > q_{\max}$ stop, else go to Step 2.

We call this the Maximum Eigenvalue (ME) Algorithm.

Denote the true number of factors by q_0. Then we have:

Theorem 8.2.2 *Under assumptions* $1-5$ *and as* $p, n \to \infty$, $\widehat{q}$ *converges in probability to* q_0.

Proof. First we prove $\mathrm{P}(\widehat{q} > q_0) \to 0$:
Since the lemma 2.1 of Yin *et al* (1988) also holds as $y \to 0$, the Theorem 1 in Yin *et al* holds valid for $y \to 0$. By Theorem 1 of Bai (2003) $\widehat{\mathbf{f}}_t$ is a $\sqrt{p} \wedge n$—consistent estimator for $\mathbf{Q}\mathbf{f}_t$, where $\mathbf{Q}$ is some nonsingular matrix, Wlog, assume $\mathbf{Q} = \mathbf{I}$. By Theorem 2 of Bai (2003), we have that $\widehat{\lambda}_i - \lambda$ is $o_p(p \wedge \sqrt{n})$. Thus, we conclude that $\widehat{\varepsilon}_{it} - \varepsilon_{it} = o_p(\sqrt{p} \wedge \sqrt{n})$. These results hold for $q \geq q_0$. Hence, it follows that

$$\frac{1}{n}\sum_{t=1}^{n} \varepsilon_{it}\varepsilon_{jt} - \frac{1}{n}\sum_{t=1}^{n} \widehat{\varepsilon}_{i,t}\widehat{\varepsilon}_{j,t} = o_p(1) \tag{8.2.2}$$

uniformly over i, j. Since the maximum eigenvalue is the continuous function of the elements of the sample covariance matrix it follows from (8.2.2) that $\mu_{\max}^{(q_0)}$ convergence in probability to the maximum eigenvalue of $\frac{1}{n}\boldsymbol{\varepsilon}'\boldsymbol{\varepsilon}$. Thus by Theorem 3.1 in Yin *et al* (1988), we have $\widehat{\mu}_{\max}^{(j)} < b$ for $j = q_0$ in probability as $n, p \to \infty$. This guarantees that $\mathrm{P}(\widehat{q} > q_0) \to 0$.

Then we need show that $\mathrm{P}(\widehat{q} < q_0) \to 0$. Suppose that $j < q_0$. Let $\boldsymbol{\lambda}_i = (\lambda_{1,i}, \dots \lambda_{q_0,i})'$. For all $1 \le j < q_0$, we have asymptotically

$$y_{i,t}^{(j)} = \lambda_{j+1,i} f_{j+1,t} + \dots \lambda_{q_0,i} f_{q_0,t} + \varepsilon_{i,t}.$$

Note that this is the case since $\widehat{\lambda}_{j,i}, j < q_0$ enjoys the asymptotic properties given in above even if it is estimated with fewer than q_0 factors, since orthogonality of estimate factors. Thus the population covariance matrix of $y_{i,t}^{(j)}$ is given by

$$\left(\Lambda^{(j+1)}\right)' \Sigma^{(j+1)} \Lambda^{(j+1)} + \Sigma_\varepsilon \tag{8.2.3}$$

where $\Lambda^{(j+1)} = (\lambda_i^{(j+1)}, \dots \lambda_p^{(j+1)}), \lambda_i^{(j+1)} = (\lambda_{j+1}, \lambda_0)'$ and $\Sigma^{(j+1)}$ is the covariance matrix of $f_{j+1,t}, \dots, f_{q_0,t}$. The maximum eigenvalue of the covariance matrix in (8.2.3) tends to infinity at rate p as $p \to \infty$. Hence, by Corollary 1 of Bai & Silverstein (1998), the maximum eigenvalue of the sample covariance matrix of $y_{i,t}^{(j)}$ will exceed b almost surely and thus in probability. Thus, we showed that Algorithm will stop at step 3 when and only when $j = q_0$ in probability.

When Assumption 4 is relaxed to

Assumption 7. $\boldsymbol{\varepsilon}_t = (\varepsilon_{i,1}, \dots \varepsilon_{i,t})' = \mathbf{T}_p^{1/2}\mathbf{v}_t$, where $\mathbf{v}_t$ is an $p \times 1$ vector of cross sectionally independently distributed and martingale difference random variables $v_{i,t}$ satisfying $\mathrm{E}(v_{i,t}) = 0, \mathrm{E}(v_{i,t}^2) = 1, \mathrm{E}(v_{i,t}^8) < \infty$ and $\mathbf{T}_p = [\sigma_{ij}], \sigma_{ii} = 1$, and $\sigma_{i,j}, 0 \ne |i-j| < M < \infty$ are iid random variables with $\mathrm{E}(\sigma_{i,j}) = 0$, such that the elements of square root factorization of $\mathbf{T}_p$ are bounded between $-\frac{1}{M^\beta}$ and $\frac{1}{M^\beta}$, where $\beta > 1 - \frac{1}{\ln M}$ and 0 otherwise.

Then we have:

Theorem 8.2.3 *Under Assumptions* $1-3, 5-7$, *as* $p, n \to \infty$, $\widehat{q}$ *converges in probability to* q_0.

The proof can be found in Kapetanios (2004).

Alexei. Onatski's Method:

A. Onatski (2005) impose some structures on the idiosyncratic components of the data to distinguish the increasing without bound eigenvalues of the covariance matrix with bounded sequence. Under the structures, he shows that the first q largest eigenvalues of sample covariance matrix are larger than u while the $q+1$-st largest eigenvalue converges almost surely to u, where q is the true number of factors and u is some positive number which will be specified in the following propositions. The model is

$$\mathbf{X}_t = \boldsymbol{\Lambda}\mathbf{F}_t + \mathbf{e}_t \tag{8.2.4}$$

$\mathbf{X}_t$ is $p \times 1$ vector of the cross-sectional observations at time period t. $\boldsymbol{\Lambda}\mathbf{F}_t$ and $\mathbf{e}_t$ are unobserved systematic and idiosyncratic components, $\boldsymbol{\Lambda}$ is $p \times q$ factor loading, $\mathbf{F}_t$ is $q \times 1$ vector of factors. We want to estimate the unknown number q of factors.

The largest eigenvalue of the covariance matrix for idiosyncratic vectors is bounded, whereas all eigenvalues of the covariance matrix of the systematic part $\boldsymbol{\Lambda}\mathbf{F}_t$ tend to infinity.

Assuming the true number q of factors does not exceed q_{max}, the small integer larger than $\min(n^\alpha, p^\alpha)$, where $0 < \alpha < 1$. A couple of assumptions are needed:
Assumption 1: $n, p \to \infty, p/n \to y$ where $y \in (0, \infty)$.
Assumption 2: $\lambda_{\min}$ of $\boldsymbol{\Lambda}'\boldsymbol{\Lambda} \to \infty$, the eigenvalues μ of $\frac{1}{n}\sum_{t=1}^n \mathbf{F}_t\mathbf{F}_t'$ is bounded, *i.e.*

$$B_1 < \mu < B_2 < \infty.$$

Assumption 3: There exists a $p \times p$ random matrix $\mathbf{S}_p$, such that

$$\mathbf{e}_t = \mathbf{S}_p \boldsymbol{\varepsilon}_t$$

where $\boldsymbol{\varepsilon}_t = (\varepsilon_{1t}, \ldots, \varepsilon_{pt})$, $\mathrm{E}\varepsilon_{it} = 0$, $\mathrm{E}\varepsilon_{it}^2 = 1$, $\mathrm{E}\varepsilon_{it}^4 < \infty$, ε_{it} iid, and $\mathbf{S}_p$ and ε_t are independent. ($\mathrm{Cov}(\boldsymbol{\varepsilon}_t) = \mathbf{S}_p\mathbf{S}_p'$)
Assumption 4:
(1) $F^{\mathbf{S}_p'\mathbf{S}_p} \xrightarrow{D} H$ a.s. H is a fixed CDF with bounded support.
(2)$u(F^{\mathbf{S}_p'\mathbf{S}_p}) \to u(H)$ a.s, where $u(\cdot)$ be the upper bound of the support.
(3)$y \int \frac{t^2 dH(t)}{(u(H)-t)^2} > 1$, if the integral exists.
Assumption 5: $\left\| F^{\frac{1}{n}\mathbf{ee}'} - G \right\| \overset{P}{\sim} n^{-\beta_1}$, $|\mu_1 - u| \overset{P}{\sim} p^{-\beta_2}$, $\beta_1, \beta_2 \in (0, 1]$, when μ_1 is largest eigenvalue of $\frac{1}{n}\mathbf{ee}'$, where $\mathbf{e} = (\mathbf{e}_1, \cdots, \mathbf{e}_n)$.

The estimator of Onatski is as follows: When p moves rewrite (8.2.4) as

$$\mathbf{X} = \boldsymbol{\Lambda}\mathbf{F} + \mathbf{e} \tag{8.2.5}$$

$\mathbf{X}$ is $p \times n$, $\mathbf{F}$ is $q \times n$, $\mathbf{e}$ is $p \times n$ matrices, λ_i is i-th largest eigenvalues of $\frac{1}{n}\mathbf{XX}'$. Define a family of estimators:

$$\widehat{q}_\delta = \#\{i \le p;\ \lambda_i > (1 + \delta)\widehat{u}\}$$

where

$$\widehat{u} = w\lambda_{r_{max}+1} + (1 - w)\lambda_{2r_{max}+1}$$

and $w = \alpha^{2/3}/(\alpha^{(2/3)} - 1)$ and δ will be suggested by simulation.

Then we have

Proposition 8.1. *Under Assumptions* $1 - 4$, *we have*
(1) The spectral distribution of $\frac{1}{n}\mathbf{ee}' \xrightarrow{w} G$ *with bounded support a.s.*

(2) For any i such that $\frac{i}{p} \to 0, p \to \infty$, *the i-th eigenvalue* μ_i *of* $\frac{1}{n}\mathbf{ee}'$ *a.s* $\to u$ *(the support of G).*
(3) The spectral distribution of $\frac{1}{n}\mathbf{XX}' \xrightarrow{w} G$ *a.s.*
(4) For any $i > q$ *such that* $i/p \to 0, p \to \infty$, *the i-th eigenvalue of* $\frac{1}{n}\mathbf{XX}', \lambda_i \to u$, *a.s.*

Proof. c.f. Chapter 2.

Thus from conclusion *(4)* we get $\widehat{u} \to u$ a.s. Then we get:

Proposition 8.2. *Under Assumptions* $1-4$, *for any fixed* $\delta > 0$, *we have*

$$\widehat{q}_\delta \to q \ a.s, \quad p \to \infty.$$

Proof. See Onatski.
Define

$$g(\alpha, \beta) = \begin{cases} \frac{4}{3}(1-\alpha) & \text{if} \quad \frac{5}{3}(1-\alpha) < \beta < 1 \\ \beta - \frac{1}{3}(1-\alpha) & \text{if} \quad 1-\alpha < \beta \le \frac{5}{3}(1-\alpha) \\ \frac{2}{3}\beta & \text{if} \quad 0 < \beta \le (1-\alpha) \end{cases}$$

and

$$h(\alpha, \beta) = \frac{2}{3}\min\{\beta, 1-\alpha\}$$

We also have:

Proposition 8.3. *Let Assumptions* $1-5$ *hold, then:*
(1)$\widehat{u} - \mu_1 = o_P(n^{-\min\{g(\alpha_1 \beta_1), \beta_2\}})$
(2)$\lambda_{q_{\max}+1} - \mu_1 = o_P(n^{-\min\{h(\alpha_1 \beta_1), \beta_2\}})$

Proof. See Onatski.

The proposition suggests that the optimal choice of α in $q_{max} \sim \min(n^\alpha, p^\alpha)$ depends on β_1, β_2 when $\frac{4}{5}\beta_1 \le \beta_2, \alpha = 1 - \frac{3}{5}\beta_1$, when $\frac{2}{3}\beta_1 \le \beta_2 < \frac{4}{5}\beta_1$. α can be any number in $[1 - 3(\beta_1 - \beta_2), 1 - \frac{3}{4}\beta_1]$ when $\beta_2 < \frac{2}{3}\beta_1$, any α from the segment $[0, 1 - \frac{3}{4}\beta_2]$ is optimal in the sense of optimizing the rate of the convergence of $\widehat{u} - \mu_1 \to 0$.

When $\mathbf{S}_p$ is identify, $\boldsymbol{\varepsilon}_t$ iid normal, it is known $\beta_2 = 2/3$. The conjecture for β_1 by Bai (1999) is 1.
Based on Proposition 8.3, we get finally:

Proposition 8.4. *Let Assumptions* $1-5$ *hold, then* $\widehat{q}_\delta$ *is consistent for* q *when* $\delta \sim p^{-\gamma}$, *for any* γ *such that* $0 \le \min\{g(\alpha, \beta_1), \beta_2\}$.

The Monte Carlo simulation suggests that the best choice for δ is $\delta = \max\{p^{-2/5}, n^{-2/5}\}$.

Matthew C. Harding's work

He took a structural approach and showed how to improve the estimation of factor model by incorporating the economic assumptions which allow for arbitrary parametric forms of heteroscedasticity and autocorrelation. The key to identify the number of latent factors lies in correctly understanding the structure of noise (idiosyncratic effects) in the data. Once one can split the estimated eigenvalues associated with a large factor model into those due to the latent structure and those due to the noise, they could estimate the number of factors consistently. He mentioned that even though the main factors play a key role in the effect of factors, many other factors may exist in the data and contain potentially valuable economic information.

He developed in details on this issue.

8.3 Some Application in Finance of Factor Model

8.3.1 Inflation Forecasting

One main aim of monetary policy is maintaining price stability over the medium term. Inflation forecasting is important for financial policy decisions and investment strategy. Two approaches for filtering a price index are adopted. One is on the cross section dimension and the other is on the time series dimension of the aggregate price index series. Kapetanios (2004) proposed method of estimating dynamic factor models that exploits both the cross section and the time series dimensions. It can also treat the case when the number of variables exceeds the number of observations. Bryan and Cecchetti (1993) did propose the dynamic factor index model to treat the US CPI (Consumer Price Index) using Kalman filter. For core inflation's measurement c.f. Wynne (1999) for a review. Bryan *et al*'s model are difficult to estimate when the dimension is large.

Consider model

$$\mathbf{x}_t = \mathbf{C}\mathbf{f}_t + \mathbf{D}\mathbf{u}_t$$

$$\mathbf{f}_t = \mathbf{A}\mathbf{f}_{t-1} + \mathbf{B}\mathbf{u}_{t-1} \qquad t = 1, \ldots, n$$

where $\mathbf{f}_t$ are factors and $\mathbf{u}_t$ is white noise.
Assuming non-singularity of $\mathbf{D}$, in the state space representation we get

$$\mathbf{X}_t^f = \mathbf{O}\mathbf{K}\mathbf{X}_t^P + \varepsilon\mathbf{E}_t^f = \mathbf{O}\mathbf{f}_t + \varepsilon\mathbf{E}_t^f$$

where

$$\mathbf{X}_t^f = (\mathbf{X}_t', \mathbf{X}_{t+1}', \ldots)'$$

$$\mathbf{X}_t^p = (\mathbf{X}_{t-1}', \mathbf{X}_{t-2}', \ldots)'$$

$$\mathbf{E}_t^f = (\mathbf{u}_t', \mathbf{u}_{t+1}', \ldots)'$$

f stands for future, p stands for past respectively.

$$\mathbf{O} = [\mathbf{C}', \mathbf{A}'\mathbf{C}', (\mathbf{A}^2)'\mathbf{C}', \ldots]'$$

$$\mathbf{K} = [\overline{\mathbf{B}}, (\mathbf{A} - \overline{\mathbf{B}}\mathbf{C})\overline{\mathbf{B}}, (\mathbf{A} - \overline{\mathbf{B}}\mathbf{C})^2\overline{\mathbf{B}}, \ldots]$$

where $\overline{\mathbf{B}} = \mathbf{BD}^{-1}$ and

$$\begin{pmatrix} \mathbf{D} & 0 & \cdots & 0 \\ \mathbf{CB} & \ddots & \ddots & \vdots \\ \mathbf{CAB} & \ddots & \ddots & 0 \\ \cdots & & \cdots \mathbf{CB} & \mathbf{D} \end{pmatrix}$$

The best linear predictor of future is $\mathbf{OKX}_t^p$. We need an estimator of $\mathbf{K}$. Due to infinity of dimension, in real practice, only finite sample approximations are available and denoted by

$$\mathbf{X}_{s,t}^f = (\mathbf{x}_t', \ldots \mathbf{x}_{t+s-1}')'$$

$$\mathbf{X}_{q,t}^p = (\mathbf{x}_{t-1}', \ldots, \mathbf{x}_{t-q}')'.$$

Their sample covariance matrices are denoted by $\widehat{\mathbf{T}}^f$ and $\widehat{\mathbf{T}}^p$. To estimate $\mathbf{F} = \mathbf{OK}$ by regressing $\mathbf{X}_{s,t}^f$ on $\mathbf{X}_{q,t}^p$, we get $\widehat{\mathbf{F}}$. Then find the singular value decomposition $\widehat{\mathbf{U}}\widehat{\mathbf{S}}\widehat{\mathbf{V}}'$ of $\left(\widehat{\Gamma}^f\right)^{-\frac{1}{2}} \widehat{\mathbf{F}} \left(\widehat{\Gamma}^p\right)^{\frac{1}{2}}$. The desired estimator of $\mathbf{K}$ is given by

$$\widehat{\mathbf{K}} = \widehat{\mathbf{S}}_m^{\frac{1}{2}} \widehat{\mathbf{V}}_m' \left(\widehat{\Gamma}^p\right)^{-\frac{1}{2}}$$

where $\widehat{\mathbf{V}}_m$ is the first m columns of $\widehat{\mathbf{V}}$, $\widehat{\mathbf{S}}_m$ is the heading $m \times m$ submatrix of $\widehat{\mathbf{S}}$, $\widehat{S}$ contains the singular values of $\left(\widehat{\Gamma}^f\right)^{-\frac{1}{2}} \widehat{\mathbf{F}} \left(\widehat{\Gamma}^p\right)^{\frac{1}{2}}$ in decreasing order. The factor estimators are given by $\widehat{\mathbf{K}}\mathbf{X}_t^P$.

For discussion of consistency, wlog assume $\widehat{\Gamma}^f$ and $\widehat{\Gamma}^p$ are identities. The lag truncation parameter q is increasing at a rate greater than $\ln(n^\alpha)$, for some $\alpha > 1$, depends on the maximum eigenvalue of $\mathbf{A}$ and is slower than $n^{1/3}$. The lead truncation parameter s is required to satisfy $sp \geq m$.

Theorem 8.3.1 *Denote* $\mathbf{X}_p = (X_{p,1}^p, \ldots, X_{p,t}^p)', \widehat{\mathbf{f}}_t = \widehat{\mathbf{K}}\mathbf{X}_{p,t}^p$. *If* $\mathbf{u}_t$ *are iid* $(0.\Sigma_u)$ *and have finite fourth moments,* $p_1 \leq p \leq p_2$, $p_1 = O(n^{1/r})$, *for some* $r > 3$, $p_2 = o(n^{1/3})$, *the elements of* $\mathbf{C}$ *are bounded and* $|\lambda_{\max}(\mathbf{A})| < 1$ $|\lambda_{\min}(\mathbf{A})| > 0$, λ *stands for eigenvalue of the indicated matrix in brackets, then* $\widehat{\mathbf{f}}_t$ *converges in probability to the space spanned by the true factors.*

Proof. See Kopetanios *et al* (2003).

In the case where $p \to \infty$, with the additional assumption $p_p = o(n^{1/3}), p = O(n^{1/r}), r > 3$ and matrix $\mathbf{OK}$ has k_1 singular values tending to infinity as $p \to \infty$ and k_2 non-zero finite singular values, we have the following theorem.

Theorem 8.3.2 *Under assumptions of Theorem 8.3.1 and the above, if* $p = o(n^{1/3-1/r})$, *then when* $p, n \to \infty$, $\widehat{\mathbf{f}}_t = \widehat{\mathbf{K}}\mathbf{X}_t^p$ *converges in probability to the space spanned by the true factors.*

Proof. See Kopetanros *et al* (2003).

Headline inflation is denoted by $\pi_t = 100 \ln P_t/P_{t-12}$, where P_t is a price index measure. Define p as the number of subcomponents of the price measure, and w_i for $i = 1$ to p as the weights associated with the i-th subcomponent, it follows that $P_t = \sum_{i=1}^{p} w_i P_{i,t}$, $P_{i,t}$ is the price index for subcomponent i at time t, based on dynamic factor model

$$\mathbf{X}_t = (x_{1,t}, \ldots x_{n,t})' \qquad x_{i,t} = 100 \ln(P_{i,t}/P_{i,t-12})$$

$i = 1, 2, \ldots n$. The measure of underlying inflation is the factor estimate of $\widehat{\mathbf{F}}\mathbf{X}_t^P$. Camba-Mendez and Kapetanios compared their estimation with other measure like excluding measure, trimmed mean measure and Quah and Vahey (1995)'s VAR system method. It is reported that the dynamic factor model is applicable against other traditional measures.

As forecasting methods related to factor models is concerned, Forni *et al* (2003) discussed in details about improvement to the method proposed by Stock and Watson (2002). They proposed to estimate the common components by projecting onto the static principal components of the data, but failed to exploit the potential crucial information contained in the leading-lagging relations between the elements of the panel. Forni. *et al* using their early work in 2000 to obtain estimates of common and idiosyncratic covariance matrices at all leads and lags as inverse Fourier transforms of the corresponding estimated spectral densities of various matrices. Then they use their estimates in the construction of the contemporaneous linear combinations of the observations having smallest idiosyncratic-common variance ratio. The resulting aggregates can be obtained by generalized PC. They also proved that, as p and $n \to \infty$, this forecast is a consistent estimator of the optimal h-step ahead forecast.

8.3.2 Leading and Coincident Index

It is well known that fixed-income market are directly linked to interest rates. The stock Market is tied to corporate profits, inflation and interest rates. Federal Reserve in U.S. or Central Bank in China determines the monetary policy. Their policy influence the money supply through open market operations and interest rates. Although they can determine short-term interest rates but long-term rates are basically determined by economy itself. To understand the trend of economic development, different indices and indicators are designed and announced to public.

The index of leading economic indicators (LEI), a composite of several different indicators, is designed to predict future aggregate economic activity. Historically, the LEI reaches peaks and throughs earlier than the underlying turns in the economy and therefore is an important tool for forecasting and planning. In general, turning points in the economy are signaled by three consecutive monthly LEI changes in the same direction. In U.S., manufacturing, initial unemployment claims, new orders for consumer goods, vendor performance, plant and equipment orders, building permits, change in unfilled orders-durable, sensitive material prices, stock prices, like S&P500, Real M_2, index of consumer expectations, among others, are the major components of the indices of LEI. The Commerce Department publishes also the coincident indicators & Lagging indicators. The rise or fall in coincident indicators suggest economic expansion or contraction in the given month. The coincident index consists of components like: employees on nonagricultural payrolls, personal income less transfer payments, index of industrial production and business sales. The lagging index indicators the up and down 4-9 months after the economy begins to emerge from or enter into a recession. They include average duration of unemployment; ratio of business inventories to sales; index of unit labor costs for manufacturing; average prime rate charged by banks; commercial and industrial loans; and ratio of consumer installment debt to personal income.

The problem we meet is the construction of coincident or leading index. According Stock and Watson (1989), five equations are

$$\begin{aligned}
&\Delta \mathbf{X}_t = \beta + \gamma(L)\Delta \mathbf{C}_t + \mathbf{u}_t && (1)\\
&\mathbf{D}(L)\mathbf{u}_t = \boldsymbol{\varepsilon}_t && (2)\\
&\Phi(L)\Delta \mathbf{C}_t = \delta_t \boldsymbol{\eta}_t && (3)\\
&\Delta \mathbf{C}_t = \boldsymbol{\mu}_{\mathbf{C}} + \lambda_{\mathbf{CC}}(L)\Delta \mathbf{C}_{t-1} + \lambda_{\mathbf{CY}}(L)\mathbf{Y}_{t-1} + \nu_{\mathbf{C}t} && (4)\\
&\mathbf{Y}_t = \boldsymbol{\mu}_{\mathbf{Y}} + \lambda_{\mathbf{YC}}(L)\Delta \mathbf{C}_{t-1} + \lambda_{\mathbf{YY}}(L)\mathbf{Y}_{t-1} + \nu_{\mathbf{Y}t} && (5)
\end{aligned}$$

where $\mathbf{C}_t$ is "the state of the economy" and unobserved. It is assuming that the co-movements of observed coincident time series at all leads and lags arise solely form movements in $\mathbf{C}_t$. Let $\mathbf{X}_t$ denote an $p \times 1$ vector of the logarithms of macroeconomic variables that are hypothesized to move contemporaneously with overall economic conditions. $\mathbf{u}_t$ represent idiosyncratic movements in the series and measurement error. By empirical study, the coincident variables used appear to be integrated but not co-integrated (*c.f.* Engle and Granger 1987), so, the model is specified in terms of $\triangle \mathbf{X}_t$ and $\triangle \mathbf{C}_t$, L is the lag operator, $\Phi, \gamma, \mathbf{D}, \lambda_{\mathbf{CC}}, \lambda_{\mathbf{CY}}, \lambda_{\mathbf{YC}}, \lambda_{\mathbf{YY}}$ are all lag polynomials of scalar vectors and/or matrices, $\boldsymbol{\varepsilon}_t, \boldsymbol{\eta}_n$ are disturbances. Some uncorrelated conditions are assumed. The proposed CEI is computed as the minimum MSE (minimum mean squared error) linear estimate of this single common factor, $\mathbf{C}_{t|t'}$, produced by applying the Kalman filter to the estimated system. In other words, $\mathbf{C}_{t|t'}$ is a linear combination of current and past logarithms of coincident variables.

Given CEI, the proposed LEI is the estimate of the growth of the factor: $\mathbf{C}_{t+6|t'} - C_{t|t'}$. Hence, the sum of CEI and LEI is $\mathbf{C}_{t+6|t'}$, a forecast of the log level of the CEI in six months.

The LEI is constructed by modeling the leading variables ($\mathbf{Y}_t$) and the state ($\mathbf{C}_t$) as a vector autoregressive system with some modifications: eliminating higher lags of variables in all equations expect one for coincident variable. These are the above equations (4) and (5), where ν's are serially uncorrelated error terms. $\lambda_{..}(L)$'s orders of lag polynomials are determined empirically by using the statistical criteria.

The parameters of coincident and leading models are estimated in two steps. First, we estimate parameters given in equations (1)-(3) by ML (Maximum Likelihood) and Kalman filter (by evaluating the likelihood function). Secondly, the leading model is estimated conditioning on the estimated parameters of the coincidence. Technically, the equations (1),(2),(4),(5) form a state space model with $\Delta \mathbf{C}_t$ and its lags as elements of state vector. Using EM algorithm and MLE to estimate parameters in equations (4) and (5), conditioning on the estimation of parameters given in (1) and (2). The coincident index is a weighted average of $\Delta \mathbf{X}_t$. The leading model is as a projection of $\Delta \mathbf{C}_{t|t'}$ onto leading variables in VAR framework.

Since Burns and Mitchell (1946), people make inference about the state of the economy. The traditional coincident index constructed by the Department of Commerce is a combination of four representative monthly variables on total output, income, employment and trade. The indices constructed by The Conference Board (TCB) are used widely due to their ready availability. As described in Issler and Vahid (2001), they used the information content in the NBER Business Cycle Dating Committee decisions to construct index of economic out of noisy information in the coincident series. They also take into account of measurement error in the coincident series by using instrument-variable methods. The resulting index is a simple linear combination of the four coincident series originally proposed by Burns and Mitchell (1946). Start and end form a cycle, but in a higher level.

8.3.3 Financial Crises Warning

The Asian financial crises in 1990's swept through East Asia. Massive capital inflows had been pouring in the form of bank landing. New currency crisis models are motivated. The importance of debt and liquidity management was emphasized by IMF to prevent external crises. External Vulnerability indicators are reported by IMF. It is important to monitor a number of vulnerability indicators like the short term external debt ratio to foreign reserves. Early warning system (EMS) suggest whether they could be a leading indicator of crisis. Cipollini and Kapetanios use factor model to estimate common shock interpreted as a vulnerability indicator. Then, warning of crises are possible in advance. Due to the external debt data provided by Bank for International Settlements(BIS) are available at low frequency, the number of cross sections exceed the time series. Thus the method by Stock and Watson (2002) is useful to extract factor. An Exchanger Market Pressure index (EMP) are used to proxy of stress. An increase in the value of EMP indicates that the net demand for a country is weakening (see Tanner (2002)). The idiosyncratic shock for different countries are estimated as the residuals of each country's specific EMP on the estimated static factor. Cipollini *et al* also get the out of sample probability forecasting through simulation of dynamic factor model. Accuracy could be compared by quadratic probability score and other scores. They pool the set of information provided by different vulnerability indicators in each country to obtain measure in that region's vulnerability.

The model discussed is

$$\mathbf{X}_{n,t} = \mathbf{A}_n \mathbf{f}_t + \varepsilon_{nt}$$

$$\mathbf{f}_t = \mathbf{D}\mathbf{f}_{t-1} + \mathbf{R}\mathbf{u}_t \qquad (*)$$

where $\mathbf{X}_{nt}$ is p-dimensional dataset of balance sheet data, $\mathbf{f}_t$ is the vector of static factors, $\mathbf{A}_n$ the $n \times r$ matrix of factor loadings. $\mathbf{u}_t$ is the common shock (interpreted as the regional vulnerability indicator). $\mathbf{R}$ measures the impact of the common shock on the static factors. First, they estimate $\mathbf{f}_t$ by PC analysis, then an OLS estimation is used to retrieve the $\varepsilon_t = \mathbf{R}\mathbf{u}_t$ in $(*)$. The $\mathbf{R}$ is obtained by using eigenvalue-eigenvector decomposition of covariance matrix of ε_t, *i.e.* $\mathbf{R} = \mathbf{KM}$, M is the diagonal matrix having the square roots of q largest eigenvalues of $\text{Cov}(\varepsilon_t)$. The matrix $\mathbf{K}$ is an $r \times q$ matrix whose columns are the corresponding eigenvectors. After these, through OLS regression, β is estimated from

$$\text{EM}P_{i,t} = \beta \mathbf{f}_t + \nu_t.$$

The simulation forecasting is obtained from equation

$$\text{EM}P_{i,t+1} = \beta(\mathbf{D}\mathbf{f}_t + \mathbf{R}\mathbf{u}_{t+1}) + \nu_t$$

where ν_{t+1} measures the idiosyncratic shock(country specific shock).
The empirical study and simulation tell us the evidence for contagion can be explained by the large short-term external borrowing from a common lender country, which provided valueable Lessons for Asia Crisis in 1997.

References

1. Agrell, E., Eriksson, T., Vardy, A. and Zeger, K. (2002) Closest point search in lattices. *IEEE Trans. Inf. Theory*, 48(8): 2201–2214.
2. Albeverio, S., Pastur, L. and Shcherbina, M. (2001) On the $1/n$ expansion for some unitary invariant ensembles of random matrices. Dedicated to Joel L. Lebowitz. *Comm. Math. Phys.*, 224(1): 271–305.
3. Angls d'Auriac, J. Ch. and Maillard, J. M. (2003) Random matrix theory in lattice statistical mechanics. Statphys-Taiwan-2002: Lattice models and complex systems (Taipei/Taichung). *Phys. A*, 321(1-2): 325–333.
4. Arnold, L. (1967) On the asymptotic distribution of the eigenvalues of random matrices. *J. Math. Anal. Appl.*, 20: 262–268.
5. Arnold, L. (1971) On Wigner's semicircle law for the eigenvalues of random matrices. *Z. Wahrsch. Verw. Gebiete*, 19: 191–198.
6. Aubrun, Guillaume (2005) A sharp small deviation inequality for the largest eigenvalue of a random matrix. *Sminaire de Probability XXXVIII*, p. 320–337, Lecture Notes in Math., 1857, Springer, Berlin.
7. Bai, J. (2003), Inferentral Theory For Factor Models of Large Dimensions. *Econometrica*, 71: 135–173.
8. Bai, J. and Ng, S.(2002), Determining the number of factors in approximate factor models. *Econometrica*, 70: 191–221.
9. Bai, J. and Ng, S. (2007) Determing the number of primitive shocks in factor models. Journal of Business & Economic Statistics, 25(1): 52–60.
10. Bai, Z. D. (1999) Methodologies in spectral analysis of large dimensional random matrices, A review. *Statistica Sinica*, 9(3): 611–677.
11. Bai, Z. D. (1997) Circular law. *Annals of Probab.*, 25(1): 494–529.
12. Bai, Z. D. (1995) Spectral analysis of large dimensional random matrices. *Journal of Chinese Statistical Association*, 33(3): 299–317.
13. Bai, Z. D. (1993a) Convergence rate of Expected spectral distributions of large random matrices. Part I. Wigner Matrices. *Ann. Probab.*, 21(2): 625–648.
14. Bai, Z. D. (1993b) Convergence rate of expected spectral distributions of large random matrices. Part II. Sample Covariance Matrices. *Ann. Probab.*, 21(2): 649–672.
15. Bai, Z. D., Liu, H. X. and Wong, W. K. (2008) Enhancement of the applicability of markowitz's portfolio optimization by utilizing random matrix theory. To appear *Mathematical Finance*.

16. Bai, Z. D., Miao, B. Q. and Jin, B. S. (2007) On limit theorem for the eigenvalue of product of two random matrices. *J. Multivariate Anal.*, 98: 76–101.
17. Bai, Z. D., Miao, P. Q. and Pan, G. M. (2007) On asymptotics of eigenvectors of large sample covariance matrix. *Ann. Probab.*, 35(4): 1532–1572.
18. Bai, Z. D. Miao, B. Q. and Tsay, J. (2002) Convergence rates of the spectral distributions of large Wigner matrices. *International Mathematical Journal*, 1(1): 65–90.
19. Bai, Z. D., Miao, B. Q. and Tsay, J. (1999) Remarks on the convergence rate of the spectral distributions of Wigner matrices. *J. Theoret. Probab.*, 12(2): 301–311.
20. Bai, Z. D. Miao, B. Q. and Tsay, J. (1997) A note on the convergence rate of the spectral distributions of large dimensional random matrices. *Statist. & Probab. Letters*, 34(1): 95–102.
21. Bai, Z. D., Miao, B. Q. and Yao, J. F. (2003) Convergence rates of spectral distributions of large sample covariance matrices *SIAM J. Matrix Anal. Appl.*, 25(1): 105–127.
22. Bai, Z. D. and Saranadasa, H. (1995) Effect of High Dimension Comparison of Significance Tests for a High Dimensional Two Sample Problem. To appear in *Statistica Sinica.*
23. Bai, Z. D. and Silverstein, J. W. (2007) On the Signal-to-Interference-Ratio of CDMA Systems in Wireless Communications. *Annals of Applied Probability*, 17(1): 81–101.
24. Bai, Z. D. and Silverstein, J. W. (2006) *Spectral analysis of large dimensional random matrices.* 1st Edition, Science Press, Beijing. ISBN: 7-03-017766-5.
25. Bai, Z. D. and Silverstein, J. W. (2004) CLT for linear spectral statistics of large–dimensional sample covariance matrices. *Ann. Probab.*, 32(1): 553–605.
26. Bai, Z. D. and Silverstein, J. W. (1999). Exact separation of eigenvalues of large dimensional sample covariance matrices. *Ann. Probab.*, 27(3): 1536–1555.
27. Bai, Z. D. and Silverstein, J. W. (1998). No eigenvalues outside the support of the limiting spectral distribution of large dimensional sample covariance matrices. *Ann. Probab.*, 26(1): 316–345.
28. Bai, Z. D., Silverstein, J. W. and Yin, Y. Q. (1988) A note on the largest eigenvalue of a large dimensional sample covariance m *J. Multiv. Anal.*, 26: 166–168.
29. Bai, Z. D. and Yao, J. F. (2005) On the convergence of the spectral empirical process of Wigner matrices. *Bernoulli*, 11(6): 1059–1092.
30. Bai, Z. D. and Yin, Y. Q. (1993) Limit of the smallest eigenvalue of large dimensional covariance matrix. *Ann. Probab.*, 21(3): 1275–1294.
31. Bai, Z. D. and Yin, Y. Q. (1988a) A convergence to the semicircle law. *Ann. Probab.*, 16(2): 863–875.
32. Bai, Z. D. and Yin, Y. Q. (1988b) Necessary and sufficient conditions for the almost sure convergence of the largest eigenvalue of Wigner matrices. *Ann. Probab.*, 16(4): 1729–1741.
33. Bai, Z. D. and Yin, Y. Q. (1986). Limiting behavior of the norm of products of random matrices and two problems of Geman–Hwang. *Probab. Th. Rel. Fields*, 73: 555–569.
34. Bai, Z. D., Yin, Y. Q. and Krishnaiah, P. R. (1987) On the limiting empirical distribution function of the eigenvalues of a multivariate F matrix. *Theory Probab. and Its Appl.*, 32: 537–548.

35. Bai, Z. D., Yin, Y. Q. and Krishnaiah, P. R. (1986) On limiting spectral distribution of product of two random matrices when the underlying distribution is isotropic. *J. Multiv. Anal.*, 19: 189–200.
36. Bai, Z. D. and Zhang, L. X. (2006) Semicircle Law for Hadamard Products. *SIAM J. Matrix Analysis and Applications*, 29(2): 473–495.
Baik, J. and Silverstein, J. (2006) Eigenvalues of Large Sample Covariance Matrices of Spiked Population Models. *Journal of Multivariate Analysis*, 97(6): 1382–1408.
37. Barry, R. P. and Pace, R. K. (1999) Monte Carlo estimates of the log determinant of large sparse matrices. Linear algebra and statistics (Istanbul, 1997). *Linear Algebra Appl.*, 289(1-3): 41–54.
38. Baum, K. L. Thomas, T. A. Vook, F. W. and Nangia, V. (2002) Cyclic-prefix CDMA: an improved transmission method for broadband DS-CDMA cellular systems. *Proc. of IEEE WCNC'2002*, Vol. 1, pp. 183-188, March 2002.
39. Beheshti, S., Isabelle, S. H. and Wornel, G. W. (1998) Joint intersymbol and multiple access interference suppression algorithms for CDMA systems. *Euporean Trans. Telecommun.*, 9: 403–418.
40. Bekakos, M. P. and Bartzi, A. A. (1999) Sparse matrix and network minimization schemes. *Computational methods and neural networks*, 61–94, Dynamic, Atlanta, GA.
41. Bellman, R. E. (1957) *Dynamic Programming.* Princeton University Press, Princeton. N.J.
42. Ben-Hur, A., Feinberg, J., Fishman, S. and Siegelmann, H. T. (2004) Random matrix theory for the analysis of the performance of an analog computer: a scaling theory. *Phys. Lett. A*, 323(3-4): 204–209.
43. Biely, C. and Thurner, S. (2006) Ramdom matrix ensemble of time-lagged correlation matrices: derivation of eigenvalue spectra and analysis of financial time-series. *arXiv: physcs/0609053.* I, 7.
44. Biglieri, E. Caire, G. and Taricco, G. (2001) Limiting performance of block-fading channels with multiple antennas, *IEEE Trans. on Inf. Theory*, 47(4): 1273–1289.
45. Biglieri, E. Proakis, J. and Shamai, S. (1998) Fading channels: Information-theoretic and communications aspects. *IEEE Trans. on Inf. Theory*, 44(6): 2619–2692.
46. Billingsley, P. (1968) *Convergence of Probability Measures.*, Wiley, New York.
47. Black, F. (1972) Capital Market Equilibrium with Restricted Borrowing. *Journal of Business*, 45, July 444–454.
48. Bleher, P. and Kuijlaars, A. B. J. (2004) Large n limit of Gaussian random matrices with external source. I. *Comm. Math. Phys.*, 252(1-3): 43–76.
49. Bliss, D. W. Forsthe, K. W. and Yegulalp, A. F. (2001) MIMO communication capacity using infinite dimension random matrix eigenvalue distributions. *Thirty-Fifth Asilomar Conf. on Signals, Systems and Computers*, 2: 969–974.
50. Boley, D. and Goehring, Todd (2000) LQ–Schur projection on large sparse matrix equations. Preconditioning techniques for large sparse matrix problems in industrial applications (Minneapolis, MN, 1999). *Numer. Linear Algebra Appl.*, 7(7-8): 491–503.
51. Botta, E. F. F. and Wubs, F. W. (1999) Matrix renumbering ILU: an effective algebraic multilevel ILU preconditioner for sparse matrices. Sparse and structured matrices and their applications (Coeur d'Alene, ID, 1996). *SIAM J. Matrix Anal. Appl.*, 20(4): 1007–1026.

52. Breitung, J. and Eickmeier, S. (2005) Dynamic factor models. Discussion Paper Series 1: Economics Studies. N038/2005. Deutsche Bundes bank.
53. Bryan, M. F. (1994) *Measuring Core inflation*, in Monetary Policy. Ed. by N. G. Mankiw, Chicago Univ. Press.
54. Bryan, M. F. and Cecchetti, S. G. (1993) The Consumer Price Index as a Measure of Inflation. *Federal Reserve Bank of Cleveland Ecnomic Review*, pp. 15–24.
55. Burda, Z. and Jurkiewicz, J. (2004) Signal and noise in financial correlation matrices. *Phys. A*, 344(1-2): 67–72.
56. Burkholder, D. L. (1973) Distribution function inequalities for martingales. *Ann. Probab.*, 1: 19–42.
57. Bernanke, B. S., Boivin, J. and Eliasz, P. (2004). Measuring the Effects of Monetary Policy: A Factor-Augmented Vector Autoregressive (FAVAR) Approach. *NBER Working Paper*, 10220.
58. Burns A. F. and Mitchell W. C. (1946) Measuring Business Cycle. *New York National Burean of Ecnomic Research.*
59. Caire, C., Taricco, G. and Biglieri, E. Optimal power control over fading channels. *IEEE Trans. on Inf. Theory*, 45(5): 1468–1489, July 1999.
60. Caire, G., Müller, R. R. and Tanaka, T. Iterative multiuser joint decoding: optimal power allocation and low-complexity implementation. *IEEE Trans. Inf. Theory*, 50(9): 1950–1973, Sept. 2004.
61. Camba-Méndez, G. and Kapetanios, G. (2004) Forecasting Euro Area Inflation using Dynamic Factor Measures of underlying inflation. *Working paper series*, 402.
62. Campbell, J. Y. Lo, Andren, W. and Macknlay, A. Craig. (1997) The Econometrics of Finacial Markets. Princeton Univ. Press.
63. Chamberlain, G. and Rothschild, M. (1983), Aribitrage Factor Structure and Mean-Variance Analysis on Large Asset Markets. *Econometrica*, 51: 1281–1324.
64. Caselle, M. and Magnea, U. (2004) Random matrix theory and symmetric spaces. *Phys. Rep.*, 394(2-3): 41–156.
65. Chan, A. M. and Wornell, G. W. A class of block-iterative equalizers for intersymbol interference channels: fixed channel results. *IEEE Trans. on Commun.*, 49(11), 1966–1976, Nov. 2001.
66. Chan, A. M. and Wornell, G. W. (2002) Approaching the matched-filter bound using iterated-decision equalization with frequency-interleaved encoding. *Proc. of IEEE Globecom'02*, Vol. 1, pp. 17–21.
67. Chuah, C. N., Tse, D., Kahn, J. M. and Valenzuels, R. A. (2002) Capacity scaling in MIMO wireless systems under correlated fading. *IEEE Trans Inform Theory*, 48: 637–650.
68. Cioffi, J. M. (2003) *Digital Communications*, CA: Stanford Univ., Course Reader.
69. Cipollini, A. and Kapetanios, G. (2008) Forecasting Financial Crises and Contagion in Asia Using Dynamic Factor Analysis. Working Paper, 538. Queen Mary. University of London.
70. Conlon, T., Ruskin, H. J. and Crane, M. (2007) Random matrix theory and fund of fund s portfolio optimisation. *Physica A: Statistical Mechanics and its applications*, 382(2): 565–576.
71. Cover, T. and Thomas, J. (1991) *Elements of Information Theory*, New York: Wiley.

72. Damen, M. O., Gamal, H. E. and Caire, G. (2003) On maximum-likelihood detection and the search for the closest lattice point. *IEEE Trans. on Inf. Theory*, 49(10): 2389–2402.
73. Debbah, M., Hachem, W., Loubaton, P. and Courvill, M. (2003) MMSE analysis of certain large isometric random precoded systems. *IEEE Trans. Inf. Theory*, 49(5): 1293–1311.
74. Deift, P. (2001) Four lectures on random matrix theory. Asymptotic combinatorics with applications to mathematical physics (St. Petersburg, 2001), 21–52, *Lecture Notes in Math.*, 1815, Springer, Berlin, 2003.
75. Del Corso, G. M. and Romani, F. (2001). Heuristic spectral techniques for the reduction of bandwidth and work–bound of sparse matrices. In memory of W. Gross. *Numer. Algorithms*, 28(1-4): 117–136.
76. Dembo, A., Guionnet, A. and Zeitouni, O. (2003) Moderate deviations for the spectral measure of certain random matrices. *Ann. Inst. H. Poincaré Probab. Statist.*, 39(6): 1013–1042.
77. Dempster, A. P. (1958). A high dimensional two sample significance test. *Ann. Math. Statis.*, 29: 995–1010.
78. des Cloizeaux, J. and Mehta, M. L. (1973) Asymptotic behavior of spacing distributions for the eigenvalues of random matrices. *J. Mathematical Phys.*, 14: 1648–1650.
79. Diaconis, P. and Evans, S. N. (2001) Linear functionals of eigenvalues of random matrices. *Trans. Amer. Math. Soc.*, 353(7): 2615–2633.
80. Diaconis, P. and Evans, S. N. (2002) A different construction of Gaussian fields from Markov chains: Dirichlet covariances. *Annales of Institute of Henry Poincare (B) Probability and Statistics*, 38(6): 863–878.
81. Dilworth, S. J. (1993). Some probabilistic inequalities with applications to functional analysis. *Banach Spaces: Contemporary Mathematics*, 144: AMS, Providence 53–67.
82. Dudley, R. M. (1985) An extended Wichura theorem, definitions of Donsker class, and weighted empirical distributions. Probability in Banach spaces. *Lecture Notes in Math.*, 1153: V: 141–178, Springer, Berlin.
83. Dumitriu, I. and Edelman, A. (2002) Matrix models for beta ensembles. *J. Math. Phys.*, 43(11): 5830–5847.
84. Efron, B. (1979) Bootstrap methods: another look at the jackknife. *Ann. Statist.*, 7(1): 1–26.
85. Elderman, A. (1997) The probability that a random real Gaussian matrix has k real eigenvalues, related distributions, and the circular law. *Journal of Multivariate Analysis*, 60(2): 203–232.
86. Engle, R. F., and Gearger, C. W. J. (1987) Co-integration and error correction representaton, estimation and testing. *Econometrica*, 55(2): 251–276.
87. Fan, K. (1951). Maximum properties and inequalities for the eigenvalues of completely continuous operators. *Proc. Nat. Acad. Sci. U.S.A.*, 37: 760–766.
88. Form, M. and Lippi, M. (1999). Aggregation of linear dynamic microeconomic models. *Journal of Mathematical Ecnomics*, 31: 131–158.
89. Forni, M., Giamond, D., Lippi, M. and Reichlin, L. (2004) Opening the Black Box: Structural Factor Models Versus Structural VARS. *C.E.P.R. Discussion Papers in its series CEPR Discussion Papers* with number 4133.
90. Forni, M., Hallin, M., Lippi, M. and Reichlin, L. (2000) The generalized dynamic factor model: identification and estimation. *The Review of Economics and Statistics*, 82: 540–554.

91. Forni, M., Hallin, M., Lippi, M. and Reichlin, L. (2003) Do Financial Variables Help Forecasting Inflation and Read Activity in the Euro Area. *Journal of Monetary Ecnomics*, 50: 1243–1255.
92. Forni, M., Hallin, M., Lippi, M. and Reichlin, L. (2001) Coincident and leading indicators for the Euro area. *The Economic Journal* 111: 62–85.
93. Forni, M., Hallin, M., Lippi, M. and Reichlin, L. (2003) The Generalized Dynamic Factor Model One-side Estimation and Forecasting. *LEM Working Paper Series.*
94. Forni, M., Hallin, M., Lippi, M., and Reichlin, L. (2004) The generalized dynamic factor model: consistency and rates. *Journal of Econometrics*, 119: 231–255.
95. Forni, M and Reichlin, L. (1998) Let's get real: a factor analytical approach to disapggregated business cycle dynamic. *Review of Economic Studies*, 65: 453–473.
96. Foschini, G. J. (1996) Layered space-time architechure for wireless communications in a fading environment when using multi-element antennas. *Bell Labs Technical Journal*, 1(2): 41–59,
97. Foschini, G. J. and Gans, M. (1998) On limits of wireless communications in fading environment when using multiple antennas. *Wireless Personal Communi.*, 6(3): 315–335.
98. Gadzhiev, Ch. M. (2004) Determination of the upper bound for the spectral norm of a random matrix and its application to the diagnosis of the Kalman filter. (Russian) *Kibernet. Sistem. Anal.*, 40(2): 72–80.
99. Gamburd, A., Lafferty, J. and Rockmore, D. (2003) Eigenvalue spacings for quantized cat maps. Random matrix theory. *J. Phys. A*, 36(12): 3487–3499.
100. Geman, S. (1980) A limit theorem for the norm of random matrices. *Ann. Probab.*, 8: 252–261.
101. Geman, S. (1986) The spectral radius of large random matrices. *Ann Probab.*, 14(4): 1318–1328.
102. Gerlach, D. and Paulraj, A. (1994) Adaptive transmitting antenna arrays with feedback, *IEEE Signal Process. Lett.*, 1(10): 150–152.
103. Ginibre, J. (1965) Statistical ensembles of of complex, quaterion and real matrices. *J. Math. Phys.*, 6: 440–449.
104. Girko, V. L. (1984a) Circle law. *Theory Probab. Appl.*, 4: 694–706.
105. Girko, V. L. (1984b) On the circle law. *Theory Probab. & Math. Statist.*, 28: 15–23.
106. Girko, V. L., Kirsch, W. and Kutzelnigg, A. (1994) A necessary and sufficient conditions for the semicircular law. *Random Oper. and Stoch. Equ.*, 2: 195–202.
107. Girshick, M. A. (1939) On the sampling theory of roots of determinantal equations. *Ann. Math. Statistics*, 10: 203–224.
108. Gnedenko, B. V. and Kolmogorov, A. N. (1954) *Limit distributions for sums of independent random variables*, Addison–Wesley.
109. Goldberger, A. S. (1972) Maximum-Likelihood Estimation of Regression Containing Un-observable Independent Variables. *International Economic Review*, 13(1): 1–15.
110. Goldberg, G. and Neumann, M. (2003) Distribution of subdominant eigenvalues of matrices with random rows. *SIAM J. Matrix Anal. Appl.*, 24(3): 747–761.
111. Goldsmith, A. J. and Varaiya, P. P. (1997) Capacity of fading channels with channel side information. *IEEE Trans on Inf. Theory*, 43(6): 1986–1992.

112. Grant A. J. and Alexander, P. D. (1998) Random sequence multisets for synchronous code-division multiple-access channels. *IEEE Trans. Inf. Theory*, 44(7): 2832–2836.
113. Grenander, U. and Silverstein, J. W. (1977) Spectral analysis of networks with random topologies. *SIAM J. Appl. Math.*, 32(2): 499–519.
114. Götze, F. and Tikhomirov, A. (2004) Limit theorems for spectra of positive random matrices under dependence. *Zap. Nauchn. Sem. S.-Peterburg. Otdel. Mat. Inst. Steklov. (POMI)*, 311(7): *Veroyatn. i Stat.*, 92–123,
115. Grönqvist, J., Guhr, T. and Kohler, H. (2004) The k-point random matrix kernels obtained from one-point supermatrix models. *J. Phys. A*, 37(6): 2331–2344.
116. Guionnet, A. (2004) Large deviations and stochastic calculus for large random matrices. *Probab. Surv.*, 1: 72–172 (electronic).
117. Guo, D., Verdú, S. and Rasmussen, L. K. (2002) Asymptotic normality of linear multiuser receiver outputs. *IEEE Trans. Inf. Theory*, 48(12): 3080–3095.
118. Guo, D. and Verdú, S. (2005) Randomly spread CDMA: Asymptotics via statistical physics. *IEEE Trans. Inf. Theory*, 51(6): 1982–2010.
119. Gupta, A. (2002) Improved symbolic and numerical factorization algorithms for unsymmetric sparse matrices. *SIAM J. Matrix Anal. Appl.*, 24(2): 529–552.
120. Khorunzhy, A. and Rodgers, G. J. (1997) Eigenvalue distribution of large dilute random matrices. *J. Math. Phys.*, 38: 3300–3320.
121. Hanly, S. and Tse, D. (2001) Resource pooling and effective bandwidths in CDMA networks with multiuser receivers and spatial diversity. *IEEE Trans. Inform. Theory*, 47: 1328–1351.
122. Hara, S. and Prasad, R. (1997) Overview of multicarrier CDMA. *IEEE Commun. Magazine*, 35(12): 126–133.
123. Haykin, S. (2005) Cognitive radio: Brain-empowered wireless communications. *IEEE J. Selected Areas in Commun.*, 23(2): 201–220.
124. Honig, M. L. and Ratasuk, R. (2003) Large-system performance of iterative multiuser decision-feedback detection. *IEEE Trans. Commun.*, 51(8): 1368–1377.
125. Honig, M. L. and Xiao, W. M. (2001) Performance of reduced-rank linear interference suppression. *IEEE Trans. Inf. Theory*, 47(5): 1928–1946.
126. Horn, R. A. and Johnson, C. R. (1985) *Matrix Analysis*, Cambridge Univ. Press.
127. Horn, R. A. and Johnson, C. R. (1991) *Topics in Matrix Analysis*, Cambridge Univ. Press.
128. Hsu, P. L. (1939) On the distribution of roots of certain determinantal equations. *Ann. Eugenics*, 9: 250–258.
129. Huber, P. J. (1964) Robust estimation of a location parameter. *Ann. Math. Statist.*, 35: 73–101.
130. Hwang, C. R. (1986) A brief survey on the spectral radius and the spectral distribution of large dimensional random matrices with iid entries. *Random Matrices and Their Applications: Contemporary Mathematics*, 50: AMS, (ed. M. L. Mehta) Providence, 145–152.
131. Hoeffding, W. (1963) Probability inequalities for sums of bounded random variables. *J. Amer. Statist. Assoc.*, 58: 13–30.
132. Ingersol, J. (1984). Some results in the theory of arbitrage pricing. *The Journal of Finance*, 39: 1021–1039.

133. Issler, J. V. and F. Vahid (2001) Common Cycles and the Importance of Transitory Shocks to Macroeconomic Aggregates. *Journal of Monetary Economics*, 47: 449–475.
134. Janik, R. A. (2002) New multicritical random matrix ensembles. *Nuclear Phys. B*, 635(3): 492–504.
Jiang, T. (2005) Maxima of entries of Haar distributed matrices. *Probab. Theory Related Fields*, 131(1): 121–144.
135. Jiang, T. (2004) The limiting distributions of eigenvalues of sample correlation matrices. *Sankhyā*, 66(1): 35–48.
136. Johansson, K. (1998) On fluctuations of eigenvalues of random Hermitian matrices. *Duke Math. J.*, 91(1): 151–204.
137. Johansson, K. (2000) Shape fluctuations and random matrices. *Comm. Math. Phys.*, 209: 437–476.
138. Johnstone, I. M. (2001) On the distribution of the largest eigenvalue in principal components analysis. *Ann. Statist.*, 29(2): 295–327.
139. Johansson, K. (2002) Non-intersecting paths, random tilings and random matrices. *Probab. Theory Related Fields*, 123(2): 225–280.
140. Jonsson, D. (1982) Some limit theorems for the eigenvalues of a sample covariance matrix. *J Multivariate Anal.*, 12: 1–38.
141. Kapetanios, G. (2004) A Note on modelling Core inflation for the UK using a new dynamic factor estimation method and alarge disaggregated price index dataset. *Economics Letters*, 85: 63–69.
142. Kapetanios, G. and Marcellino, M. (2006) A Parametric Estimation method for Dynamic Factor Models of Large Dimensions, CEPR Discussion Paper No. 5620 Available at SSRN: http://ssrn.com/abstract=913392.
143. Keating, J. P. (2000) Random matrices and the Riemann zeta-function. *Highlights of mathematical physics*, (London), 153–163, Amer. Math. Soc., Providence, RI, 2002.
144. Keating, J. P. and Mezzadri, F. (2004) Random matrix theory and entanglement in quantum spin chains. *Comm. Math. Phys.*, 252(1-3): 543–579.
145. Keeling, K. B. and Pavur, R. J. (2004) A comparison of methods for approximating the mean eigenvalues of a random matrix. *Comm. Statist. Simulation Comput.*, 33(4): 945–961.
146. Khorunzhy, A. and Rodgers, G. J. (1998) On the Wigner law in dilute random matrices. *Reports on Mathematical Physics*, 42: 297–319.
147. Khorunzhy, O., Shcherbina, M. and Vengerovsky, V. (2004) Eigenvalue distribution of large weighted random graphs. *J. Math. Phys.*, 45(4): 1648–1672.
148. Krzanowski, W. (1984) Sensitirily of principal components. *J Royal stats*, Soc B, 46 3: 558–563.
149. Kuhn, K. W. and Tucker, A. W. (1951) Nonlinear Programming. *Proceedings of the second Berkeley Symposium on Mathematical Statistics and Probability*, edited by Jerzy Neymann. University of California Press, Berkeley and Los Angoles, pp. 481–492.
150. Lahiri, I. K. and Moore. G. H. eds. (1991) *Leading Ecnomic Indicators. New Approches and Forecastig Records.* Cambridge: Cambridge Univ. Press, pp. 63–89.
151. Laloux, L., Cizeau, P., Bouchaud, J. P. and Potters. M. (1999) Noise Dressing of Financial Correlation Matrices. *Phy. Rev. Lett.* 83: 1467–1470.

152. Laloux, L., Cizeau, P. Bouchaud, J. P. and Potters, M. (1999) *Random Matrix Theory Applied to Portfolio Optimization in Japanese Stock Market*, Risk 12(3): 69.
153. Laloux, L., Cizeau, P., Potters, M. and Bouchaud, J. (2000) Random matrix theory and financial correlations. *Int. J. Theor, Appl. Finance*, 3: 391–397.
154. Latała, R. (2005) Some estimates of norms of random matrices. *Proc. Amer. Math. Soc.*, 133(5): 1273–1282.
155. Le Car, G. and Delannay, R. (2003) The distributions of the determinant of fixed-trace ensembles of real-symmetric and of Hermitian random matrices. *J. Phys. A*, 36(38): 9885–9898.
156. Letac, G. and Massam, H. (2000) The normal quasi-Wishart distribution. Algebraic methods in statistics and probability (Notre Dame, IN, 2000), 231–239, *Contemp. Math., 287, Amer. Math. Soc., Providence, RI, 2001.*
157. Li, L., Tulino, A. M. and Verdú, S. (2004) Design of reduced-rank MMSE multiuser detectors using random matrix methods. *IEEE Trans. Inf. Theory*, 50(6): 986–1008.
158. Li, P., Paul, D., Narasimhan, R. and Cioffi, J. (2006) On the distribution of SINR for the MMSE MIMO receiver and performance analysis. *IEEE Trans. Inf. Theory*, 52(1): 271–286.
159. Liang, Y. C. (2006) Asymptotic performance of BI-GDFE for large isometric and random precoded systems. *Proc. IEEE VTC-Spring*, Stocholm, Sweden.
160. Liang, Y. C., Chen, H. H., Mitolla, J., Mahonen, III, P. Kohno, R. and Reed, J. H. (2008) Guest Editorial, Cognitive Radio: Theory and Application. *IEEE J. Selected Areas in Commun.*, 26(1): 1–4.
161. Liang, Y. C., Cheu, E. Y., Bai, L. and Pan, G. (2008) On the relationship between MMSE-SIC and BI-GDFE receivers for large multiple-input multiple-output (MIMO) channels, to appear in *IEEE Trans. Signal Process.*
162. Liang, Y. C. and Chin, F. (2001) Downlink channel covariance matrix (DCCM) estimation and its applications in wireless DS-CDMA systems with antenna array. *IEEE J. Selected Areas in Commun.*, 19(2): 222–232.
163. Liang, Y. C., Chin, F. and Liu, K. J. R. (2001) Downlink beamforming for DSCDMA mobile radio with multimedia services. *IEEE Trans. Commun.*, 49(7): 1288–1298.
164. Liang, Y. C., Pan, G. M. and Bai, Z. D. (2007) Asymptotic performance analysis of MMSE receivers for large MIMO systems using random matrix theory. *IEEE Trans. Inf. Theory*, 53(11): 4173–4190.
165. Liang, Y. C., Sun, S. and Ho, C. K. (2006) Block-iterative generalized decision feedback equalizers (BI-GDFE) for large MIMO systems: algorithm design and asymptotic performance analysis. *IEEE Trans. Signal Process.*, 54(6): 2035–2048.
166. Liang, Y. C., Zhang, R. and Cioffi, J. M. (2006) Subchannel grouping and statistical water-filling for vector block fading channels. *IEEE Trans. Commun.*, 54(6): 1131–1142.
167. Liang, Y. C. *et al.* (2005) System description and operation principles for IEEE 802.22 WRANs, Available at IEEE 802.22 WG Website http://www.ieee802.org/22/.
168. Liang, Y. C., Zeng, Y., Peh, E. and Hoang, A. T. (2008) Sensing-throughput tradeoff for cognitive radio networks. *Proc. IEEE International Conference on Communications (ICC) 2007*, pp. 5330–5335, Glasgow, UK.

169. Liang, Y. C., Zeng, Y., Peh, E. and Hoang, A. T. (2008) Sensing-throughput tradeoff for cognitive radio networks. *Communications, 2007. ICC '07. IEEE International Conference on Publication Date: 24–28 June 2007 On page(s)*: 5330–5335.
170. Li, L., Tulino, A. M. and Verdú S. (2004) Design of reduced-rank MMSE multiuser detectors using random matrix methods. *IEEE Trans. Inform. Theory*, 50(6): 986–1008.
171. Lin, Y. (2001) Graph extensions and some optimization problems in sparse matrix computations. *Adv. Math.*, (China) 30(1): 9–21.
172. Lintner J. (1965) Security Prices, Risk and Maximal Gains from Diversification. *Journal of Finance*, 20: 578–615.
173. Linusson, S. and Wstlund, J. (2004) A proof of Parisi's conjecture on the random assignment problem. *Probab. Theory Related Fields*, 128(3): 419–440.
174. Loève, M. (1977) *Probability Theory:* 4 ed. Springer–Verlag, New York.
175. Malgorzata, S. and Jakub, K. (2006) Automatic trading agent, RMT based Portfolio Theory and Portfolio Selection. *ACTA Physica Polonica B*, 37: 3145–3160.
176. Marčenko, V. A. and Pastur, L. A. (1967) Distribution for some sets of random matrices. *Math. USSR–Sb.*, 1: 457–483.
177. Markowitz, H. (1952) Portpolio Selection *The Journal of Finance*, VII(1): 77–91.
178. Markowitz, H. (1956) The Optimization of a Quadratic Function Subject to linear Constraints. *Naval Research Logistics Quarterly*, III: 111–133.
179. Markowitz, H. (1991) *Portfolio Selection, Efficient Diversification of investiments* 2rd Edition. Blackwell. Publishing.
180. Martegna, R. N. (1999) Hierarchical structure in finacial market. *European Physical Journal B*, 11: 193–197.
181. Marti R., Laguna, M., Glover, F. and Campos, Vicente (2001) Reducing the bandwidth of a sparse matrix with tabu search. Financial modelling. *European J. Oper. Res.*, 135(2): 450–459.
182. Masaro, J. and Wong, C. S. (2003) Wishart distributions associated with matrix quadratic forms. *J. Multivariate Anal.*, 85(1): 1–9.
183. Matthew. C. H. (2007) Essays in Ecnometrics and Random Matrix Theory. Ph.D. Theses, MIT.
184. Mayya, K. B. K. and Amritkar, R. E. (2006) Analysis of delay correlation matriceo. *oar: arXiv.org. cond-mat/0601279.*
185. McKenzie, B. J. and Bell, T. (2001) Compression of sparse matrices by blocked Rice coding. *IEEE Trans. Inform. Theory*, 47(3): 1223–1230.
186. Mehta, M. L. (2004) *Random matrices.* Third edition. Pure and Applied Mathematics (Amsterdam), 142. Elsevier/Academic Press, Amsterdam, 2004. xviii+688 pp. ISBN: 0-12-088409-7 82-02.
187. Mehta, M. L. (1991) *Random Matrices.* 2nd Edition. Academic Press, New York.
188. Mehta, M. L. (1960) On the statistical properties of the level-spacings in nuclear spectra. *Nuclear Phys.*, 18: 395–419.
189. Mendez, C., Kapetanios, G, G., Smith, R. J. and Weale, M. R. (2001) An Automatic Leading Indicator of Economic Activity: Forecasting GDP Growth for European Countries. *Ecnometrics Journal*, 4: S56–S90.
190. Merton, R. (1973) Antertemporal Capital Asset Pricing Model. *Econometrica*, 41: 867–887.

191. Mitola, J. and Maguire, G. Q. (1999) Cognitive radios: making software radios more personal. *IEEE Personal Commun.*, 6(4): 13–18.
192. Müller, R. R. (1998) *Power and Bandwidth Efficiency of Multiuser Systems with Random Spreading,* PhD Thesis, Universtät Erlanger-Nürnberg, Erlangen, Germany.
193. Naulin, Jean–Marc (2002) A contribution of sparse matrices tools to matrix population model analysis. Deterministic and stochastic modeling of biointeraction (West Lafayette, IN, 2000). *Math. Biosci.*, 177/178: 25–38.
194. von Neumann, J. and Morgenstern, O. (1953) *Theory of Games and Economic Behavior.* 3rd edition, Princeton University Press, Princeton. N. J.
195. Ndawana, M. L. and Kravtsov, V. E. (2003) Energy level statistics of a critical random matrix ensemble. Random matrix theory. *J. Phys. A*, 36(12): 3639–3645.
196. Neub, N. (2002) A new sparse-matrix storage method for adaptively solving large systems of reaction-diffusion-transport equations. *Computing*, 68(1): 19–36.
197. Onatski, A. (2008) Determining the Number of Factors from Empirical Distribution of Eigenvalue. *Columbia University, Department of Economics in its series Discussion Papers* with number 0405-19.
198. Ormerod, P. (2004) Extracting information from noisy time series data. Vol terra Consulting ltd.
199. Oyman, O., Nabar, R. U., Bolcskei, H. and Paulraj, A. J. (2002) Tight lower bounds on the ergodic capacity of Rayleigh fading MIMO channels, *Proc. of Globecom'02*, pp. 1172–1176.
200. Pan, G. M., Guo, M. and Liang, Y. C. (2007) Asymptotic performance of reduced-rank linear receivers with principal component filter. *IEEE Trans. Inf. Theory*, 53(3): 1148–1151.
201. Papp, G., Sz. Pafka, Nowak, M. A. and Kondor, I. (2005) Random matrix filtering in Portfolio wptimazation. *Acta Physica Polonica B*, 36(9): 2757–2765.
202. Pastur, L. A. (1972) On the spectrum of random matrices. *Teoret. Mat. Fiz.*, 10: 102–112, (Teoret. Mat. Phys.) 10: 67–74.
203. Pastur, L. A. (1973) Spectra of random self–adjoint operators. *Uspekhi Mat. Nauk*, 28(1): 4–63, (Russian Math. Surveys 28(1): 1–67.
204. Paulraj, A., Nabar, R. and Gore, D. (2003) *Introduction to Space-Time Wireless Communications*, Cambridge UK, Cambridge University Press.
205. Petz, D. and Rffy, J. (2004) On asymptotics of large Haar distributed unitary matrices. *Period. Math. Hungar.*, 49(1): 103–117.
206. Plerou, V., Gopikrishnan, P., Rosenow, B., Amasel, L. A. N. and Stanley, H. E. (1999) Universal and Non-universal Properties of Cross-correlations in Financial Time Series. *Phys. Rev, Lett.* 83: 1471.
207. Plerou, V. Gopikrishnan, P. Rosenow, B. Amasel, L. A. N. and Stanley, H. E. (2000) A Random Matix THeory Approach to Financial Cross-correlations. *Physica A*, 287: 374–382.
208. Potters, M., Bouchaud, J.-P. and Laloux, L. (2005) Financial Application of random matrix theory: Old laces and New pieces. *Acta Physica Polonica B*, 36,(9): 2767–2784.
209. Proakis, J. G. (1995) *Digital Communications*, Third Edition, New York: McGraw-Hill.
210. Quah, D. and Vahey, S. P. (1995) Measuring Core Inflation. *Ecnomic Journal*, 105: 1130–1140.

211. Rao, M. B. and Rao, C. R. (1998) *Matrix Algebra and Its Applications to Statistics and Econometrics*, 1st Ed. World Scientific, Singapore.
212. Rashid-Farrokhi, F., Liu, K. J. R. and Tassiulas, L. (1998) Transmit beamforming and power control for cellular wireless systems. *IEEE J. Select. Areas Commun.*, 16(8): 1437–1449.
213. Rashid-Farrokhi, F., Tassiulas, L. and Liu, K. J. R. (1998) Joint optimal power control and beamforming in wireless networks using antenna arrays. *IEEE Trans. Commun.*, 46(10): 1313–1324.
214. Ratnarajah, T., Vaillancourt, R. and Alvo, M. (2004/2005) Eigenvalues and condition numbers of complex random matrices. *SIAM J. Matrix Anal. Appl.*, 26(2): 441–456.
215. Robinson, P. M. (1974) Indentification, Estimation and Large Sample Theory for Regressions Containing Unobservable Variables. *Internatinal Economic Review*, 15(3): 680–692.
216. Rojkova, V. and Kantardzic, M. (2007) Analysis of inter-domain traffic correlations: random matrix theory approach. *arXiv.org:cs/0706.2520*.
217. Rojkova, V. and Kantardzic, M. (2007) Delayed Correlation in Inter-Doman Network Traffic. *arXiv:0707.10838 v1*. [cs.NI], 7.
218. Rosenthal, H. P. (1970) On the subspaces of L^p $(p > 2)$ spanned by sequences of independent random variables. *Israel J. Math.*, 273–303.
219. Ross. S. (1976) The Arbitrage Theory of Capital Asset Pricing. *Journal of Economic Theory*, 13: 341–360.
220. Eickmeier, S. and Ziegler, C. (2006) How good are dynamic factor models at forecasting output and inflation Ameta-analysic approach. *Discussion Paper Series 1: Economics Studies*, 42, Dentsche Burdesbank.
221. Saranadasa, H. (1993) Asymptotic expansion of the misclassification probabilities of D-and A-criteria for discrimination from two high dimensional populations using the theory of large dimensional random matrices. *J. Multiv. Anal.*, 46: 154–174.
222. Semerjian, G. and Cugliandolo, L. F. (2002) Sparse random matrices: the eigenvalue spectrum revisited. *J. Phys. A.* 35(23): 4837–4851.
223. Sharifi, S., Crane, M., Shamaie, A. and Ruskin, H. (2004) Random matrix theory for portfolio optimization: a stability approach. *Phys. A*, 335(3-4): 629–643.
224. Sharpe, W. (1964) Capital Asset Prices: A Theony of Market Equilibrium under Conditions of Risk. *Journal of Finance*, 19: 425–442.
225. Sinai, Ya. and Soshnikov, A. (1998) Central limit theorem for traces of large random symmetric matrices with independent matrix elements. *Bol. Soc. Brasil. Mat. (N.S.)*, 29(1): 1–24.
226. Silverstein, J. W. (1995) Strong convergence of the eimpirical distribution of eigenvalues of large dimensional random matrices *J. Multivariate Anal.*, 5: 331–339.
227. Silverstein, J. W. (1990) Weak convergence of random functions defined by the eigenvectors of sample covariance matrices. *Ann. Probab.*, 18: 1174–1194.
228. Silverstein, J. W. (1989a) On the eigenvectors of large dimensional sample covariance matrices. *J. Multiv. Anal.*, 30: 1–16.
229. Silverstein, J. W. (1989b) On the weak limit of the largest eigenvalue of a large dimensional sample covariance matrix *J. Multiv. Anal.*, 30: 307–311.
230. Silverstein, J. W. (1985a) The limiting eigenvalue distribution of a multivariate F matrix. *SIAM J. Math. Anal.*, 16(3): 641–646.

231. Silverstein, J. W. (1985b) The smallest eigenvalue of a large dimensional Wishart Matrix. *Ann. Probab.*, 13(4): 1364–1368.
232. Silverstein, J. W. (1984a) Comments on a result of Yin, Bai and Krishnaiah for large dimensional multivariate F matrices. *J. Multiv. Anal.*, 15: 408–409.
233. Silverstein, J. W. (1984b) Some limit theorems on the eigenvectors of large dimensional sample covariance matrices. *J. Multiv. Anal.*, 15: 295–324.
234. Silverstein, J. W. (1981) Describing the behavior of eigenvectors of random matrices using sequences of measures on orthogonal groups. *SIAM J. Math. Anal.*, 12: 174–281.
235. Silverstein, J. W. (1979) On the randomness of eigenvectors generated from networks with random topologies. *SIAM J. Applied Math.*, 37: 235– 245.
236. Silverstein, W. J. and Bai, Z. D. (1995) On the empirical distribution of eigenvalues of a class of large dimensional random matrices. *J. Multiv. Anal.*, 54: 175–192.
237. Silverstein, W. J. and Choi, S. I. (1995) Analysis of the limiting spectral distribution of large dimensional random matrices. *J. Multiv. Anal.*, 54: 295–309.
238. Silverstein, J. W. and Combettes, P. L. (1992) Signal detection via spectral theory of large dimensional random matrices. *IEEE ASSP*, 40: 2100–2104.
239. Skorohod, A. V. (1956) Limit theorems for stochastic processes. (Russian) *Teor. Veroyatnost. i Primenen.*, 1: 289–319.
240. Soshnikov, A. (2004) Poisson statistics for the largest eigen-values of Wigner random matrices with heavy tails. *Electron. Comm. Probab.*, 9: 82–91.
241. Soshnikov, A. (2002) A note on universality of the distribution of the largest eigenvalues in certain sample covariance matrices. Dedicated to David Ruelle and Yasha Sinai on the occasion of their 65th birthdays. *J. Statist. Phys.*, 108(5-6): 1033–1056.
242. Stariolo, D. A., Curado, E. M. F. and Tamarit, F. A. (1996) Distributions of eigenvalues of ensembles of asymmetrically diluted Hopfield matrices. *J. Phys. A.*, 29: 4733–4739.
243. Stock, J. H. and Watson, M. W. (1989) New Indexes of Coincident and leading Ecnomic Indicators. *NBER Macroecnomics Annual.* pp 351–394.
244. Stock, J. H. and Watson, M. W. (1999) Diffusion Indexes. Manuscript, Economics Department, Harvard University.
245. Stock, J. H. and Watson, M. W. (2002) Macroecnomic Forecasting Vsing diffusion indexes. *Journal of Business and Ecnomic Statistic*, 20: 147–162.
246. Stock, J and Watson, M. (2003) Implicatons of Dynamic Factor Models for VAR Analysis, mimeo.
247. Tanaka, T. (2002) A statistical-mechanics approach to large-system analysis of CDMA multiuser detectors. *IEEE Trans. Inf. Theory*, 48: 2888–2910.
248. Tao, T. and Vu, V. (2010) Random matrices: Universality of ESD and the Circular Law (with appendix by M. Krishnapur), *Annals of Probability*, 38(5): 2023–2065.
249. Telatar, E. (1999) Capacity of multi-antenna Gaussian channels. *Euro. Trans. Telecommun.*, 10(6): 585–595.
250. Teodorescu, R., Bettelheim, E., Agam, O., Zabrodin, A. and Wiegmann, P. (2005) Normal random matrix ensemble as a growth problem. *Nuclear Phys.*, 704(3): 407–444.
251. Titchmarsh, E. C. (1939) *The Theory of Functions, Second Edition.* Oxford University Press, London.

252. Thomas, G. and Bernd, K. (2002) A New Method to Estimate the Nosie in Financial Correlation Matrices. *arXiv:cond-mat/0206577, v1, cond-mat-mech.* Submitted to *J. Phys. A*: Math Gen.
253. Toth, B. and Kertesz, J. (2007) On the origin of the Epps effect. *arXiv: Physics/0701110 v3.* [physics. soc-ph], 21.
254. Toth, B., Toth, B. and Kertesz, J. (2007) Modeling the Epps effect of cross-correlations in asset prices. *arXiv: 0704.3798 v1.* [physics. data-an].
255. Tracy, C. A. and Widom, H. (1994) Level-spacing distributions and the Airy kernel. *Comm. Math. Phys.*, 159: 151–174.
256. Tracy, C. A. and Widom, H. (1996) On orthogonal and symplectic matrix ensembles. *Comm. Math. Phys.*, 177(3): 727–754.
257. Tracy, C. A. and Widom, H. (1998) Correlation functions, cluster functions, and spacing distributions for random matrices. *J. Statis. Phys.*, 92: 809–835.
258. Tracy, C. A. and Widom, H. (2002) Distribution functions for largest eigenvalues and their applications. *Proceedings of the International Congress of Mathematicians*, 1: 587–596, Higher Ed. Press, Beijing, 2002.
259. Trichard, L. G., Evans, J. S. and Collings, I. B. (2002) Large system analysis of linear multistage parallel interference cancellation. *IEEE Trans. Commun.*, 50(11): 1778–1786.
260. Tse, D. and Hanly, S. (1999) Linear multiuser receivers: effective interference, effective bandwidth and user capacity. *IEEE Trans. Inf. Theory*, 47(2): 641–657.
261. Tse, D. and Zeitouni, O. (2000) Linear multiuser receivers in random environments. *IEEE Trans. Inf. Theory*, 46(1): 171–188.
262. Tulino, A. M. and Verdú, S. (2004) *Random matrix theory and wireless communications: Foundations and Trends in Commun. and Inf.*, Now Publishers Inc.
263. Vassilevski, P. S. (2002) Sparse matrix element topology with application to AMG(e) and preconditioning. Preconditioned robust iterative solution methods, PRISM '01 (Nijmegen). *Numer. Linear Algebra Appl.*, 9(6-7): 429–444.
264. Verdú, S. and Shamai, S. (1999) Spectral efficiency of CDMA with random spreading. *IEEE Trans. Inf. Theory*, 45(3): 622–640.
265. Wachter, K. W. (1978) The strong limits of random matrix spectra for sample matrices of independent elements. *Ann Probab.*, 6(1): 1–18.
266. Wachter, K. W. (1980) The limiting empirical measure of multiple discriminant ratios. *Ann. Stat.*, 8: 937–957.
267. Wang, K. and Zhang, J. (2003) MSP: a class of parallel multistep successive sparse approximate inverse preconditioning strategies. *SIAM J. Sci. Comput.* 24(4): 1141–1156.
268. Wang, X. and Poor, H. V. (1999) Iterative (turbo) soft interference cancellation and decoding for coded CDMA. *IEEE Trans. Commun.*, 47(7): 1046–1061.
269. Wiegmann, P. and Zabrodin, A. (2003) Large scale correlations in normal non-Hermitian matrix ensembles. Random matrix theory. *J. Phys. A*, 36(12): 3411–3424.
270. Wigner, E. P. (1955) Characteristic vectors bordered matrices with infinite dimensions. *Ann. of Math.*, 62: 548–564.
271. Wigner, E. P. (1958) On the distributions of the roots of certain symmetric matrices. *Ann. of Math.*, 67: 325–327.

272. Witte, N. S. (2004) Gap probabilities for double intervals in Hermitian random matrix ensembles as τ-functions-spectrum singularity case. *Lett. Math. Phys.*, 68(3): 139–149.
273. Wolfe, P. (1957) A Simplex Method for Quadratic Programming, Privately circulated as RAND p-12.5.
274. Wynne, M. A. (1999) Core Inflation: A review of some conceptual issues. *European Central Bank, Working paper No. 5.*
275. Xia, M. Y., Chan, C. H., Li, S. Q., Zhang, B. and Tsang, L. (2003) An efficient algorithm for electromagnetic scattering from rough surfaces using a single integral equation and multilevel sparse–matrix canonical–grid method. *IEEE Trans. Antennas and Propagation*, 51(6): 1142–1149.
276. Yin, Y. Q. (1986) Limiting spectral distribution for a class of random matrices. *J. Multiv. Anal.*, 20: 50–68.
277. Yin, Y. Q., Bai, Z. D. and Krishnaiah, P. R. (1988) On the limit of the largest eigenvalue of the large dimensional sample covariance matrix. *Probab. Th. Rel. Fields*, 78: 509–521.
278. Yin, Y. Q., Bai, Z. D. and Krishnaiah, P. R. (1983) Limiting behavior of the eigenvalues of a multivariate F matrix *J. Multiv. Anal.*, 13: 508–516.
279. Yin, Y. Q. and Krishnaiah, P. R. (1985) Limit theorem for the eigenvalues of the sample covariance matrix when the underlying distribution is isotropic. *Theory Probab. Appl.*, 30: 861–867.
280. Yin, Y. Q. and Krishnaiah, P. R. (1983) A limit theorem for the eigenvalues of product of two random matrices. *J. Multiv. Anal.*, 13: 489–507.
281. Yonina, C. and Chan, A. M. (2003) ON the asymptotic performance of the decorrelator. *IEEE Trans Inform Theory*, 49: 2309–2313.
282. Zabrodin, A. (2003) New applications of non-Hermitian random matrices. *Ann. Henri Poincaré*, 4(suppl. 2): S851–S861.
283. Zellner, A. (1970) Estimation of Regression Retotionships Containing Unobservable Independent Variables. *International Economic Review*, 11(3): 441–454.
284. Zeng, Y. H. and Liang, Y. C. (2006) Eigenvalue based sensing algorithms, in doc.: *IEEE* 802.22-06/0119r0.
285. Zeng, Y. H. and Liang, Y. C. (2007) Maximum-minimum eigenvalue detection for cognitive radio, *Proc. IEEE Intern. Symp. on Personal, Indoor and Mobile Radio Commun. (PIMRC).*
286. Zeng, Y. H. and Liang, Y. C. (2007) Eigenvalue based spectrum sensing algorithms for cognitive radio, submitted to *IEEE Trans. Commun.*
287. Zeng, Y., Koh, C. L. and Liang, Y. C. (2008) Maximum eigenvalue detection: theory and application, to appear in *Proc. IEEE ICC'2008*, Beijing, China.
288. Zhang, J., Chong, E. and Tse, D. (2001) Output MAI distributions of linear MMSE multiuser receivers in DS-CDMA systems. *IEEE Trans. Inf. Theory*, 47(3): 1028–1144.
289. Zheng, S. R. (2012) Central Limit Theorem for Linear Spectral Statistics of Large Dimensional F-Matrix. *Annales de l'Institut Henri Poincaré-Probabiliteset Statistiques*, 48(2): 444–476.

Index

Printed in the United States
By Bookmasters